C. J. Dunnicliff
REGISTERED PROFESSIONAL ENGINEER
PROVINCE OF ONTARIO

HUMBER

D0810539

ECONOMIC ANALYSIS
for
ENGINEERS and
MANAGERS

161904179

THE CANADIAN CONTEXT

J. C. Sprague
J. D. Whittaker

Prentice-Hall Canada, Inc.

Canadian Cataloguing in Publication Data

Sprague, J. C.
    Economic analysis for engineers and managers
    ISBN 0-13-224049-1

    1. Engineering economy — Canada.   2. Managerial
    economics — Canada.   I. Whittaker, J. D.   II. Title.

    TA177.4.S68 1986   658.1'5   C85-098576-5

Editorial/production supervision and interior design: Fred Dahl
Cover design: Photo Plus Art
Manufacturing buyer: Rhett Conklin

ISBN   0-13-224049-1   01

© 1986 by Prentice-Hall Canada Inc.
A Division of Simon & Schuster, Inc.
Scarborough, Ontario

ALL RIGHTS RESERVED

No part of this book may be reproduced in any form without permission in writing from the
publisher.

Prentice-Hall Inc., Englewood Cliffs, *New Jersey*
Prentice-Hall International, Inc., *London*
Prentice-Hall of Australia, Pty., Ltd., *Sydney*
Prentice-Hall of India Pvt., Ltd., *New Delhi*
Prentice-Hall of Japan Inc., *Tokyo*
Prentice-Hall of Southeast Asia (Pte.) Ltd., *Singapore*
Editora Prentice-Hall do Brasil Ltda., *Rio de Janeiro*
Prentice-Hall Hispanoamericana, S.A., *Mexico*
Whitehall Books Ltd., Wellington, *New Zealand*

ISBN 0-13-224049-1

Contents

PART III

ECONOMIC ANALYSIS OF ALTERNATIVES

CHAPTER SIX

Basis for Comparison of Investment Alternatives, 151

CHAPTER TEN
Replacement Analysis, 315

CHAPTER ELEVEN
Economic Analysis of Public Projects, 359

CHAPTER FOURTEEN
The Budgeting Process, 467

PART V
COMPLETING AN ECONOMIC ANALYSIS

CHAPTER FIFTEEN
Presentation, Implementation, and Tracking Investment Decisions, 505

APPENDIXES, 513

APPENDIX A
Notation, 515

APPENDIX B
Summary of Formulas, 518

APPENDIX C
Glossary of Terms, 522

APPENDIX D
Selected References, 527

APPENDIX E
Answers to Selected Problems, 531

APPENDIX F
Tables, 534

Index, 569

Preface

Making decisions is the principal function of the work of an engineer or manager. Economic analysis is fundamental to this decision-making process.

This book is directed to people who are concerned with the economic analysis associated with the decision-making process. It is designed for the practicing engineer, technologist, and manager. The mathematics have been simplified. Derivations and proofs have been included for the benefit of those interested but are not essential to the use of the book.

The book has also been planned for use as a text for universities, colleges, and technical institutes for such courses as engineering economy, managerial economics, and business analysis. No prerequisite courses are essential to its use for a one-term course. However, the book covers more material than can be discussed in one term, allowing the instructor a good selection of special topics.

The book is structured to permit maximum flexibility in the choice of material and the sequence in which it is presented. Chapters 1 through 7 are essential to the person acquiring the necessary fundamentals associated with economic analysis and a familiarization with the notation used throughout the book. A good working knowledge of Chapter 6 represents the foundation to a comprehensive understanding of the remaining chapters.

Depreciation and income tax considerations are included in the economic analysis in Chapter 4. This early introduction greatly enhances the transition from a before-tax to an after-tax analysis. This transition allows the reader to develop, at an early stage, an appreciation for the impact of taxation on the investment decision.

Several important considerations have been introduced that are worthy of mention. Chapter 2, Managing the Corporation, stresses the value of developing an organized comprehensive financial package, including financial statements and cash flow projections as a basis for obtaining financial assistance. In addition, the important topics of working-capital management and ratio analysis are discussed.

A major feature of the text is the emphasis on annual equivalent revenue requirements as a decision criterion. This measure, which shows the before-tax cash flows that must be generated to cover costs, taxes, and return on investment, is perhaps the easiest to understand of the discounted cash flow measures.

An entire chapter has been devoted to the topic of inflation. Measuring the impact of inflation on investment decisions is mandatory to the economic health of an organization. Chapters 9 and 10 discuss the difficult topic of replacement analysis and place the subject in a framework that is easy to follow and straightforward in its application. Chapter 11 is devoted specifically to the analysis of public financing.

Chapter 12 deals with the treatment of risk and uncertainty. Different methods and strategies, ranging from risk aversion to Monte Carlo simulation of probabilistic data, are developed and discussed. Chapter 13 discusses cost estimating, an essential ingredient in every economic analysis. Chapter 14 gives a comprehensive coverage of the budgeting system and the important topic of lease financing. The problems of preparing and presenting an economy study are addressed in Chapter 15.

Many example problems are included throughout the body of the text and a large number of problems are included at the end of each chapter.

Our objective is to present to the student, practicing engineer, technologist, and manager a reasonably complete and ready-reference source of material that may be applied when economic feasibility studies are called for.

Special thanks are given to the many students and practitioners who have assisted us in clarification and expansion of many areas throughout the text. We also gratefully acknowledge the editorial and typing assistance provided by the computing center and office staff.

J. C. SPRAGUE

J. D. WHITTAKER

CHAPTER ONE

Introduction

Decision making involves choosing between alternatives. *Economic decision making* involves choosing between alternatives where the differences are measured in terms of money. The fundamental questions asked of any project, proposal, process, equipment, or venture are: "How much will it cost?" and "Will it make money?". This text provides methods of answering these questions for major capital investments.

Capital investment decisions are generally made at the highest level in an organization. They are the preserve of the board, or owner, or chief executive officer. The reason that they are given this importance is because of the high cost attached to making the wrong decision. An error in staffing levels can be corrected by hiring or laying-off people. An error in ordering supplies can be corrected by expediting and adjusting inventory levels. A plant designed around the wrong process, or in the wrong location, is an error that the company may have to live with. A mistaken capital investment decision is often irredeemable in that the cost of correcting it is often greater than the cost of bearing it.

Economic decision making combines the vision of the developer, the organizing talent of the manager, the analytical ability of the economist, and the technical capability of the engineer, together with the mathematics of finance. A project must be well conceived, appropriate to the circumstances, technologically feasible, and in the final analysis it must be profitable.

1

Although the final decision is made at the top level, it is essential that all parties to the design and development process be aware of the criteria by which the investment will be judged. Only then can the company rationally make the many mini-decisions that are part of the design process. An engineer who is not aware of the economic ramification of the material selection, or process selection, is not a competent engineer. A financer who is not well informed on the technical feasibility of the investments will fail. A project is an entity and although it may be considered from different aspects, such as technical, economic, market, and financial, all aspects must be considered as an integral system for the project to succeed.

The best way to introduce economic analysis is by example. Therefore, we have selected a fairly commonplace situation to introduce some of the techniques and methods and to illustrate some of the issues.

A commonly held opinion is that money spent on rent is money "down the drain," whereas money spent on home ownership is an investment in the future. The following example, comparing rental with ownership, illustrates some of the principles of economic analysis and some of the problems with opinions.

Example 1.1

Problem: A young couple in their late twenties with two children wish to compare renting a centrally located, three-bedroom apartment with owning a three-bedroom suburban house.

Basic Information

1. The cost of the house is $85,000. It can be purchased with a $20,000 down payment and a 20-year mortgage at 12% interest. Mortgage and legal fees associated with the purchase are $2,000. The current yearly rental cost of the apartment is $8,400 ($700 per month). A $700 damage deposit is required, but the landlord is required to pay 6% interest on the deposit.

2. Current taxes on the house are $1,000 per year. The taxes on the apartment are included in the rent.

3. Parking costs $600 ($50 per month) at the apartment. The house has a garage.

4. The house has a large fenced yard; the apartment has an exercise room and a swimming pool.

5. Appliances come with the apartment. The homeowners must purchase them and can expect to have to replace their appliances every 10 years at a cost of $2,000.

6. The homeowners require fire, theft, and liability insurance at a cost of $500 per year. The apartment dwellers' insurance on their possessions is $200 per year.

7. House utilities (heat, light, water, and phone) cost $150 per month. The apartment phone costs are $30 per month.

8. The suburban homeowner has a higher commuting cost, to and from work and shopping. The difference is an extra $50 per month.

9. Household maintenance costs average about $1,200 per year.

10. The homeowners have the freedom to alter their home, subject only to local building codes. The apartment dwellers cannot change their surroundings, although the landlord will repaint the apartment every 5 years.

11. The inflation rate will average 5% per year over the next 20 years, and the value of the house plus rent, taxes, maintenance, utilities, and so on, will inflate at that rate.

12. The couple can invest any surplus funds in an account that earns 10% annual interest.

Data Summary: For simplicity we shall neglect interest payments made on periods of less than 1 year. Thus the utility cost of $150 per month will be treated as $12 \times 150 = $1,800$ per year. However, the appliance cost of $2,000 every 10 years is not $2,000 \div 10 = 200 per year. If $125.40 is invested each year in a deposit account earning 10%, the balance in the account at the end of 10 years would be $2,000. Thus the annual appliance cost is taken as $125.40.

Repayment of a $65,000, 12%, 20-year mortgage requires annual payments of $8,703.50. This payment is not subject to inflation and will remain constant over the 20 years.

Summarizing these cash flows, we have:

	House	*Apartment*	*Difference*
Initial Costs			
Deposit	$ 0.00	$ 700.00	
Fees (mortgage, etc.)	2,000.00	0.00	
Down payment	20,000.00	0.00	
Appliances	2,000.00	0.00	
Total	$24,000.00	$ 700.00	$23,300.00
Annual costs			
Taxes	$ 1,000.00	$ 0.00	
Rent	0.00	8,400.00	
Parking	0.00	600.00	
Appliances	125.40	0.00	
Insurance	500.00	200.00	
Utilities	1,800.00	360.00	
Transportation	600.00		
Maintenance	1,200.00		
Total inflating items	$ 5,225.40	$ 9,560.00	$ 4,334.60
Mortgage payment	$ 8,703.50	$ 0.00	

Thus, in the first year, the apartment option would provide $8,703.50 − $4,334.60 = $4,368.90 that could be invested. In the second year, the annual costs increase by 5%, so the extra amount available to the apartment dweller is reduced to

$$\$8,703.50 - \$4,334.60(1.05) = \$4,152.17$$

This trend would continue and in the twentieth year, the amount calculates to

$$\$8,703.50 - \$4,334.60(1.05)^{20} = \$-2,797.48$$

which means that the annual cash difference is in favor of the houseowner.

Solution: To compare these two alternatives, it is necessary to look at all the associated cash flows. All money differences will be assumed invested. That is, the apartment option has a deposit of $23,300 earning 10% interest, plus $700 earning 6% interest for 20 years, plus the annual differences that go into the 10% account. Considering all the cash flows involved, at the end of 20 years the position is as follows:

1. The houseowner owns a house that, because of inflation, is worth $225,500.
2. The apartment dweller has a deposit account with a balance of $304,290.

Thus, on a strictly economic basis, the analysis favors the apartment dweller. However, what value do you place on the freedom to sit in your own yard; or conversely, the luxury of not having to mow the lawn or shovel snow?

This example is useful because it raises many relevant points about economic decision making. First, it demonstrates that some common opinions, such as renting being less economic than buying, are not always substantiated by the figures. Second, it opens the question of financing. Does the couple have the $20,000 down payment, and will a mortgage company finance the purchase of the home? What alternative uses do people have for money, and what return can they generate? Finally, it illustrates that the economic analysis is often only a part of the overall decision analysis. The choice of renting versus buying involves more than just economic factors, and the decision is usually made on the basis of both economics and a choice of life-style.

1.1 APPLICATIONS OF ECONOMIC ANALYSIS

Any decision situation where the outcomes can be measured in monetary terms is a candidate for economic analysis. It is applicable both to corporate and individual situations, and to government-sponsored public works.

Some of the more common industrial–commercial applications are:

1. *Design optimization:* What method, structure, or material should be used?
2. *Equipment selection:* What balance of capacity, initial cost, and operating costs are correct?
3. *Equipment replacement:* When are existing operating costs or capacity limitations such that new equipment should be acquired?
4. *Capacity decisions:* What mix of equipment is best for the forecasted capacity requirements?
5. *Location studies:* Should the plant be located near the material supplies, near the market, or where the best labor force is available?
6. *Process choices:* What method of process is the best for a given situation?
7. *Resource development:* How, and at what rate, should resources be developed, extracted, and marketed?
8. *Setting prices:* What price provides the balance between competitiveness and rate of return?
9. *Setting rates:* What rate should a public utilities board allow a utility to charge its customers?
10. *Acquisitions, mergers, and new ventures:* What price should be paid to acquire a new or existing company?

All these questions require the careful application of the techniques of economic analysis. Only with analysis can one hope to guard the long-term health of the organization.

Government-sponsored projects have acceptance criteria that are slightly different, but the economic analysis techniques are the same. Commercial and industrial projects are judged on the basis of net benefit to the owners or shareholders, whereas government projects are judged on the basis of net benefit to a community as a whole. Some areas where economic analysis is used to determine these net benefits are:

1. *Regional development programs:* Are some regions more in need of special development programs than others?
2. *Incentives, subsidies, and grants:* Does some industrial sector, or area need help, encouragement, or stimulation?
3. *Choice of method:* What method would bring the greatest result for each dollar expended?
4. *Timing of new needs:* When should a highway be upgraded, a park built, a new social insurance benefit introduced? What is the cost and what is the effect?

5. *Evaluation of benefit-cost ratios:* A common method for comparing and accepting major public works is on the basis of the benefit-cost ratio. Detailed consistent economic analysis is necessary to calculate benefit-cost ratios that are accurate and comparable.

When people, both privately and publicly through the government, invest money, they wish to choose the projects that will provide the best return. The techniques of economic evaluation are the method of measuring and comparing these expected returns. Thus we find that the techniques employed by an individual entrepreneur when assessing a small franchising operation are the same as those used by the World Bank when evaluating a major development loan which will alter the economic situation of an entire country.

1.2 THE ALL-PERVASIVE TECHNIQUES

Because their application is so broad and their use so essential, the techniques of economic analysis are not specific to any discipline. These techniques are obviously a necessary component of the education of an engineer, an accountant, an economist, or a business major. The lawyer must understand them for corporate law, the political scientist needs them to follow the government allocative process, the sociologist to assess benefits and impact, the computer scientist to understand the economic computer models, and the list goes on. Some of the more common names or areas under which the basic principles are taught are:

Engineering economy
Managerial economics
Capital budgeting
Financial management
Economic evaluation
Benefit-cost analysis
Financial accounting
Economic feasibility analysis
Project analysis

Regardless of the name, the principles and techniques are consistent. They currently represent the most valid method of assessing and analyzing a decision situation.

1.3 THE CORRECT ANALYSIS

Discouraging as it may sound, it is highly unlikely that an economic analysis is right or correct in the true sense of the words. An economic analysis requires predicting the future, which hopefully we can do with some degree of accuracy. However, past performance indicates that significant errors in forecasting can result.

Economic analyses completed in the 1960s forecasted low stable energy prices, an assumption destroyed by the OPEC cartel of the 1970s. Plant feasibility studies undertaken in the 1970s assumed an exponential growth in the cost of energy—an assumption negated by the stable and falling prices of the 1980s. Still, if one is preparing today an economic analysis of a plant or process that will have a productive life of 20 to 30 years, it is necessary to make some forecast or estimate of materials, supplies, labor and energy costs, and markets over that period. The forecast will probably be in error, but with reasoned analysis it is usually possible to estimate the probable magnitude of the error and to assess the possible effects.

A correct economic analysis is not one that manages to predict the future accurately. Rather, a correct analysis is one that makes the best use of the information that is available at the time the decision is made. The techniques of economic analysis are the means whereby the available information is analyzed and assessed for its impact on the project's success.

1.4 A PROJECT IS AN ENTITY

A project, from inception to commissioning, involves many different participants. Figure 1.1 shows some of the forces that interact during the development of a project. Each participant, from the owners to promoters, engineers, and government agencies, has a particular viewpoint and special interest. The economic analyst must integrate all these disparate forces into a single analysis. To ignore one, or to focus too heavily on another, is to prejudge and bias the analysis. At the outset no one knows for certain which factors will cause the success or demise of a venture.

History gives us many examples of projects that became overnight successes or failures through a change in government policy, a major political event, or a highly publicized defect. The analyst must be sensitive to all, recognizing that a project is a single entity and that any of its factors could have a major effect on the outcome.

1.5 CHECKLIST FOR EVALUATION

The scope of economic analysis is revealed by the following checklist. The list is intended both to highlight the factors included in an analysis and to

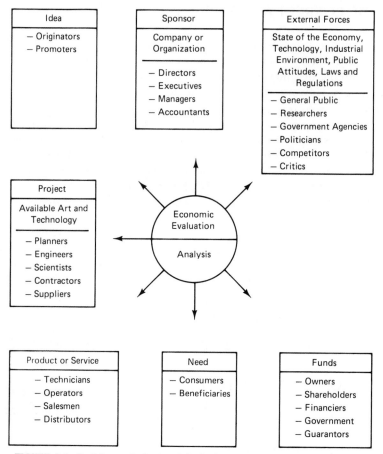

FIGURE 1.1 Participants in Project Life Cycle. (Adapted from W. B. Magyar, "Economic Evaluation of Engineering Projects," *The Engineering Economist,* Vol. 13, No. 2, pp. 67–85, 1968.)

serve as a guide for preparation of an evaluation. The start is a technically sound project or proposal, with clearly defined objectives directed toward satisfying a perceived need. Then the following items must be considered.

Environmental Aspects

1. What, and who, is the competition?
2. What are the short- and long-term industry trends (infant, young, mature)?
3. What is the economic outlook for the industry and the nation (inflation/recession)?
4. How, and where, is the labor market?

5. What changes in government policy, regulation, or legislation would help or hinder the project?

Technical–Economic Aspects

1. What alternatives are, or could be, available?
2. Have fixed and variable costs been estimated?
3. Are supplies and suppliers available?
4. What is the project life?
5. What is the possibility of technical obsolescence?
6. What is the salvage or end value?

Financial Aspects

1. What is the cost of raising the money?
2. What is the income tax rate?
3. How stable is the firm financially?
4. Is the firm able to finance the project?
5. Are any special tax concessions, grants, or incentives available?

Administrative

1. What is the nature of decision making within the firm? Who authorizes a project, and on what basis?
2. On what criteria will the project be judged?
3. What is the size of the project versus the firm's normal project size?
4. What are the legal ramifications (licenses, copyrights, patents, contracts)?
5. What are the deadlines?
6. What method of distribution is appropriate?

1.6 OUTLINE OF THE TEXT

Economic evaluation cuts across many traditional disciplines. A project is evaluated with respect to the firm's policies. Sound accounting and income tax principles, by which the firm measures its success, are relevant. The high technical component of capital investment requires engineering judgment and assessment. Future sales, economic conditions, technical obsolescence, labor rates, and government policy are all part of the project evaluation, and hence a knowledge of economics, forecasting, marketing, econometrics, and finance is necessary. All of these must be considered in the evaluation process, so the analyst must possess some understanding of all these areas.

This text is designed primarily for the person concerned with economic feasibility in the decision-making process. It concentrates, therefore, on developing and explaining the economic rationale for project selection.

Chapters 2 to 6 deal with the basic skills, techniques, and principles that are fundamental to all analyses. The accounting, financial, and tax principles by which a firm's performance is measured. The concept of discounted cash flow is introduced and applied to individual projects. The conventional measures of present value, rate of return, annual cost, and revenue requirement are then presented and related to the firm's performance.

The remainder of the book develops those special topics that support or extend the basic evaluation core. Cost estimating (Chapter 13) and risk and uncertainty (Chapter 12) are supporting areas. Break-even analysis and replacement analysis are applications of the principles of economic analysis to crucial operational areas.

Ideally, the market should determine the success or failure of a project. However, many economically sound proposals have never been initiated because they were not well presented, explained, or defended. Chapter 15 discusses methods of developing and presenting an economy study.

PROBLEMS

1.1 A $1 million investment in inventory has certain factors in common with a $1 million investment in a parking ramp.
   a. What factors are common to the two investments?
   b. What factors differ between these two investments?

1.2 What nonmonetary factors would you consider critical in your decision to rent versus buy a home?

1.3 Your company is planning to locate a steel mill in western Canada. What factors would you consider relevant to the location you select?

CHAPTER TWO

Managing the Corporation

In every decision and action, management must consider economic performance. It can justify its existence and its authority only by the economic results it produces. There are always noneconomic factors, such as the happiness of the members of the enterprise and the contribution to the welfare or culture of the community to be considered. However, any benefits to be derived by these factors can result only if economic performance is maintained.

The economic resources entrusted to management are usually organized within the framework known as the company or the corporation. The corporation is an entity in itself which acts as the legal framework under which management carries on the day-to-day activities necessary to maintain economic health. The primary function of this text is to present the financial structure within which the corporation functions and to demonstrate the financial tools with which an effective manager must be familiar in order to utilize the economic resources entrusted to management.

We should, at the outset, realize that management and economic performance must be an active ingredient at all levels within the organization. The success of the organization depends on such people as the engineer, shop supervisor, and sales manager performing their independent functions as part of the total management team. Therefore, as individual members of the team, we should understand the total structure of the organization and our role as individuals in maintaining the economic health of the total entity.

A firm's success, or failure, is measured by its books of accounts. In this chapter we outline basic accounting concepts and describe the principal accounting instruments. Problems involving working capital are a common cause of corporate distress. Factors affecting working capital are monitored by comparison with similar firms and industry standards. The means of comparison is financial ratios. The principal financial ratios are introduced and described below.

2.1 THE CORPORATE STRUCTURE

When a company is formed, financial assets are acquired from such sources as individual shareholders and the bank. To carry on the day-to-day business of the company, these financial assets are partially converted into fixed assets such as land, buildings, and equipment. The remaining financial assets are transformed into current assets, such as bank accounts, inventories, lease payments, and insurance policies.

To ensure that the company maintains economic health, management must exercise sound business practices at every stage in the decision-making process. Management must evaluate:

1. *Market feasibility:* Enter a competitive market that indicates a reasonable probability of success.
2. *Financial feasibility:* Search for financial assets from reputable sources at the best possible interest rate.
3. *Asset feasibility:* Do the necessary cash flow analysis to determine if the fixed assets should be leased or purchased, and constantly monitor the status of such current assets as accounts receivable and inventory.

2.2 THE COST OF CAPITAL

All capital has a cost. The user—an individual or a company—of capital must satisfy the profit motive of the supplier (e.g., shareholder and/or bank) of the capital. Management, to whom the capital is entrusted, must be fully aware of this capital cost and should therefore use this capital in the most effective manner possible.

There are two important sources of external capital to the firm:

1. Debt capital (e.g., bank loans, bonds)
2. Equity capital (shareholders)

If management goes to the bank in search of financial assets (capital), the banker will ask at least three important questions:

1. How much money do you need?
2. What are you going to use the money for?
3. How are you going to repay the loan?

Provided that management can answer these questions to the bank manager's satisfaction, the company will probably be able to borrow the money. The bank's profit (interest on the loan) becomes a debt or cost of borrowed capital to the company. The interest on the loan is normally at a fixed rate or tied directly to the bank's prime rate of interest.

The shareholders, the owners of the company, also expect a fair rate of return (profit) on their investment. This return is usually referred to as *net income* or *return to equity*. The shareholder receives this return in the form of dividends and/or an increase in the market value of each individual share. Thus the profits to the equity holder represent a cost of equity capital to the company. If no profits exist for an extended period of time, management has failed in its economic performance and the company may cease to exist.

In summary, the company must generate sufficient cash flow in revenues to meet (1) the day-to-day operating expenses: wages, materials, insurance, property taxes, and so on; (2) repayment of the investment in assets; (3) the profit requirements of both the debt and equity capital generated; and (4) last but not least, income tax payments.

Figure 2.1 is a simplified diagram that outlines the flow of funds within the corporation. A thorough study of this diagram is essential to a complete understanding of the concepts outlined in this chapter and the chapters to follow. The flow of cash within a corporation is somewhat analogous to the flow of fluid in a pipeline with various branches off the main stream.

2.3 ACCOUNTING SYSTEMS

Before proceeding with a discussion of the steps recommended in organizing a small company and the development of the associated financial statements, it is desirable briefly to review accounting systems.

The basis of an accounting system is a set of ledgers or accounts in which all the financial transactions of the firm are listed. There is one account for each type of transaction. For example, there is one account

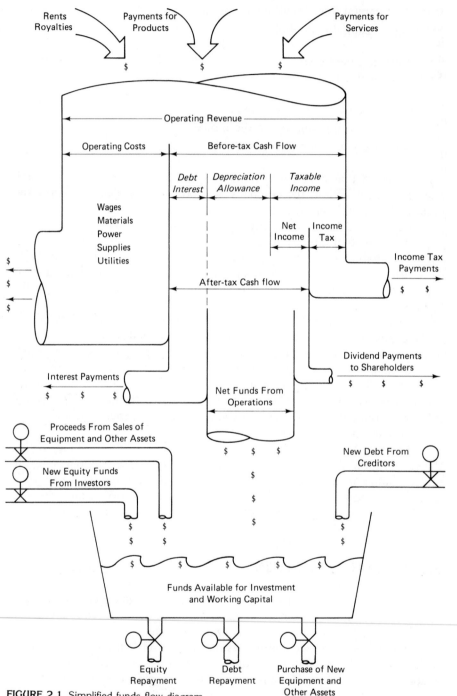

FIGURE 2.1 Simplified funds flow diagram.

for equipment, one account for bank loans, and one account for accounts receivable. The entry for an individual transaction includes the date and a description of the transaction. There are two columns, a left-hand column and a right-hand column, in which the dollar value of the transaction is recorded.

The left-hand side of any account is arbitrarily called the *debit side,* and the right-hand side is called the *credit side.* Amounts entered, or to be entered, on the left-hand side are called *debits,* and amounts entered on the right-hand side are called *credits.* The verb "to debit" means to make an entry in the left-hand side of an account, and the verb "to credit" means to make an entry in the right-hand side of the account. The words "debit" and "credit" have no other meaning in accounting.

The rule in accounting, to which there are absolutely no exceptions, is that for each transaction the debit amount (or the sum of all the debit amounts, if there are more than one) must equal the credit amount (or the sum of all the credit amounts): thus the name *double-entry accounting.* It follows that the recording of a transaction in which debits do not equal credits is incorrect. It also follows that for the accounts combined, the sum of the debit balances must equal the sum of the credit balances; otherwise, something has been done incorrectly. Thus the debit and credit arrangement used in accounting provides a useful means of checking the accuracy with which the work has been done.

The *balance sheet* is the principal accounting instrument for showing the financial condition of a company. All the asset accounts (cash, inventory, equipment, buildings, land, etc.) are listed on the left-hand side. All the liability accounts (loans, etc.) and equity accounts are listed on the right-hand side. This arrangement follows from the accounting system, where an increase in an asset is entered in the left-hand column (a debit) and an increase in a liability is listed in the right-hand column (a credit). With this method the sum of the asset column equals the sum of the liability and equity column and the accounts are said to *balance.* This rationalization can be expressed by the simple equation

$$\text{Assets} = \text{liabilities} + \text{owners' equity}$$

One must also take into account the fact that expenses and revenue are subdivisions of owners' equity, since the rules of debits and credits are merely an expansion of the algebraic relationships that follow from this equation. The following five rules may be used:

1. Increases in assets are debits; decreases are credits.
2. Increases in liabilities are credits; decreases are debits.
3. Increases in owners' equity are credits, decreases are debits.

4. Increases in expense are debits; decreases are credits.
5. Increases in revenue are credits; decreases are debits.

The normal accounting procedure is as follows:

1. Post all entries initially to the appropriate journal accounts.
2. Summarize these journal accounts by periodically posting to ledger accounts.
3. A *trial balance* is simply a list of the account names and the balances in each account as of a given time, with debit balances in one column and credit balances in another column. The preparation of the trial balance serves two principal purposes:
   a. It shows whether the equality of debits and credits has been maintained.
   b. It provides a convenient transcript of the ledger record as a basis for making adjusting and closing entries or in the preparation of financial statements.
4. As indicated above, the financial statements are normally prepared from the trial balance.

2.4 GLOBAL INDUSTRIES

Assume that you have decided to start your own small material-handling company called Global Industries. You have decided to enter the business on a small scale by initially manufacturing and marketing a small hand truck to which a plastic bag can be readily attached for collecting grass cuttings, garbage, and so on. You have done sufficient homework to believe that this market is substantial, and several community hardware stores have agreed to carry your product. You plan to finance this operation with $15,000 of your own money and by borrowing the balance required from the bank.

Your first task is to develop the necessary cash flow statements as a basis for discussions with the banker. Equipment requirements are shown in Table 2.1 and the cash flows are shown for the year in Table 2.2. The cash flows are expected to be identical during January through August. From September through December, the only major expenses will be the rent of $500 per month and your salary of $1,500 per month. Note that these cash flows do not include payments of principal or interest on your loan.

Table 2.3 can now be developed to determine the amount of cash needed each month (working-capital requirements) to carry on the day-to-day expenses. Table 2.3 indicates that in April the working-capital require-

TABLE 2.1 Equipment summary for Global Industries.

Type	Installed Cost
MIG welder	$ 5,000
Cutting torch and tanks	500
Jigs	1,000
Painting equipment	1,000
Grinders	500
Air compressor	500
Pipe-bending machine	1,000
Shear	2,000
Hand tools	500
Delivery truck	6,000
Office furniture	2,000
	$20,000

TABLE 2.2 Cash budget for Global Industries.

	Jan.–Apr.	May–Aug.	Sept.–Dec.
1. Material	$3,000	$ 3,000	
2. Direct labor	1,500	1,500	$1,500
3. Manufacturing overhead			
a. Rent $500			500
b. Heat, light, and			
insurance $200			
c. Maintenance $300	1,000	1,000	
4. Selling expense	700	700	
5. Administrative expense	300	300	
Total expenses/month	$6,500	$ 6,500	$2,000
Total income/month ($25/unit)	$ 0	$17,750	$ 0

TABLE 2.3 Estimated consolidated cash position for Global Industries during 1985.

Month	Previous Month's Balance	Current Month's Receipts	Operating Costs	Combined Month-End Balance
January	$ 0	$ 0	$ 6,500	$ (6,500)
February	(6,500)	0	6,500	(13,000)
March	(13,000)	0	6,500	(19,500)
April	(19,500)	0	6,500	(26,000)
May	(26,000)	17,750	6,500	(14,750)
June	(14,750)	17,750	6,500	(3,500)
July	(3,500)	17,750	6,500	7,750
August	7,750	17,750	6,500	19,000
September	19,000	0	2,000	17,000
October	17,000	0	2,000	15,000
November	15,000	0	2,000	13,000
December	13,000	0	2,000	11,000
		$71,000	$60,000	

ments will reach $26,000, probably much more than you expected. However, you are now in a position to negotiate a loan with your banker.

The banker agrees to loan you the additional $5,000 required for equipment, to be repaid at $200 per month, and will arrange for a working-capital drawing account of up to $30,000. The interest rate on these loans will be 1% per month. The cash flows associated with the bank loans during the first year are shown in Table 2.4.

TABLE 2.4 Repayment schedule to the bank for the $5,000 loan during 1985.

	Loan Principal Remaining Beginning of Period	Loan Payment on Principal	Interest on the Principal Remaining	Working Capital Req'd	Interest on the Working Capital	Month-End Payment
Jan.	$5,000	$ 200	$ 50	$ 6,500 + 250	$ 67.50	$ 317.50
Feb.	4,800	200	48	13,000 + 248	132.48	380.48
Mar.	4,600	200	46	19,500 + 246	197.46	443.46
Apr.	4,400	200	44	26,000 + 244	262.44	506.44
May	4,200	200	42	14,750 + 242	149.92	391.92
June	4,000	200	40	3,500 + 240	37.40	277.40
July	3,800	200	38	0		238.00
Aug.	3,600	200	36	0		236.00
Sept.	3,400	200	34	0		234.00
Oct.	3,200	200	32	0		232.00
Nov.	3,000	200	30	0		230.00
Dec.	2,800	200	28	0		228.00
		$2,400	$468		$847.20	$3,715.20

Total payments:

	Principal	$2,400.00
	Interest	1,315.20
		$3,715.20

Note: For estimating purposes only. Detailed calculations would consider interest payments on the interest. The difference is not significant.

2.5 THE ACCOUNTING FUNCTION

The fundamental statement on which the accounting function is based is

$$\text{Assets} = \text{liabilities} + \text{owners' equity}.$$

The transactions that were undertaken during the year by Global Industries are listed below.

1. Supply Global Industries with personal capital	$15,000
2. Negotiate a bank loan for $5,000 (long-term assets)	5,000
3. Purchase equipment, furniture, etc.	20,000
4. In the first year of operation:	

a. Manufacturing costs equal:
    (1) Direct labor   18,000
    (2) Direct material   24,000
    (3) Manufacturing overhead (excluding consumption of capital)   10,000

                Total manufacturing cost (the manufacturing cost becomes $52,000
                a debit to inventory)

b. Selling expense $ 5,600

c. Administrative expense   2,400

d. Bank interest on the $5,000 loan   468

e. Bank interest on the working-capital loan   847

5. Recognize the partial consumption of capital (equipment) with the passing of time (debit to inventory)

6. Sell for cash the hand trucks produced (2,840 units) at $25 per unit

7. Pay accounts

8. Determine and pay income tax (25% of taxable income)

9. Pay bank loan

The accountant will first enter these transactions in the appropriate journal. The transactions are entered in the order in which they occur throughout the year.

		Debit	Credit
1.	Cash	$15,000	
	Equity		$15,000
2.	Cash	5,000	
	Bank loan		5,000
3.	Equipment	20,000	
	Cash		20,000
4. a.	Inventory	52,000	
	Accounts payable		52,000
b.	Selling expense	5,600	
	Accounts payable		5,600
c.	Administrative expense	2,400	
	Accounts payable		2,400
d.	Bank interest	1,315	
	Accounts payable		1,315
5.	Inventory	5,000	
	Equipment		5,000
6.	Cash	71,000	
	Inventory		57,000
	Selling expense		5,600
	Administrative expense		2,400
	Bank interest		1,315
	Equity		4,685
7.	Accounts payable	61,315	
	Cash		61,315
8.	Income tax	1,171	
	Cash		1,171
9.	Bank loan	2,400	
	Cash		2,400

The nine entries are posted into *T-accounts* or *ledger accounts*. For example:

Cash				Inventory			Equipment	
(1) $15,000	$20,000 (3)		(4) $52,000	$57,000 (6)		(3) $20,000	$ 5,000 (5)
(2) 5,000	61,315 (7)		(5) 5,000				
(6) 71,000	1,171 (8)						
	2,400 (9)						

Note: The financial statements to follow assume that no accounts are paid during the year. In practice, payments are made on a continuing basis throughout the year. The balance sheet represents a snapshot of the company's financial position at a given time. Therefore, the balance sheet to follow assumes that all payments occurred after December 31.

The *balance sheet* and *statement of income* can now be developed.

Global Industries
Balance Sheet
December 31, 1985

Current Assets		*Current Liabilities*	
Cash	$71,000	Accounts payable	$61,315
Inventory	0	Estimated tax liability	1,171
		Bank loan due	2,400
Total current assets	$71,000	Total current liabilities	$64,886
Fixed Assets		*Long-Term Liabilities*	
Equipment	$20,000	Bank loan payable	$ 2,600
Less accumulated			
depreciation	5,000		
Total fixed assets	$15,000	Total long-term liability	$ 2,600
		Total liabilities	$67,486
		Stockholders' Equity	
		Common stock	$15,000
		Retained earnings	3,514
		Total liabilities	
Total assets	$86,000	+ shareholders' equity	$86,000

Global Industries
Statement of Income
for Year Ending December 31, 1985

Sales (revenues)			$71,000
Cost of sales			
Material		$24,000	
Direct labor		18,000	
Manufacturing overhead			
Rent	$6,000		
Heat, light, insurance	4,000	10,000	
Provision for depreciation		5,000	57,000
Gross margin			$14,000
Expenses			
Selling expense	$5,600		
Administrative expense	2,400		
Bank interest	1,315		
			$ 9,315
Net income before provision for taxes			4,685
Provision for income tax			1,171
(25% of net income before taxes)			
Net income			$ 3,514

One additional statement of importance is the *funds flow statement.* Its format is as follows:

Global Industries
Funds Flow Statement
for Year Ending December 31, 1985

Source of funds	
Operating revenues − operating costs − income tax =	
after-tax cash flow ($71,000 − $60,000 − $1,171)	$ 9,829
Less: Interest on debt	1,315
Less: Dividends on equity capital	
Plus: Proceeds from disposal of equipment	
Plus: New debt or equity capital	20,000
	$28,514
Application of funds	
Increase in working capital during the period	$ 6,114
Repayment of debt or equity capital	$ 2,400
Capital expenditures for new plant and equipment	$20,000
	$28,514

In addition to these statements, a *schedule of the permitted depreciation allowance* is usually included with the financial statements. The declining-balance method was used to arrive at the values stated.

Global Industries
Schedule of Depreciation Allowance
for Year Ending December 31, 1985

Item	Account	Book Value	Depr. Rate %	Annual Depr.	Accrued Depr.
Factory machines	1	$11,500	20	$2,300	$2,300
Vehicles	2	6,000	30	1,800	1,800
Tools	3	500	100	500	500
Furniture	4	2,000	20	400	400
					$5,000

2.6 WORKING-CAPITAL MANAGEMENT

It can be seen from the Global Industries example that although the overall plan forecasts a surplus of money, there are times during the year when the actual cash position is negative. This is because many of the costs of production, including the payment of workers and suppliers, precede the sale and collection of money. The money necessary to bridge this time gap is called *working capital*.

Ensuring that an adequate level of working capital is maintained within the organization is one of the prime functions of top management. The major portion of a financial manager's time is devoted to the day-to-day internal operations of the firm, which may be classified as working-capital management. A shortage of working capital usually results from:

1. Failure to calculate and account for this capital requirement initially
2. Failure to collect receivables, which is one of the largest single causes of bankruptcy

No organization wants to carry too much working capital because this means that the money is not available for investment elsewhere. On the other hand, a shortage of working capital leads to an inability to pay bills and a poor credit rating. Once this situation occurs, bankruptcy usually results.

2.6.1 Determining Working-Capital Requirements

The definition of working capital (WC) is

WC = current assets − current liabilities

    = (cash + accounts receivable, including sales tax + inventories
      + prepaid expenses + miscellaneous current assets)
     − (accounts payable + wages + sales tax payable
      + income tax liability + current debt liability + dividends
      + miscellaneous current liabilities)

This statement is true in a broad sense but requires some clarification to be of meaningful value to a manager.

The amount of working capital required within a company is directly dependent on the following criteria:

1. The size of the cash flows
2. The timing of these cash flows

The working-capital requirements within a company consist of two types. These may be described as:

1. Long-term working capital
2. Current working capital

Long-term working capital represents the working capital that should be supplied through long-term financing. *Current working capital* represents that portion of the total working capital that can be financed by a line of credit with the bank and is necessary to meet the month-to-month fluctuations in the working-capital requirements. This line of credit in turn depends on such factors as management credit ratings and previous business experience.

To assess fully the amount of working capital required as a long-term investment and to meet fluctuations in current operating expenses, cash flow budgets should be generated. Cash flow budgets are also excellent management control measures that can be used to compare actual cash flows to budgeted cash flows. Undesirable increases in such factors as accounts receivable, inventories, and costs of goods sold can be readily detected, initiating studies to determine the cause of these discrepancies.

Table 2.5 presents the first 6 months of a cash flow budget to establish working-capital requirements for the Flexigrow Manufacturing Company. The maximum working capital required during the period January–June was $1,280,000.

A cash flow budget of this nature should be developed annually by management to determine the amount of working capital required. This capital requirement is represented by the average monthly cumulative net cash flow balance after inventories have stabilized (Inventories have stabilized at approximately $500,000). As indicated in Table 2.5, the average level of working capital is approximately $1,300,000. This value represents the average long-term working-capital requirement for the level of sales, inventories, and expenditures indicated. Fluctuations in accounts receivable, inventory levels, and cost of sales can conceivably account for the working-capital requirements in any one month reaching $1,500,000. On the basis of budgeted cash flows, management is now in a position to go to the bank and establish a working line of credit and thereby establish the initial capital requirement in working capital.

In the past, banks have generally not been in favor of financing any portion of long-term working capital. However, this policy appears to be changing. For management with a good payment record, the bank may set the upper limit on working capital at two-thirds of average outstanding

TABLE 2.5 Projected cash budget to estimate working capital requirements for the Flexigrow Manufacturing Company (thousands of dollars).

	Jan.	Feb.	Mar.	Apr.	May	June
Net sales	1,000	1,500	1,300	1,150	1,850	1,540
Cash balance beginning of period	350	210	0	0	0	0
Cash Receipts						
Accounts receivable collected	1,175	1,140	1,270	1,285	1,335	1,499
Miscellaneous receipts	25	30	40	20	25	25
Total receipts	1,200	1,170	1,310	1,305	1,360	1,524
Total cash available	1,550	1,380	1,310	1,305	1,360	1,524
Disbursements						
Operating expenses	400	800	400	500	650	625
Material purchases	350	425	425	320	450	200
Administrative costs	100	120	110	110	120	110
Selling costs	120	180	150	145	175	160
Fringe benefits	150	190	160	155	185	165
Taxes	60	50	85	65	60	100
Equipment purchases	50	150	200	45	100	70
Dividends	110			110		
Total disbursements	1,340	1,915	1,530	1,450	1,740	1,430
Cash balance end of month before bank loans	210	(535)	(220)	(145)	(380)	94
Bank loans to cover working capital		535	220	145	380	
Cash balance end of month	210	0	0	0	0	94
Cumulative working-capital requirement	0	535	755	900	1,280	

accounts receivable plus 50 to 75% of the average inventory balance. This policy will normally cover the current working capital plus part of the long-term working capital. However, many new companies in which the managers have no previous experience in operating their own business have been fortunate to establish a sufficient working line of credit to handle fluctuations above average in monthly payables. For the Flexigrow Manufacturing Company the bank line of credit could vary from zero to approximately $1,300,000 (two-thirds of average accounts receivable + 75% of inventory), depending on the company's previous record.

The development of an annual cash flow budget such as that outlined in Table 2.5 is invaluable to management whether for the analysis of a new company or an ongoing concern. To generate a cash flow budget of this nature, revenues and costs must be analyzed carefully and completely. This analysis results in a systematic plan for the year's operations that can be used by management as an excellent control measure requiring a minimum of effort.

2.6.2 An Equitable Balance between Long-Term and Current Working-Capital Requirements

In developing any working-capital policy, the question of flexibility, cost, and risk involved must be considered. (*Note:* Any working-capital policy adopted must be monitored continuously.) If a firm finances long-term assets with permanent capital and short-term assets with short-term capital, its financial risk is lower than if long-term assets are financed with short-term debt or short-term assets are financed with long-term debt. Figure 2.2 represents the situation where a firm finances all its fixed assets with long-term capital but part of its permanent current assets with short-term financing. The lower the dashed line is positioned, the more aggressive is the policy adopted by the firm. That is, the firm is more vulnerable to loan-renewal problems and fluctuations in interest rates.

Alternatively, as in Figure 2.3, the dashed line could be drawn above the line, designating permanent current assets, indicating that permanent capital is being used to meet seasonal demands. In this case the firm uses a small amount of short-term credit to meet its peak seasonal requirements, but it also meets a part of its seasonal needs by "storing liquidity" in the form of marketable securities during the off-season. The humps above the dashed line represent short-term financing; the troughs below the dashed line represent short-term security holdings.

Short-term financing certainly increases the financing flexibility. If the need for funds is seasonal or cyclical, the firm may not want to commit itself to long-term debt. Such debt can be refunded, provided that the loan agreement includes a call or prepayment provision, but even so, prepayment penalties can be expensive. Accordingly, if a firm expects its needs for funds to diminish in the near future, or if it thinks there is a good chance that such a reduction will occur, it may choose short-term debt for the

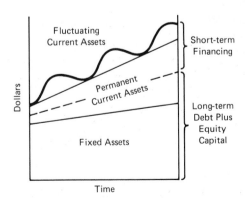

FIGURE 2.2 Capital structure.

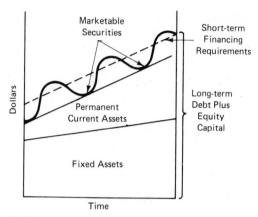

FIGURE 2.3 Capital structure.

flexibility it provides. As indicated earlier, the cash flow budget may be used to analyze the flexibility aspect of projected cash flows.

Interest rates are frequently lower on short-term debt than on long-term debt. This factor can result in significant savings to the firm over a period of time, but subjects the firm to more risk than does long-term financing. The risk effect occurs for two reasons:

1. If a firm borrows on a long-term basis, its interest costs will be relatively stable over time, but if it borrows on a short-term basis, its interest costs will fluctuate widely, at times going excessively high. For example, government Treasury bills may rise and drop as much as 3 to 4 points over a 90-day period.
2. If a firm borrows heavily on a short-term basis, it may find itself unable to repay this debt, or it may be in such a shaky financial position that the lender will not extend the loan: thus the firm could be forced into bankruptcy.

2.6.3 Factors Affecting Working-Capital Requirements

Accounts Receivable

Let us assume that the turnaround of the accounts receivable for a company with average monthly sales of $1,000,000 can be changed as follows:

	Present	Proposed Change
Paid in cash	10%	20%
Paid within 1 month	40	50
Paid within 2 months	30	30
Paid within 3 months	20	
	100%	100%

The present average turnaround period equals:

$$(0.1)(0) + (0.4)(1) + (0.3)(2) + (0.2)(3) = 1.6 \text{ months}$$

The proposed averaged turnaround period equals:

$$(0.2)(0) + (0.5)(1) + (0.3)(2) = 1.1 \text{ months}$$

This reduction in the turnaround period reduced accounts receivable and thereby the working capital required by

$$(\text{Average sales})(0.5) = \$1,000,000(0.5) = \$500,000$$

which is a significant saving if this reduction in the turnaround period can be accomplished with little or no increase in costs.

Another major advantage in reducing the turnaround on accounts receivable is portrayed in Figure 2.4. The older an overdue account, the less chance there is of collecting that account. Figure 2.4 is derived from government statistics and discussions with financial managers.

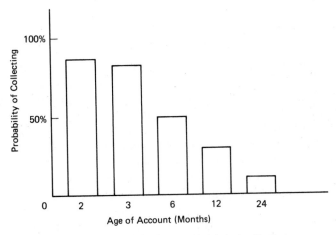

FIGURE 2.4 Age of account versus probability of collecting.

Accounts Payable

The effect of the structure of the accounts payable is just the opposite of the effect of the structure of accounts receivable. It is good business to take advantage of the full turnaround available on accounts payable; however do not run the risk of losing a valuable supplier. One must also consider the value in taking advantage of discounts available.

Inflation

An inflationary economy usually has an adverse effect on working-capital requirements. Working-capital requirements usually increase in an inflationary economy, for the following reasons:

1. Accounts receivable are often not fully responsive to inflationary trends in labor and material costs. At the very minimum there will normally be a lag time between increased wages and material costs, and receivables.
2. Individual company inventories will tend to fluctuate to a greater degree. The shortage of supply of raw materials may force a company to order larger quantities and to accept delivery of materials earlier than desirable to ensure production requirements.
3. Price breaks on quantity buying are often discontinued when demand exceeds supply.
4. Raw-material suppliers tend to shorten their turnaround time on accounts receivable when demand exceeds supply.
5. Inflation usually results in a tight-money situation, which may result in no increase in the line of credit at the bank to offset increased cash flows due to inflation.

Inventories

In almost any organization inventories represent a major percentage of the dollars associated with working capital. The development of a fully integrated inventory system is essential to maintain an acceptable level of working capital. Perhaps the most critical parameter is inventory turnover. Does someone in your organization know what the inventory turnover is for most products that you handle? The return on investment will vary significantly for different products within an individual company. Perhaps a study of the existing system regarding such factors as handling costs and losses may indicate a very beneficial course of action.

A company that is subject to seasonal influences may have very high inventory requirements for a major portion of the year. These high inventories will contribute directly to a high-working-capital requirement. In extreme cases it may be necessary to accumulate inventories for 11 months and receive profits for only 1 month. In a case such as this, working-capital requirements should be calculated for a point in time when inventories are at a peak, provided that the company's line of credit at the bank is not flexible enough to meet these fluctuations in working capital required.

The Flexigrow Manufacturing Company has a total working-capital requirement of approximately $1,300,000, which represents 33% of the total capital investment of $4,000,000. This high percentage of working capital to total investment is not unusual and does signify the importance of each individual organization generating an annual cash flow budget to project, among other valuable information, the working-capital requirements for comparison with the actual cash flows that occur during the year. The organization is then in a position to assess the optimum levels with respect to such factors as turnaround on accounts receivable, inventories, and cost of goods sold.

Exercising continuous control over all revenues and expenditures within the organization through a planned cash flow budget can do a lot to help ensure a continued profitable organization.

2.7 FINANCIAL RATIOS

Careful and continuous planning is essential to a manager's success. Effective financial management of the firm requires that the manager be aware of the strengths and weaknesses within the organization. The manager can then capitalize on these strengths and take corrective action in areas of concern. Information regarding such key areas as follow is essential:

1. Are inventories too low or too high with respect to the sales demand?
2. Does the firm have an excessive investment in accounts receivable, and can this be corrected by a more aggressive collection policy?
3. Is the firm's capital structure such that it will be able to raise additional external capital when the need arises?

These and many other questions regarding the financial stability of the organization require answers on a continuing basis. *Ratio analysis* represents the most commonly used tool to analyze financial statements.

Ratios are very helpful, but they must be used with a certain amount of caution. Ratios represent a diagnostic tool that can be used as a basis on which to form a judgment. They represent a management-by-exception tool that signals that a problem may be developing in a specific area. Action can then be taken to pinpoint the problem, if indeed one does exist, and corrective action initiated.

Ratios may be classified into four fundamental types:

1. *Liquidity ratios:* measure the firm's ability to meet its maturing short-term obligations
2. *Leverage ratios:* measure the extent to which the firm has been financed by debt
3. *Activity ratios:* measure how effectively the firm is utilizing its resources

4. *Profitability ratios:* measure management's overall effectiveness as shown by the profits generated from sales and investment

The data to calculate ratios come from the financial statements. Therefore, to present and study examples of ratios, comparative balance sheets and condensed income statements are presented for Buildwell Industries. Buildwell manufactures and markets material-handling equipment.

Buildwell Industries
Comparative Balance Sheets
for Years Ending December 31
(Thousands of Dollars)

	1980	1981	1982	1983	1984	1985
Assets						
Cash	680	560	400	460	250	330
Accounts receivable	2,200	2,600	2,800	2,800	3,000	3,100
Inventories	4,500	4,550	5,660	6,000	6,100	5,600
Prepaid expenses	140	150	180	140	100	80
Land and buildings	1,050	1,500	1,480	1,440	1,400	1,360
Plant and equipment	920	1,400	1,350	1,290	1,240	1,190
Total assets	9,490	10,760	11,870	12,130	12,090	11,660
Liabilities						
Accounts payable	1,200	1,380	1,430	1,520	1,750	1,780
Accrued expenses	270	320	350	330	420	425
Notes payable (bank)	—	—	—	200	300	300
Taxes payable	740	830	860	750	300	—
Dividends payable	250	250	250	125	125	125
Long-term debt	—	300	290	275	260	245
Common stock	5,000	5,000	5,000	5,000	5,000	5,000
Retained earnings	2,030	2,680	3,690	3,930	3,935	3,785
Total liabilities and equity	9,490	10,760	11,870	12,130	12,090	11,660

Buildwell Industries
Condensed Income Statements
for Year Ending December 31
(Thousands of Dollars)

	1980	1981	1982	1983	1984	1985
Sales	20,000	23,000	24,000	23,800	25,500	26,500
Discounts and allowances	570	540	575	760	860	1,190
Net sales	19,430	22,460	23,425	23,040	24,640	25,310
Cost of goods sold	14,330	16,650	17,550	17,890	18,760	20,070
Gross profit on sales	5,100	5,810	5,875	5,150	5,880	5,240
Operating expenses	3,550	4,040	4,240	4,530	5,300	5,380
Profit before taxes	1,550	1,770	1,635	620	580	(140)
Income taxes (at 48%) (refund)	744	850	785	298	278	(67)
Net profit (loss)	806	920	850	322	302	(73)

Financial data are presented for a 6-year period. Two years are sufficient to show how the ratios are calculated, but several years' data are necessary to detect and analyze trends in a meaningful manner.

2.7.1 Liquidity Ratios

Liquidity ratios represent one of the first concerns of the financial manager. These ratios measure the ability of the firm to meet its short-term obligations. The two most commonly used liquidity ratios are presented here.

Current Ratio

The *current ratio* is calculated by dividing current liabilities into current assets. Current assets normally include cash, marketable securities, accounts receivable, and inventories. Current liabilities consist of accounts payable, short-term notes payable, current maturities of long-term debt, accrued income taxes, dividends once declared, and other accrued expenses (principally wages). The current ratio is the most commonly used measure of short-term solvency. It indicates the extent to which the claims of short-term creditors are covered by assets that may be converted to cash in a period roughly corresponding to the maturity of the claims.

The calculation of the current ratio for the Buildwell company in 1980
is

$$\text{Current ratio} = \frac{\text{current assets}}{\text{current liabilities}}$$

$$= \frac{\$7,520,000}{\$2,460,000} = 3.1$$

The industry average, derived from published reports, is close to 2.0.

The current ratio is significantly higher than the industry average. Buildwell Industries could liquidate its current assets at 30% of book value and still pay off current creditors in full. However, the fact that the ratio is well above the industry average may be an indication that accounts receivable and inventory levels are too high.

The industry average is not a magic number that all firms should strive to maintain. However, if values vary significantly from this average, it is a signal to the manager that a check should be made.

Comparison of ratios internal to the firm from period to period are usually more meaningful than a comparison with industry averages. An individual firm's ratios may vary significantly from the industry average and still present no reason for concern. These differences may result from such factors as:

1. Rental versus ownership of assets
2. Inventory valuation methods
3. Marketing strategies (e.g., marketing in a local area versus marketing over a large geographic area with a low margin)
4. Methods of financing, both long and short term
5. The state of maturity of the organization

A comparison of ratios, both internal to the firm and with the industry average, offers the most meaningful strategy to produce positive results.

Quick Ratio or Acid Test Ratio

The *quick ratio* is calculated by deducting inventories from current assets and dividing the difference by current liabilities. Inventories are typically the least liquid of a firm's current assets and the assets on which losses are most likely to occur in the event of liquidation. Therefore, this measure of the firm's ability to pay off short-term obligations without relying on the sale of inventories is important. The quick ratio for the Buildwell company in 1980 is

$$\text{Quick, or acid test, ratio} = \frac{\text{current assets} - \text{inventory}}{\text{current liabilities}}$$

$$= \frac{\$3,020,000}{\$2,460,000} = 1.2$$

The industry average is close to 1.0. Thus Buildwell Industries would appear to be well secured in the short term with respect to meeting creditors' demands.

2.7.2 Leverage Ratios

Leverage ratios provide measures of the contribution of the owners (equity capital) versus the financing provided by the creditors (bonds, bank loans, etc.). Firms with a low leverage ratio (low debt ratio) have a lower risk of losses when the economy is in a recession, but they also have lower expected returns when the economy is buoyant. In other words, high leverage runs the risk of high profits with a chance of large losses. The financial manager must maintain a desirable balance between debt and equity funds. Two of the more common leverage ratios are discussed next.

Debt-to-Total Assets Ratio
The debt-to-total assets ratio, usually referred to as the *debt ratio,* measures the percentage of total funds that have been supplied by creditors. Debt includes current liabilities and all bonds. This value is calculated for Buildwell Industries in 1980.

$$\text{Debt ratio} = \frac{\text{total debt}}{\text{total assets}}$$

$$= \frac{\$2,210,000}{\$9,490,000} = 23\%$$

The industry average for manufacturing in general is about 25%. Since Buildwell Industries ratio is close to the industry average, the company should be cautious regarding the borrowing of additional funds. However, there may be no serious problem created by increasing the debt ratio. It is important to remember that no single ratio should be considered in isolation. In addition, several years' ratios are essential to give a useful picture of the company's financial stability.

Times-Interest-Earned Ratio
The *times-interest-earned ratio* measures the extent to which earnings can decline without resulting in financial embarrassment to the firm because

of an inability to meet annual interest costs. Failure to meet this obligation can bring legal action by the creditors, possibly resulting in bankruptcy. Interest expense is deducted to arrive at taxable income; therefore, the ability to pay current interest is not affected by income taxes.

This ratio is determined by dividing earnings before interest and taxes by the interest charges. For the year 1981 this ratio is

$$\text{Times interest earned} = \frac{\text{profit before taxes} + \text{interest charges}}{\text{interest charges}}$$

$$= \frac{\$1,770,000 + \$30,000}{\$30,000^*} = 60$$

The industry average is approximately 10. This ratio indicates that Buildwell Industries has a good margin of safety in meeting interest charges.

2.7.3 Activity Ratios

Activity ratios measure how effectively the firm employs the resources at its command. These ratios involve comparisons between the level of sales and the investment in various asset accounts. The activity ratios presume that a "proper balance" should exist between sales and the various asset accounts, such as inventories, accounts receivable, and fixed assets. Four of these ratios are discussed below for Buildwell Industries.

Inventory Turnover Ratio

The *inventory turnover ratio* will provide not only a useful indication of the liquidity of inventories, and thereby a measure of short-term debt paying ability, but also some insights into managerial efficiency. This ratio is calculated by dividing cost of goods sold by average inventory. Typically, the denominator is estimated by taking the average of the beginning and ending inventories for the period.

This ratio is calculated for Buildwell Industries for 1981.

$$\text{Inventory turnover} = \frac{\text{cost of goods sold}}{\text{inventory}}$$

$$= \frac{\$16,650,000}{\$4,525,000} = 3.7$$

An inventory turnover of 3.7 times annually is equivalent to inventory turning over every 99 days.

* $300,000 in long-term debt at 10%.

This value should be used with caution. If an unusually busy or slow period occurs just before or after the year end, some distortion is inevitable. However, if the manager has been examining the firm over a period of several years, a declining ratio may indicate overstocking and the buildup of obsolete stock, whereas an increasing ratio may indicate an unusually high inventory turnover, resulting in too low a level of inventories and frequent costly stock outs.

Average Collection Period Ratio

The *average collection period ratio* indicates the number of days of credit sales outstanding and uncollected. The credit sales value is often not available and the total sales figure is used. This ratio is calculated by dividing receivables times days in the year by annual credit sales. For Buildwell Industries this ratio is calculated for 1980.

$$\text{Average collection period ratio} = \frac{(\text{receivables})365}{\text{net sales}}$$

$$= \frac{(\$2,200,000)365}{\$19,430,000} = 41 \text{ days}$$

The industry average is on the order of 60 days.

A ratio that is high relative to the industry standards, or that has increased significantly from previous ratios for the company, may suggest a lax collection problem and could result in an increase in bad debts. A ratio that is low relative to industry standards may be the result of excessively restrictive credit policies, which have the effect of decreasing sales.

Fixed-Asset Turnover Ratio

The *fixed-asset turnover ratio* measures the turnover of book value of fixed assets, and is a ratio of sales to fixed assets. For 1980 for Buildwell this ratio is

$$\text{Fixed-asset turnover} = \frac{\text{net sales}}{\text{fixed assets}}$$

$$= \frac{\$19,430,000}{\$1,970,000} = 9.9$$

The industry average is on the order of 5 to 6. A ratio of 9.9 indicates that Buildwell Industries is probably making good use of its fixed assets. It may also be an indication that additional capital investment is in order. In any event, it is a signal that indicates further investigation.

Total Asset Turnover Ratio

The *total asset turnover ratio* measures the turnover of all the firm's assets. It is calculated by dividing sales by total assets. This ratio for Buildwell Industries in 1980 is

$$\text{Total asset turnover} = \frac{\text{net sales}}{\text{total assets}}$$

$$= \frac{\$19,430,000}{\$9,490,000} = 2.0$$

The industry average is close to 2.0. Buildwell industries average of 2.0 indicates that the company is making good use of its assets.

2.7.4 Profitability Ratios

The profitability of a company is due to the result of many policies and decisions generated throughout the total organization. *Profitability ratios* represent the final proof of how well management is managing the total resources of the firm. Three of these ratios will now be discussed.

Profit Margin on Sales Ratio

The *profit margin on sales ratio* is calculated by dividing net profit after taxes by sales. This ratio gives the profit per dollar of sales. This ratio for Buildwell Industries in 1980 was

$$\text{Profit margin} = \frac{\text{net profit after taxes}}{\text{net sales}}$$

$$= \frac{\$806,000}{\$19,430,000} = 4.1\%$$

The industry average is close to 2%. The profit margin of 4.1% for Buildwell Industries indicates that the firm's prices may be relatively high and/or costs are relatively low.

Return on Total Assets Ratio

The *ratio of net profit to total assets* measures the return on total investment (debt + equity) in the firm. For 1980 this value is

$$\text{Return on total assets} = \frac{\text{net profit after taxes}}{\text{total assets}}$$

$$= \frac{\$806,000}{\$9,490,000} = 8.5\%.$$

The industry average is close to 5.0%. Buildwell Industries average is significantly higher than the industry average. This results from the high profit margin on sales and the high turnover of total assets.

Return on Net Worth Ratio

The *ratio of net income after taxes to net worth* measures the rate of return on the shareholders' (equity) investment. This ratio for Buildwell Industries in 1980 is

$$\text{Return on net worth} = \frac{\text{net income after taxes}}{\text{net worth}}$$

$$= \frac{\$806,000}{\$7,030,000} = 11.5\%$$

The industry average is approximately 9%. Buildwell Industries average of 11.5% indicates that the firm is performing very well relative to the industry.

Trend Analysis

Ratios for any given year overlook any trends that may be developing within the organization. Table 2.6 shows the 6-year trend for Buildwell Industries.

The trend analysis indicates that Buildwell Industries performance was reasonably good until 1982. In 1982 the profitability ratios and the times interest earned ratio showed signs of deterioration, indicating that the company may be headed for financial problems. In 1983 the company profits dropped very significantly. If management had taken corrective action in 1982, serious problems probably would have been avoided.

Ratios represent a very useful management-by-exception tool which assists the manager in tracking the financial stability of the company. Attaching a ratio analysis summary to the financial statements for circulation to management should be of significant value in monitoring and maintaining a healthy company.

2.8 SUMMARY

The prime function of management within the corporation is to maintain a level of economic performance that is acceptable to all individuals and/or groups associated with the corporation. A clear picture of the manner in which funds flow into and out of a corporation is essential to everyone interested in fully understanding the basic components that enter into maintaining economic health.

A sound economic base can be maintained only if proper planning

TABLE 2.6 Ratio trend analysis for Buildwell Industries, 1980–1985.

Ratio	1980	1981	1982	1983	1984	1985
Liquidity Ratios						
1. Current ratio $= \dfrac{\text{current assets}}{\text{current liabilities}}$	$\dfrac{7{,}520}{2{,}460} = 3.1$	$\dfrac{7{,}860}{2{,}780} = 2.8$	$\dfrac{9{,}040}{2{,}890} = 3.1$	$\dfrac{9{,}400}{2{,}925} = 3.2$	$\dfrac{9{,}450}{2{,}895} = 3.3$	$\dfrac{9{,}110}{2{,}630} = 3.5$
2. Quick ratio $= \dfrac{\text{current assets} - \text{inventory}}{\text{current liabilities}}$	$\dfrac{3{,}020}{2{,}460} = 1.2$	$\dfrac{3{,}310}{2{,}780} = 1.2$	$\dfrac{3{,}380}{2{,}890} = 1.2$	$\dfrac{3{,}400}{2{,}925} = 1.2$	$\dfrac{3{,}350}{2{,}895} = 1.2$	$\dfrac{3{,}510}{2{,}630} = 1.3$
Leverage Ratios						
3. Debt ratio $= \dfrac{\text{total debt}}{\text{total assets}}$	$\dfrac{2{,}210}{9{,}490} = 23\%$	$\dfrac{2{,}830}{10{,}760} = 26\%$	$\dfrac{2{,}930}{11{,}870} = 25\%$	$\dfrac{3{,}075}{12{,}130} = 25\%$	$\dfrac{3{,}030}{12{,}090} = 25\%$	$\dfrac{2{,}750}{11{,}660} = 24\%$
4. Times interest earned $= \dfrac{\text{profit before taxes} + \text{interest charges}}{\text{interest charges}}$	—	$\dfrac{1{,}800}{30} = 60$	$\dfrac{1{,}664}{29} = 57.4$	$\dfrac{647.5}{27.5} = 23.5$	$\dfrac{606}{26} = 23.3$	$\dfrac{(164.5)}{24.5} = (6.7)$
Activity Ratios						
5. Inventory turnover $= \dfrac{\text{cost of goods sold}}{\text{average inventory}}$	—	$\dfrac{16{,}650}{4{,}525} = 3.7$	$\dfrac{17{,}550}{5{,}105} = 3.4$	$\dfrac{17{,}890}{5{,}830} = 3.1$	$\dfrac{18{,}760}{6{,}050} = 3.1$	$\dfrac{20{,}070}{5{,}850} = 3.4$
6. Average collection period $= \dfrac{\text{receivables}(365)}{\text{net sales}}$	$\dfrac{2{,}200(365)}{19{,}430} = 41$	$\dfrac{2{,}600(365)}{22{,}460} = 42$	$\dfrac{2{,}800(365)}{23{,}425} = 44$	$\dfrac{2{,}800(365)}{23{,}040} = 44$	$\dfrac{3{,}000(365)}{24{,}640} = 44$	$\dfrac{3{,}100(365)}{25{,}310} = 45$
7. Fixed asset turnover $= \dfrac{\text{net sales}}{\text{fixed assets}}$	$\dfrac{19{,}430}{1{,}970} = 9.9$	$\dfrac{22{,}460}{2{,}900} = 7.7$	$\dfrac{23{,}425}{2{,}830} = 8.3$	$\dfrac{23{,}040}{2{,}730} = 8.4$	$\dfrac{24{,}640}{2{,}640} = 9.3$	$\dfrac{25{,}310}{2{,}550} = 9.9$
8. Total asset turnover $= \dfrac{\text{net sales}}{\text{total assets}}$	$\dfrac{19{,}430}{9{,}490} = 2.0$	$\dfrac{22{,}460}{10{,}760} = 2.1$	$\dfrac{23{,}425}{11{,}870} = 2.0$	$\dfrac{23{,}040}{12{,}130} = 1.9$	$\dfrac{24{,}640}{12{,}090} = 2.0$	$\dfrac{25{,}310}{11{,}660} = 2.2$
Profitability Ratios (%)						
9. Profit margin $= \dfrac{\text{net profit after taxes}}{\text{sales}}$	$\dfrac{806}{19{,}430} = 4.1\%$	$\dfrac{920}{22{,}460} = 4.1$	$\dfrac{850}{23{,}425} = 3.6$	$\dfrac{322}{23{,}040} = 1.4$	$\dfrac{302}{24{,}640} = 1.2$	$\dfrac{(73)}{25{,}310} = (0.3)$
10. Return on total assets $= \dfrac{\text{net profit after taxes}}{\text{total assets}}$	$\dfrac{806}{9{,}490} = 8.5\%$	$\dfrac{920}{10{,}760} = 8.5$	$\dfrac{850}{11{,}870} = 7.2$	$\dfrac{322}{12{,}130} = 2.7$	$\dfrac{302}{12{,}090} = 2.5$	$\dfrac{(73)}{11{,}660} = (0.6)$
11. Return on net worth $= \dfrac{\text{net profit after taxes}}{\text{net worth}}$	$\dfrac{806}{7{,}030} = 11.5\%$	$\dfrac{920}{7{,}680} = 12.0$	$\dfrac{850}{8{,}690} = 9.8$	$\dfrac{322}{8{,}930} = 3.6$	$\dfrac{302}{8{,}935} = 3.4$	$\dfrac{(73)}{8{,}785} = (0.8)$

Year

goes into the future activities of the company. The planning function requires that the future cash flows be predicted reasonably accurately. The accounting function is to record the actual cash flows that occur. The function of management is to ensure that the predicted cash flows are close to the actual cash flows.

PROBLEMS

2.1 a. What two major sources of external capital are available to the firm?
   b. Define working capital.
   c. What is the fundamental statement on which the accounting function is based?

2.2 Outline in a brief form the structure of (a) the balance sheet, (b) the income statement, and (c) the funds flow statement, and indicate the purpose of each.

2.3 Write the appropriate equation for each of the following questions.
   a. What is the relationship between the before-tax cash flow and taxable income?
   b. What is the relationship between after-tax cash flow and net income?
   c. What is the relationship between after-tax cash flow and dividends?
   d. What is the relationship between operating revenues and income tax?

2.4 Worldwide Industries has projected the following estimates for operations next year:

Sales (operating revenues)	$3,000,000
Cash operating costs	2,000,000
Depreciation expense	500,000

The effective tax rate is 42%.
Financial data include (1) 20,000 shares of common stock, and (2) 1,000 bonds: selling price per bond = $1,000 and the bonds pay 8% interest per annum (see Figure 2.1).
   a. What is the projected earnings per share?
   b. What are the net funds available for reinvestment if:
   (1) The company pays no dividends?
   (2) The company pays 25% of net income in dividends?

2.5 For the second year of operation (1986) Global Industries (Section 2.4) has projected cash flows based on the following data:

1. Purchased $10,000 in new factory equipment. Borrowed $5,000 from the bank at an interest rate of 10% per annum. Principal payments of $300 per month.
2. Manufactured 4,000 hand trucks (500 trucks per month for 8 months) and sell 1,000 trucks per month in the May–August period.
3. Manufacturing cost per unit (spread over 8 months):

Direct labor	$6.25
Direct material	8.50
Overhead	5.25

4. Selling expense (spread over 8 months) = $7,000.
5. Administrative expense (spread over 12 months) = $3,000.
6. Selling price per unit = $25.

a. Develop a table showing the estimated consolidated cash position for Global during 1986. Will an additional line of credit be required? (*Note:* Starting balance = net income + allocation for depreciation − loan principal payments from previous year.) (Working-capital loan at 1% per month)
b. Develop a table showing the repayment schedule to the bank.
c. Produce the following statements for 1986.
   (1) Balance sheet.
   (2) Statement of income.
   (3) Funds flow statement.
   Assume a 25% income tax rate and the same depreciation rate as that used in Section 2.5.

2.6 a. "To maximize profits, a company should minimize the ratio of current assets to current liabilities." Do you agree with this statement? Discuss.
b. What factors should be considered in the development of a working-capital policy?

2.7 Realizing that the control of accounts receivable is critical to the survival of a firm:
a. Why do companies find themselves in a severe cash flow shortage position? Discuss considering company maturity (age), bank financing, and so on.
b. Discuss how you would organize the control of accounts receivable to minimize cash flow shortages.

2.8 The Flexigrow Manufacturing Company maintains a current ratio (ratio of current assets to current liabilities) of 3.0 and current assets represent a constant 25% of sales. Flexigrow has $400,000 in long-term debt. Short-term loans are maintained at 60% of current liabilities. Sales forecasts for the next 2 years are as follows:

Year	Sales
1	$2,000,000
2	2,500,000

If the cost of short-term notes is 10%, how much interest must Flexigrow pay on its short-term notes over the next 2 years?

2.9 The structure of the accounts receivable within Quarterdeck Developments is:

1. 10% paid in cash
2. 30% paid within 1 month
3. 30% paid within 2 months
4. 30% paid within 3 months

Assume that the turnaround on accounts receivable can be reduced to the following:

1. 20% paid in cash
2. 40% paid in 1 month
3. 30% paid in 2 months
4. 10% paid in 3 months

If average monthly sales are $300,000 and the cost of money is 12%, what annual saving can Quarterdeck realize by the reduction in turnaround on accounts receivable?

2.10 Assume that cash flows for the month of July for the Flexigrow Manufacturing Company are as follows: net sales = $1,600,000, miscellaneous receipts = $25,000, and total disbursements = $1,550,000. (See Table 2.5.) Calculate the cumulative working-capital requirements in July. (Accounts receivable collected − 10% cash, 40% in 1 month, 30% in 2 months, 20% in 3 months.)

2.11 Review the ratios calculated for Buildwell Industries and discuss possible reasons for the deterioration of the firm's financial structure.

2.12 Financial information for the J-D Company is presented below.
   a. Calculate the 11 ratios shown in Table 2.6 for 1985. The interest rate on debt capital is 10% and the income tax rate is 40%.
   b. During 1985 the cash, receivables, and inventories increased by $270,000. What was the source of the greatest part of the increase in these assets?

J-D Company
Comparative Balance Sheets
for Years Ended December 31, 1984 and 1985

	Dec. 31, 1985	Dec. 31, 1984
Assets		
Cash	$ 140,000	$ 120,000
Receivables (net)	250,000	200,000
Inventories	1,200,000	1,000,000
Short-term investments	190,000	390,000
Prepaid items	50,000	75,000
Land	300,000	300,000
Building and equipment (net)	1,900,000	1,820,000
	$4,030,000	$3,905,000
Liabilities and Owners' Equity		
Accounts payable	$ 490,000	$ 405,000
Notes payable	200,000	160,000
Accrued liabilities	100,000	100,000
Bonds payable due 1990	800,000	800,000
Common stock	2,000,000	2,000,000
Retained earnings	440,000	440,000
	$4,030,000	$3,905,000

J-D Company
Comparative Income Statements
for Years Ended December 31, 1984 and 1985

	Dec. 31, 1985	Dec. 31, 1984
Sales	$4,000,000	$3,580,000
Less: Cost of goods sold	2,300,000	2,150,000
Gross profit	1,700,000	1,430,000
Less: Operating expenses	460,000	396,000
Gross operating income	1,240,000	1,034,000
Less: Depreciation and other expenses	900,000	800,000
Net income before taxes	340,000	234,000
Income taxes (at 40%)	136,000	93,600
Net income	$ 204,000	$ 140,400

2.13 Presented here are comparative condensed balance sheets for the Bobcat Company at the end of 1985 and 1984, as well as a condensed income statement for the year ended December 31, 1985. Compute 10 of the 11 ratios shown in Table 2.6 (omit the times-interest-earned ratio).

Bobcat Company
Comparative Balance Sheets
for Years Ended December 31, 1984 and 1985

	Dec. 31, 1985	Dec. 31, 1984
Assets		
Cash	$ 10,000	$ 12,000
Receivables (net)	26,000	20,000
Inventories	34,000	28,000
Unexpired insurance	1,500	1,800
Land	16,000	16,000
Plant and equipment (net)	40,500	43,200
	$128,000	$121,000
Liabilities and Owners' Equity		
Accounts payable	$ 10,000	$ 8,000
Notes payable	13,000	15,000
Taxes payable	3,400	2,600
Long-term debt payable	25,000	25,000
Common stock ($50 par value)	50,000	50,000
Retained earnings	26,600	20,400
	$128,000	$121,000

Bobcat Company
Income Statement
for Year Ending December 31, 1985

Sales	$152,500
Less: Cost of goods sold	108,500
Gross profit	44,000
Less: Operating expenses	23,233
Gross operating income	$ 20,767
Less: Depreciation and other expenses	12,500
Net income before taxes	$ 8,267
Income taxes (at 25%)	2,067
Net income	$ 6,200

CHAPTER THREE
Time Value of Money

When faced with the choice between a sum of money today or the same sum a period of time hence, say in 1 year, most people will choose to receive the money today. When asked to explain their choice, the reason they usually give is that the money can be invested and earn interest over the time period, or the money can be used to reduce a deficit or current obligation. There will be agreement that although the sums of money may be the same, the value of the money will be different at two different times. It follows, therefore, that a necessary aspect of economic analysis is the timing of cash flows. It is not only important to estimate how much will be paid or received; when it is received is also a major consideration.

In Chapter 2 we discussed the timing of accounts receivable and how if the collection cycle was reduced, the need for working capital was reduced. A simple calculation indicates that if a customer has an outstanding account of $5,000 for 1 year and the firm is paying the bank 1% per month for working capital, that $5,000 costs the firm $634.13 in interest over the year. The amount is substantial, for the effects of time on the value of money are substantial.

In evaluating alternative investments we are concerned with the economical impact of cash flows over time. In such problems we must recognize the time value of money, that is, the impact of interest on the timing of cash flows. The decision to invest is based on the prospect of future cash flows of sufficient size to repay the original investment and give the investor a reasonable profit. In this chapter we present the mathematical background for considering the time value of money. The derivations for the interest formulas are presented with examples and problems that will give the reader the initial background to do investment cash flow calculations.

3.1 INTEREST

Interest may be defined as the money paid for the use of borrowed money. The *rate of interest* paid is the ratio between the interest payable at the end of a period of time, usually a year or less, and the money borrowed at the beginning of the period. For example, assume that in order to buy an automobile you have borrowed $1,000 from your parents and promise to repay the $1,000 plus $100 interest at the end of 1 year. The interest rate is $100/\$1,000 = 10\%$ per annum. This is usually described as an interest rate of 10%, the "per annum" being understood unless some other period of time is specifically stated.

3.1.1 Simple Interest

In the example above, $1,000 represents the *prinicipal* and the $100 represents the *interest paid on the principal*. Transactions involving *simple interest* apply the interest rate only to the initial principal, regardless of the number of periods involved.

Let P represent the principal sum of money involved, i the interest rate, and n the number of time periods. The simple interest I is found by the equation.

$$I = Pni$$

In our example:

1. Assume that $n = 1$ period (year). Then

$$I = (1,000)(1)(10\%) = \$100$$

The total amount due at the end of the year equals

$$\$1,000 + \$100 = \$1,100$$

2. Assume that $n = 3$ periods (years). Then the interest payment equals

$$(\$1,000)(3)(10\%) = \$300$$

The total amount due at the end of 3 years equals

$$\$1,000 + \$300 = \$1,300$$

3.1.2 Compound Interest

Transactions involving *compound interest* apply the interest rate to the remaining balance invested during each period. Compound interest is used in most business transactions (e.g., bank loans, mortgages, leases, etc.) Using the example above we have:

Period	Amount Owing Beginning of Period	Interest Charge for Period	Amount Owing End of Period
1	$1,000.00	$100.00	$1,100.00
2	1,100.00	110.00	1,210.00
3	1,210.00	121.00	1,331.00

The critical factor to remember with respect to all calculations involving compound interest is that *the interest rate is applied to the amount owing (principal + interest) at the beginning of each period.*

Compound interest, being much more common in practice than simple interest, will be used throughout the remainder of this book.

3.2 ALTERNATIVE METHODS OF REPAYMENT

Assume that you are interested in buying a new stereo set; however, to do so you have to borrow $1,000 from your parents. They have agreed to loan you the money for a period of 5 years at a 10% interest rate and have left the decision regarding a repayment plan to you. In your analysis you have developed four alternative plans of repayment to consider:

Plan A: Repay the principal plus interest in a single lump sum at the end of the 5 years.

Plan B: Repay the principal and interest in five uniform end-of-year payments.

Plan C: Repay the principal in uniform amounts at the end of each of the next 5 years. In addition to the principal, pay the interest on the unpaid balance.

Plan D: Repay the total principal in one lump sum at the end of the 5 years. Make annual end-of-year interest payments on the unpaid balance.

The cash flows for each plan are presented in Table 3.1.

Although the verbal and tabular descriptions of the cash flows present

TABLE 3.1 Repayment schemes for a $1,000 loan at 10%.

Alternative	Year	Outstanding Balance at Beginning of Year	Interest Accrued during Year	End-of-Year Payment	Outstanding Balance at End of Year
Plan A	1	$1,000	$100	0	$1,100
	2	1,100	110	0	1,210
	3	1,210	121	0	1,331
	4	1,331	133.10	0 0	1,464.10
	5	1,464.10	146.41	1,610.51	0
Plan B	1	1,000	100	263.80	836.20
	2	836.20	83.62	263.80	656.02
	3	656.02	65.60	263.80	457.82
	4	457.82	45.78	263.80	239.80
	5	239.80	23.98	263.80	0
Plan C	1	1,000	100	300	800
	2	800	80	280	600
	3	600	60	260	400
	4	400	40	240	200
	5	200	20	220	0
Plan D	1	1,000	100	100	1,000
	2	1,000	100	100	1,000
	3	1,000	100	100	1,000
	4	1,000	100	100	1,000
	5	1,000	100	1,100	0

all the information, a more explicit form of expression is the cash flow diagram. The diagram has a horizontal time line progressing from left to right. The per period (usually years) labels apply to intervals of time rather than to points on the time scale. The arrows signify cash flows. Downward arrows represent disbursements and upward arrows represent receipts. That is, upward arrows are positive (+) cash flows and downward arrows are negative (−) cash flows.

Plan A: Repay the loan plus accrued interest at the end of the fifth year.

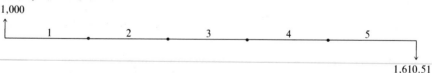

Plan B: Repay the loan plus interest in equal annual installments.

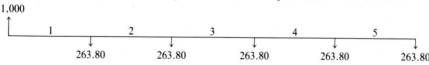

Plan C: Make annual payments of one-fifth of the principal plus accrued interest.

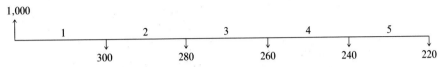

Plan D: Make annual interest payments and repay the principal at the end of year 5.

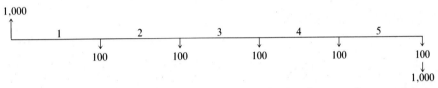

The only restrictions placed on you by your family regarding repayment is that you pay 10% interest on the remaining balance at any point in time. All plans meeting this criteria are equivalent to your parents even though the total repayment differs under each plan.

Plan	Total Payment
A	$1,610.51
B	1,319.00
C	1,300.00
D	1,500.00

However, to demonstrate equivalence, let us take the payments under plan C and deposit them to an account paying 10% interest.

Year	Interest on Balance during the Year	Deposit at the End of Year	Account Balance End of Year
1	$ 0.00	$300.00	$ 300.00
2	30.00	280.00	610.00
3	61.00	260.00	931.00
4	93.10	240.00	1,264.10
5	126.41	220.00	1,610.51

We note that we arrive at a sum identical to that of plan A. Similarly, equivalence can be demonstrated among all four plans. However, note that equivalence holds true only if the interest rate remains constant.

If you have other investment opportunities that offer a rate of return greater or less than 10%, the repayment plans are not equivalent. For example, if you cannot invest your funds above 10%, you (the borrower) will prefer plan C. Conversely, if you have many attractive investment opportunities that will earn 15%, you will prefer plan A.

In summary, the discussion above highlights at least two important points:

1. Cash flows can be moved from one point in time to any other using the same interest rate without destroying the concept of equivalence.
2. Cash flows at different times cannot be added or subtracted unless the interest rate is 0%. Therefore, to total a series of cash flows, they must be moved to a single point in time.

3.3 DERIVATION OF INTEREST FORMULAS

In this section we develop interest formulas to be used in cash flow calculations. These formulas are the basis for the discrete interest tables generated in Appendix F.

Symbols

i = interest rate per period

n = number of time periods

A = end-of-period sum in a uniform series flowing at the end of each of n periods

P = a present sum

F = a future sum flowing discretely at the end of the nth period

G = a gradient-sum, uniform period-by-period increase or decrease in cash flows

k = rate of change in a geometrically increasing or decreasing series

C = a sum discretely flowing at the end of the first period of a series of sums of C, $C(1 + k)^1$, $C(1 + k)^2$, . . ., $C(1 + k)^n$

Finding F When P Is Given

The relationship of P to F is developed as follows:

Period	Amount at Beginning of Period	+	Interest	=	Amount at End of Period
1	P	+	Pi	=	$P(1 + i)$
2	$P(1 + i)$	+	$Pi(1 + i)$	=	$P(1 + i)^2$
.					
.					
.					
n	$P(1 + i)^{n-1}$	+	$Pi(1 + i)^{n-1}$	=	$P(1 + i)^n$

The relationship between a present sum P and its equivalent future sum F is

$$F = P(1 + i)^n \qquad (3.1)$$

The quantity $(1 + i)^n$ is referred to as the *future worth of a present sum factor* and in functional notation is referred to as $(F/P, i\%, n)$. The formula for F is

$$F = P(F/P, i\%, n) \qquad (3.2)$$

Finding P When F Is Given

From equation 3.1 we can see that

$$P = F[1/(1 + i)^n] \qquad (3.3)$$

The quantity $1/(1 + i)^n$ is commonly referred to as the *present worth of a future sum factor* and in functional notation is referred to as $(P/F, i\%, n)$. The formula for P is

$$P = F(P/F, i\%, n) \qquad (3.4)$$

Example 3.1

Problem: If you deposit $1,000 today in an account that pays 8% interest each year over the next 5 years, how much money would be in the account at the end of the 5 years? Solution:

$$F = P(F/P, i\%, n) = P(F/P, 8\%, 5) = \$1,000(1.469) = \$1,469$$

The value of the $(F/P, 8\%, 5)$ factor can be calculated from $(1 + i)^n = (1.08)^5 = 1.469$ or can be read from the 8% table in Appendix F.

Problem: Assume that you would like to have $1,469 in an account 5 years from today. How much will you have to deposit today if the account pays 8% interest?

$$P = F(P/F, i\%, n) = F(P/F, 8\%, 5) = \$1,469(0.6806) = \$1,000$$

Finding F When A Is Given

If a cash flow of A dollars occurs at the end of each period for n periods at an interest rate of $i\%$ per period, the future worth, F, at the end of n periods is obtained by summing the future worths of each of the payments of amount A.

$$F = A[1 + (1 + i) + (1 + i)^2 + \cdots + (1 + i)^{n-2} + (1 + i)^{n-1}]$$

Multiply by $(1 + i)$:

$$F(1 + i) = A[(1 + i) + (1 + i)^2 + \cdots + (1 + i)^n]$$

Subtract the first line from the second:

$$Fi = A[(1 + i)^n - 1]$$
$$F = A[(1 + i)^n - 1]/i \qquad (3.5)$$

The quantity $[(1 + i)^n - 1]/i$ is referred to as the *future worth of a uniform series factor* and in functional notation is referred to as $(F/A, i\%, n)$. The equation for F is

$$F = A(F/A, i\%, n) \qquad (3.6)$$

Finding A When F Is Given

Inverting equation 3.5 gives

$$A = F\{i/[(1 + i)^n - 1]\} \qquad (3.7)$$

The quantity $i/[(1 + i)^n - 1]$ is referred to as the *uniform series amount of a future sum factor* or the *sinking-fund factor* and in functional notation is referred to as $(A/F, i\%, n)$. The equation for A is

$$A = F(A/F, i\%, n) \qquad (3.8)$$

Example 3.2

Problem: If we were to deposit $250.46 at the end of each year into an account that pays 8% interest at the end of each year for the next 5 years, how much money will be in the account at the end of the 5 years?

Solution

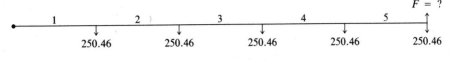

$$F = A(F/A, i\%, n) = A(F/A, 8\%, 5) = \$250.46(5.867) = \$1,469$$

Finding *P* When *A* Is Given

From Equation 3.1, $F = P(1 + i)^n$. Substituting for F in equation 3.5 results in

$$P(1 + i)^n = A[(1 + i)^n - 1]/i$$

Dividing both sides by $(1 + i)^n$ gives

$$P = A\{[(1 + i)^n - 1]/[i(1 + i)^n]\} \tag{3.9}$$

The quantity in braces is referred to as the *present worth of a uniform series factor* and in functional notation is referred to as $(P/A, i\%, n)$. The equation for P is

$$P = A(P/A, i\%, n) \tag{3.10}$$

Finding *A* When *P* Is Given

Inverting equation 3.9 gives

$$A = P\{[i(1 + i)^n]/[(1 + i)^n - 1]\} \tag{3.11}$$

The quantity in braces is referred to as the *annual worth (uniform series) amount of a present sum factor* or the *capital recovery factor* and in functional notation is referred to as $(A/P, i\%, n)$. The equation for A is

$$A = P(A/P, i\%, n) \tag{3.12}$$

Example 3.3

Problem: How much money would one have to deposit into an account today that pays 8% interest to be able to withdraw $250.46 at the end of each year for the next 5 years?

Solution

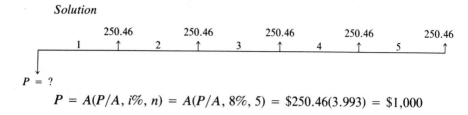

$$P = A(P/A, i\%, n) = A(P/A, 8\%, 5) = \$250.46(3.993) = \$1,000$$

3.3.1 Special Cases

Some special cases with respect to the interest formulas are worthy of note. These special cases are particularly useful in clarifying situations occurring at the limit of the applicable range of the factor values. Table 3.2 lists some extreme values for the limit as $n \rightarrow \infty$ and i is known and for $i \rightarrow 0$ and n known.

TABLE 3.2 Interest factor values for special cases.

Factor	*Limit as* $n \rightarrow \infty$ (i *is known*)	*Limit as* $i \rightarrow 0$ (n *is known*)
$P/A, i, n$	$1/i$	n
$A/P, i, n$	i	$1/n$
$P/F, i, n$	0	1
$F/P, i, n$	∞	1
$F/A, i, n$	∞	n
$A/F, i, n$	0	$1/n$

3.4 INTEREST CALCULATIONS INVOLVING A CONSTANT AMOUNT OF CHANGE PER PERIOD

Some economic analysis problems involve receipts or disbursements that can be realistically expected to increase or decrease by a uniform amount each period, thus constituting an arithmetic series. Maintenance expenses may, under certain circumstances, be portrayed in this manner.

The cash flow diagram for a series of end-of-period payments increasing by an amount, G, each period is as follows:

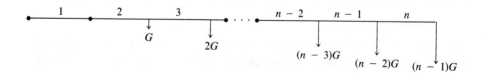

End of Period	Payment
1	0
2	G
3	$2G$
$n - 1$	$(n - 2)G$
n	$(n - 1)G$

Note that the gradient value, G, at the end of period 1 is zero. This period must be included in the calculation.

Finding F When G Is Given

$$F = G[(F/A, i, n - 1) + (F/A, i, n - 2) + \ldots + (F/A, i, 2) + (F/A, i, 1)]$$
$$= G\{[(1 + i)^{n-1} - 1]/i + [(1 + i)^{n-2} - 1]/i + \cdots + [(1 + i)^2 - 1]/i$$
$$+ [(1 + i) - 1]/i\}$$
$$= G(1/i)[(1 + i)^{n-1} + (1 + i)^{n-2} + \cdots + (1 + i)^2 + (1 + i) - n + 1]$$

Note the derivation of equation 3.5 for summing the series:

$$F = G(1/i)\{[(1 + i)^n - 1]/i - n\} \qquad (3.13)$$

The quantity to the right of G is referred to as the *future worth of a gradient series factor* and in functional notation is referred to as $(F/G, i\%, n)$. The equation for F is

$$F = G(F/G, i\%, n) \qquad (3.14)$$

Finding P When G Is Given

Multiply both sides of equation 3.13 by $1/(1 + i)^n$, the factor for the present worth of a future sum. Simplifying yields

$$P = G(1/i)\{[(1 + i)^n - 1]/[i(1 + i)^n] - [n/(1 + i)^n]\} \qquad (3.15)$$

The quantity to the right of G is referred to as the *present worth of the gradient series* factor and in functional notation is referred to as $(P/G, i\%, n)$. The equation for P is

$$P = G(P/G, i\%, n) \qquad (3.16)$$

Finding *A* When *G* Is Given

Similar to the method given immediately above for finding *P*, multiply both sides of equation 3.13 by $i/[(1 + i)^n - 1]$, the sinking-fund factor. Simplifying gives us

$$A = G\{(1/i) - n/[(1 + i)^n - 1]\} \tag{3.17}$$

The quantity in braces is referred to as the factor and in functional notation is referred to as $(A/G, i\%, n)$. The equation for *A* is

$$A = G(A/G, i\%, n) \tag{3.18}$$

Example 3.4

Problem: Maintenance costs on a certain piece of equipment are estimated to be $500 in year 1 and to increase by $100 per year. What is the annual equivalent end-of-year maintenance cost if $n = 5$ years and $i = 8\%$?

Solution

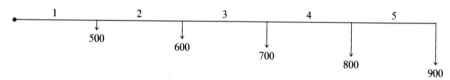

$$A = A + G(A/G, i, n) = A + G(A/G, 8\%, 5) = \$500 + \$100(1.8465)$$

$$= \$685$$

3.5 INTEREST CALCULATIONS INVOLVING A CONSTANT RATE OF CHANGE PER PERIOD

Solving engineering economy problems often involves receipts or payments that increase (or decrease) at a constant rate of growth per period (growth is exponential and the progression is geometric). Inflation may be handled in this manner, or a company may estimate that sales or profits will increase by a fixed percentage each year as a basis for projecting future sales and earnings. A model can be developed to handle such calculations. The cash flow diagram is

The formula to find the present equivalent of a series of disbursements or receipts that increase (decrease) at a constant rate of growth (decline), k, per period is

$$P = C/(1 + i) + C(1 + k)/(1 + i)^2 + \cdots + C(1 + k)^{n-1}/(1 + i)^n$$

Multiply through by $(1 + k)$:

$$P = [C/(1 + k)] [(1 + k)/(1 + i)$$
$$+ (1 + k)^2/(1 + i)^2 + \cdots + (1 + k)^n/(1 + i)^n]$$

To obtain the sum of a geometric series

$$a + ar + ar^2 + \cdots + ar^n$$

where a is the first term and r is the ratio of any term to the preceding term, we write

$$S_n = a[(1 - r^n)/(1 - r)]$$

for $r \neq 1$. Therefore, if we let $a = (1 + k)/(1 + i)$ and $r = (1 + k)/(1 + i)$, we have

$$P = \frac{[C/(1 + k)] [(1 + k)/(1 + i)]}{[1 - (1 + k)^n/(1 + i)^n]/[1 - (1 + k)/(1 + i)]}$$

For i not equal to k,

$$P = C\{[1 - (1 + k)^n/(1 + i)^n]/(i - k)\} \tag{3.19}$$

When $i = k$, equation 3.19 becomes

$$P = nC/(1 + k) \qquad \text{for } i = k \tag{3.20}$$

The quantity in braces in equation (3.19) is referred to as the *present equivalent* of a series of receipts increasing at a constant rate k per period when valued at an interest rate i or the *geometric series factor* and in functional notation is referred to as $(P/C, i\%, k\%, n)$. The equation for P is

$$P = C(P/C, i, k, n) \tag{3.21}$$

Example 3.5

Problem: You are interested in buying a small company and part of your cash flow analysis requires you to determine the present value of the net income cash flow stream over the next 5 years. Net income in year 1 is $10,000 and is expected to increase by 10% each year over the next 5 years. You require a 15% before-tax rate of return on any investments that you make. Find the present equivalent of this cash flow stream.

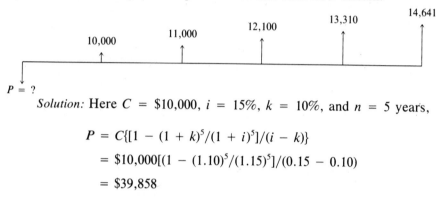

Solution: Here $C = \$10,000$, $i = 15\%$, $k = 10\%$, and $n = 5$ years,

$$P = C\{[1 - (1 + k)^5/(1 + i)^5]/(i - k)\}$$
$$= \$10,000[(1 - (1.10)^5/(1.15)^5]/(0.15 - 0.10)$$
$$= \$39,858$$

3.6 SUMMARY OF INTEREST FORMULAS

The formulas are developed from the basic compound interest relationship, $(1 + i)^n$, and an assumed pattern of cash flows. The patterns chosen are those that occur most commonly in economic analysis calculations. They are the usual assumptions made for maintenance expenditures, income growth, sales, and the like. However, one should be satisfied that the estimated cash flow pattern fits the pattern for which the formula was derived. When it does not, each cash flow can be treated separately using the single-payment, F/P and P/F, factors. The discrete interest formulas, together with their functional notation and cash flow pattern, are summarized in Table 3.3.

3.7 INTEREST TABLES AND INTERPOLATION

Prior to the advent of electronic calculators and microcomputers, interest calculations were generally done with the use of interest tables. Bankers used tables for calculating loan repayments, mortgage companies used mortgage tables, insurance companies used annuity tables, and engineering and economic analysts used present-value tables. All these tables were derived from the basic formulas developed in this chapter. They differ only in the

TABLE 3.3 Discrete interest formulas.

To Find:	Given:	Functional Notation	Formula
F	P	$F = P(F/P, i, n)$	$F = P(1 + i)^n$
P	F	$P = F(P/F, i, n)$	$P = F\left[\dfrac{1}{(1 + i)^n}\right]$
F	A	$F = A(F/A, i, n)$	$F = A\left[\dfrac{(1 + i)^n - 1}{i}\right]$
A	F	$A = F(A/F, i, n)$	$A = F\left[\dfrac{i}{(1 + i)^n - 1}\right]$
A	P	$A = P(A/P, i, n)$	$A = P\left[\dfrac{i(1 + i)^n}{(1 + i)^n - 1}\right]$
P	A	$P = A(P/A, i, n)$	$P = A\left[\dfrac{(1 + i)^n - 1}{i(1 + i)^n}\right]$
F	G	$F = G(F/G, i, n)$	$F = G\left(\dfrac{1}{i}\right)\left[\dfrac{(1 + i)^n - 1}{i} - n\right]$
P	G	$P = G(P/G, i, n)$	$P = G\left(\dfrac{1}{i}\right)\left[\dfrac{(1 + i)^n - 1}{i(1 + i)^n} - \dfrac{n}{(1 + i)^n}\right]$
A	G	$A = G(A/G, i, n)$	$A = G\left[\dfrac{1}{i} - \dfrac{n}{(1 + i)^n - 1}\right]$
P	C	$P = C(P/C, i, k, n)$	$P = C\left[\dfrac{1 - \dfrac{(1 + k)^n}{(1 + i)^n}}{i - k}\right]$

number of significant figures and in their completeness. Financial tables tend to be very complete and to many figures (eight or more digits) for financial transactions requiring precise accuracy. Economic feasibility calculations deal with estimated future cash flows, so great accuracy is inappropriate. Four- to six-digit accuracy is acceptable in these tables.

Appendix F contains a set of discrete interest tables. They are calculated from the formulas in this chapter and cover the rates and periods common to economic analysis problems. They are set out with one page per interest rate, the number of periods down the left-hand column; the number in the table corresponds to the factor listed at the head of the column.

For example, to find the present-worth factor for 8% for 10 periods,

$$(P/F, 8\%, 10)$$

refer to the 8% table for discrete compounding, reading down the P/F column and across the $n = 10$ row. The value of the factor is 0.4632. The factors for F/P, P/F, F/A, A/F, P/A, A/P, and A/G are all found on one page.

The tables do not cover all eventualities. There are gaps in the higher numbers of periods (the tables have values for $n = 35$, 40, 45, etc.) and in the higher rates of interest (15%, 20%, 25%). Intermediate values can be determined by using straight-line interpolation.

For example, to find the $(F/P, 6.5\%, 20)$ factor, we find from the discrete interest tables in Appendix F,

$$(F/P, 6.0\%, 20) = 3.207$$

$$(F/P, 7.0\%, 20) = 3.870$$

Interpolating gives us

$$(F/P, 6.5\%, 20) = 3.207 + 0.5(3.870 - 3.207)$$
$$= 3.207 + 0.3315$$
$$= 3.5385$$

The correct value is $(1.065)^{20} = 3.5236$, so for this situation the interpolation introduces an error of less than one-half of 1%, an acceptable level for economic analysis calculations.

Interpolation does not always give acceptable accuracy. The formulas contain exponential terms and so are curves, not straight lines. For larger values of i and large ranges of interpolation, the error will also be large. For example, to calculate the $(F/P, 27\%, 20)$ factor, we have, from the tables in Appendix F,

$$(F/P, 25\%, 20) = 86.736$$

$$(F/P, 30\%, 20) = 190.050$$

Interpolating, we have

$$(F/P, 27\%, 20) = 86.736 + \tfrac{2}{5}(190.050 - 86.736)$$
$$= 86.736 + 41.326 = 128.062$$

The correct value is $(1.27)^{20} = 119.1446$, so the interpolation is in error by 7.5%.

There is no hard-and-fast rule as to whether this or any other level of accuracy is correct. The accuracy depends on the nature of the problem and the accuracy of the estimates. It is sufficient to be aware that for large values of i and n and large ranges of interpolation, the error can be significant. When there is doubt, the correct value should be calculated from the formula.

3.8 PROBLEM-SOLVING METHOD

Interest problems should be solved using the following steps. This ensures completeness and facilitates checking.

> *Step 1:* Construct a cash flow diagram showing the magnitude, timing, and direction of the cash flows. Write the interest rate on the diagram.
> *Step 2:* Write the answer as an equation using the functional notation for the factors.
> *Step 3:* Substitute the table values of the factors into the equation.
> *Step 4:* Calculate the answer.
> *Step 5:* Check that the accuracy of the answer is in keeping with the accuracy of the data. Normally, interest rates will be shown to one-decimal-place accuracy and dollar values to the nearest dollar.

Example 3.6

Problem: A piece of machinery requires an overhaul today which will cost $1,000. It will require another overhaul 3 years hence which will cost $1,500. Annual operating costs are $300 per year. Find the present-worth of these costs at 6% if the machine has a life of 7 years.

Solution

Step 1: Construct a cash flow diagram

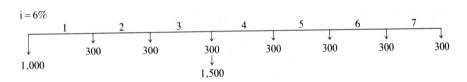

Step 2: Write the answer as an equation using functional notation:

$$P = \$1,000 + \$300(P/A, 6\%, 7) + \$1,500(P/F, 6\%, 3)$$

Step 3: Substitute table values for the factors:

$$P = \$1,000 + \$300(5.5824) + \$1,500(0.8396)$$

Step 4: Calculate the answer:

$$P = \$1,000 + \$1,674.72 + \$1,259.40 = \$3,934.12$$

Step 5: Check the accuracy.

Since the data consist of estimates, an answer to the nearest cent is inappropriate. Round the answer to the nearest dollar ($3,934).

3.9 NOMINAL AND EFFECTIVE INTEREST RATES

Assume that on a charge account invoice there is the following statement: "A service charge of 18% per annum (1.5% per month) will be added to your account each month based on your previous month's balance." What does this mean?

Suppose that I have a balance of $100, which for reasons of undue financial strain, I am unable to repay for 1 year. At the end of the first month my balance is

$$\$100(F/P, 1.5\%, 1) = \$100(1.0150) = \$101.50$$

At the end of the second month it is

$$\$100(F/P, 1.5\%, 2) = \$100(1.0302) = \$103.02$$

and at the end of 12 months it is

$$\$100(F/P, 1.5\%, 12) = \$100 \times 1.1956 = \$119.56$$

The total interest charges are $19.56. The invoice indicated that the interest rate was 18% per annum, which on $100 would amount to $18.

This example demonstrates the difference between nominal and effective interest rates. The *nominal rate* is 18%, but the *actual (effective) rate* is 19.56%. Let

i_y = effective rate of interest per year
i = rate of interest per compounding period
m = number of compounding periods per year

Then

$$im = \text{nominal interest per year}$$

If a present sum, P, is borrowed for 1 year with a stated (nominal) annual interest rate of $im\%$, the outstanding amount at the end of the year is

$$F = P(1 + i)^m$$

Thus the effective annual interest rate

$$i_y = (F - P)/P = (1 + i)^m - 1 \qquad (3.22)$$

Since $(1 + i)^n$ is the future-worth factor, the tables can be used, where applicable, for finding effective interest rates.

The nominal annual interest rate can be calculated from the effective annual rate (i_y):

$$\text{Nominal annual rate} = im = m[(1 + i_y)^{1/m} - 1] \qquad (3.23)$$

and the nominal rate per period can be calculated from the effective annual rate (i_y):

$$\text{Nominal rate per period} = i = (1 + i_y)^{1/m} - 1 \qquad (3.24)$$

For most economy studies, a year is the usual compounding period. Therefore, the effective and nominal rates are the same. However, loan transactions, savings accounts, charge accounts, and bond issues often use compounding periods of less than 1 year.

	Nominal Rate (%)	Effective Rate (%)
1% payable monthly	12.0	12.68
3% payable quarterly	12.0	12.55
6% payable semiannually	12.0	12.36

The effective interest rate increases as the number of periods increase.

When the number of compounding periods increases without limit, we have what is referred to as *continuous compounding*. As $m \to \infty$, then $i \to 0$ and $1/i \to \infty$:

$$\lim_{i \to 0} (1 + i)^{1/i} = e = 2.718 \quad \text{(base of the natural logarithm)}$$

When compounding is continuous, the effective annual interest rate (i_y) may be calculated as follows:

$$i_y = (1 + i)^m - 1$$

Rearranging the right-hand side of the equality to include i in the exponent yields

$$(1 + i)^m - 1 = (1 + i)^{im/i} - 1$$

Substituting e for $(1 + i)^{1/i}$ as $1/i \rightarrow \infty$, we have

$$i_y = e^{im} - 1 \tag{3.25}$$

Continuous compounding is developed further in Sections 3.10 and 3.11.

Example 3.7

Problem: Assume that you have the opportunity to arrange a loan at a nominal interest rate of 10% compounded monthly (alternative A) or 11% compounded quarterly (alternative B). Which arrangement would you select?

Solution: Determine the effective annual rate for each alternative.

Alternative A

$$\begin{aligned}
\text{Nominal annual rate} \quad &= 10\% \\
\text{Nominal monthly rate} &= 10\%/12 = 0.83\% \\
\text{Effective annual rate} \quad &= (1+i)^m - 1 \\
&= (1.0083)^{12} - 1 \\
&= 10.43\%
\end{aligned}$$

Alternative B

$$\begin{aligned}
\text{Nominal annual rate} \quad &= 11\% \\
\text{Nominal quarterly rate} &= 11\%/4 = 2.75\% \\
\text{Effective annual rate} \quad &= (1.0275)^4 - 1 \\
&= 11.46\%
\end{aligned}$$

Conclusion: Select alternative A.

Example 3.8

Problem: Which alternative is preferable: a loan at a nominal interest rate of 18% compounded continuously (alternative A) or a loan with an effective annual rate of 20% (alternative B)?

Solution

Alternative A

$$\begin{aligned}
\text{Nominal annual rate} &= im = 18\% \\
i_y &= e^{im} - 1 \\
&= (2.718)^{0.18} - 1 = 19.72\%
\end{aligned}$$

Conclusion: Select alternative A.

Example 3.9

Problem: You are purchasing a house for $65,000. You will pay for it with $10,000 cash that you have saved, a $48,750 first mortgage at a rate of 12% and a 20-year term, and a $6,250 second mortgage at a rate of 18% and a 10-year term. What is your monthly payment?

Solution: Although the mortgage rate is stated as an annual rate, it is actually a nominal annual rate and the actual interest period is usually less than 1 year. Since normally payments are made monthly, assume an interest period of 1 month. Thus the first mortgage is at 1% for 240 periods and the second mortgage is at 1.5% for 120 periods.

To determine the monthly payment, two calculations are necessary, one for each mortgage. For the first mortgage:

48,750

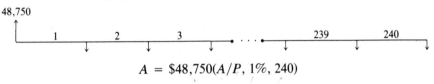

$$A = \$48{,}750(A/P, 1\%, 240)$$

For the second mortgage:

6,250

$$A = \$6{,}250(A/P, 1.5\%, 120)$$

Since the tables do not go this high, the formula must be used:

$$(A/P, 1\%, 240) = [i(1 + i)^n]/[(1 + i)^n - 1]$$

$$= [0.01(1.01)^{240}]/[(1.01)^{240} - 1] = 0.01101$$

$$A = \$48{,}750(0.01101) = \$536.74$$

$$(A/P, 1.5\%, 120) = [0.015(1.015)^{120}]/[(1.015)^{120} - 1]$$

$$= 0.018019$$

$$A = \$6{,}250(0.018019) = \$112.62$$

The monthly payment for the first 10 years equals

$$\$536.74 + \$112.62 = \$649.36$$

Example 3.10

Problem: A borrower signs a note for $15,000 and agrees to repay the loan with end-of-year payments of $5,000 each for the next 3 years. The lending agency "discounts" the note, so the borrower "prepays" $2,700 interest and actually receives $12,300. Find the actual rate of interest being paid by the borrower.

Solution

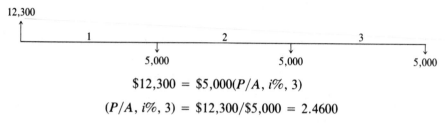

$$\$12,300 = \$5,000(P/A, i\%, 3)$$

$$(P/A, i\%, 3) = \$12,300/\$5,000 = 2.4600$$

From the tables

$$(P/A, 12\%, 3) = 2.4018$$

$$(P/A, 10\%, 3) = 2.4869$$

Interpolating gives us

$$i = 10\% + 2\%(2.4869 - 2.4600)/(2.4869 - 2.4018) = 10.7\%$$

3.10 INTEREST FORMULAS FOR CONTINUOUS COMPOUNDING WITH DISCRETE PAYMENTS

For a given nominal rate, as the frequency of compounding in a year increases, the effective interest rate also increases. Within the business world many transactions occur on at least a daily basis; therefore, compounding is occurring quite frequently. To account explicitly for the timing of these frequent cash flows, continuous compounding may be preferable. Continuous compounding means that each year is divided into an infinite number of periods. Mathematically, the single-payment compound amount factor under continuous compounding is given by

$$\lim_{m \to \infty} (1 + i)^{mn} = e^{imn}$$

where n is the number of years, m the number of compounding periods per year, and im the nominal interest rate per year.

Finding *F* When *P* Is Given

The future worth of a present sum (single-payment compound amount factor) using continuous compounding is

$$F = Pe^{imn} \qquad (3.26)$$

Using functional notation, this is

$$F = P(F/P, im, n) \qquad (3.27)$$

Example 3.11

Problem: If $1,000 is deposited in an account that pays interest at 10%, what will be accumulated in the fund at the end of 5 years using continuous compounding?

Solution

$$F = P(F/P, im, n)$$

$$= P(e^{imn})$$

$$= \$1,000(1.6487) = \$1,648.70$$

Problem: Recalculate, using daily compounding (365 days).

Solution

$$F = P(1 + i)^{mn}$$

$$= \$1,000(1.000274)^{1825} = \$1,648.69$$

Problem: Recalculate, using monthly compounding.

Solution

$$F = P(1 + i)^{mn}$$

$$= \$1,000(1.008333)^{60} = \$1,645.28$$

Problem: Recalculate, using annual compounding.

Solution

$$F = P(1 + i)^{5}$$

$$= \$1,000(1.1)^{5} = \$1,610.51$$

For most economy studies the year is the usual compounding period. However, in some banking and other business transactions, continuous or daily compounding is desirable, especially now that computers have given banks the means to calculate interest on a daily basis.

Finding P When F Is Given

The present worth of a future sum (single-payment present-worth factor) using continuous compounding is

$$P = Fe^{-imn}$$

(3.28)

Using functional notation, this becomes

$$P = F(P/F, im, n)$$

(3.29)

The common continuous compounding interest factors are summarized in Table 3.4.

TABLE 3.4 Summary of continuous compounding interest formulas.

To Find:	Given:	Discrete Payment, Continuous Compounding	Continuous Payment, Continuous Compounding
F	P	$F = Pe^{imn}$ $= (F/P, im, n)$	$F = Pe^{imn}$ $= P(F/P, im, n)$
P	F	$P = F(e^{-imn})$ $= F(P/F, im, n)$	$P = F(e^{-imn})$ $= F(P/F, im, n)$
F	A	$F = A\dfrac{(e^{imn} - 1)}{e^{im} - 1}$ $= A(F/A, im, n)$	$F = \overline{A}\left(\dfrac{e^{imn} - 1}{im}\right)$ $= \overline{A}(F/\overline{A}, im, n)$
A	F	$A = F\left(\dfrac{e^{im} - 1}{e^{imn} - 1}\right)$ $= F(A/F, im, n)$	$\overline{A} = F\left(\dfrac{im}{e^{imn} - 1}\right)$ $= F(\overline{A}/F, im, n)$
P	A	$P = A\left(\dfrac{1 - e^{-imn}}{e^{im} - 1}\right)$ $= A(P/A, im, n)$	$P = \overline{A}\left(\dfrac{e^{imn} - 1}{ime^{imn}}\right)$ $= \overline{A}(P/\overline{A}, im, n)$
A	P	$A = P\left(\dfrac{e^{im} - 1}{i - e^{-imn}}\right)$ $= P(A/P, im, n)$	$\overline{A} = P\left(\dfrac{ime^{imn}}{e^{imn} - 1}\right)$ $= P(\overline{A}/P, im, n)$

3.11 INTEREST FORMULAS FOR CONTINUOUS COMPOUNDING WITH CONTINUOUS PAYMENTS

In some situations it may be preferable to assume that cash flows taking place during the year occur uniformly throughout the year rather than at discrete periods. Some examples that may be more realistically represented by a uniform cash flow throughout the year are equipment maintenance and other operating costs. Assuming a uniform cash flow throughout the year instead of discretely on a daily, weekly, or monthly basis, it is assumed that a specific sum (e.g., $1) flows uniformly and continuously throughout the year.

The present worth of a uniform series of discrete payments is given by equation 3.9:

$$P = A\{[(1 + i)^n - 1]/[i(1 + i)^n]\}$$

Take a given sum, say \$1, and divide it into m end-of-period payments. During the year each uniform payment, referred to as $\overline{A}$, is equal to \$$1/m$ periods. Let

i = rate of interest per compounding period

m = number of compounding periods per year

im = nominal interest rate per year

n = time expressed in years

$\overline{A}$ = the uniform flow rate of money per year

Then

$$P = \frac{\$1}{m}\left[\frac{(1 + i)^m - 1}{i(1 + i)^m}\right]$$

Rearranging, we have

$$P = \frac{\$1[(1 + i)^{1/i}]^{im} - 1}{im[(1 + i)^{1/i}]^{im}}$$

As $m \to \infty$, $1/i \to \infty$ and $(1 + i)^{1/i}$ approaches the limit e. Therefore, the limiting value of P is

$$P = \$1(e^{im} - 1)/(ime^{im})$$

and the general equation for n years is

$$P = \overline{A}(e^{imn} - 1)/ime^{imn} \tag{3.30}$$

In functional notation,

$$P = \overline{A}(P/\overline{A}, im, n) \tag{3.31}$$

where $(P/\overline{A}, im, n)$ is referred to as the *continuous-flow, continuous-compounding uniform series present-worth factor.*

The difference between nominal and effective interest rates per annum is greatest when using continuous compounding. With high rates of interest, the effective rate per annum is considerably higher than the nominal annual rate *im*.

Example 3.12

Problem: What is the present worth of a uniform series of continuous cash flows totaling $1,000 per year for 10 years using an interest rate of 20% compounded continuously?

Solution: $\overline{A}$ = $1,000, *im* = 20%, and *n* = 10 years, so

$$P = \overline{A}(P/\overline{A}, 20\%, 10)$$
$$= \$1,000(e^{imn} - 1)/ime^{imn}$$
$$= \$1,000(4.3233)$$
$$= \$4,323.30$$

Example 3.13

Problem: Find the initial expenditure justified for a new piece of equipment that is expected to reduce operating costs by $5,000 per year over each of the next 5 years. *i* = 15%. Assume that the savings represent a uniform series of continuous cash flows (continuous compounding).

Solution: $\overline{A}$ = $5,000, *im* = 15%, and *n* = 5 years, so

$$P = \overline{A}(P/\overline{A}, im, n)$$
$$= \$5,000(e^{imn} - 1)/ime^{imn}$$
$$= \$5,000(3.5176)$$
$$= \$17,588$$

Problem: Assume that the savings are best represented by monthly cash flows (monthly compounding).

Solution: Each uniform payment = $5,000/12 = $416.67. *i* = 15%/12 = 1.25%, *m* = 12, and *n* = 5 years.

$$P = \frac{\$5,000}{m}\left[\frac{(1 + i)^{mn} - 1}{i(1 + i)^{mn}}\right]$$
$$= \frac{\$5,000}{12}\left[\frac{(1.0125)^{60} - 1}{0.0125(1.0125)^{60}}\right]$$
$$= \$416.67(42.0346) = \$17,514$$

3.12 SUMMARY

Two crucial aspects of an economic analysis are the magnitude and timing of cash flows. Because money has a time value, it is not possible to simply add or compare sums that occur at different points in time.

The arithmetic of compound interest makes it possible to move amounts through time while retaining their equivalent value. In this way sums occurring at different times can be combined or compared.

Formulas have been developed for most of the common cash flow patterns. These formulas, combined with the five-step solution method, make it possible to do a concise and complete calculation of any interest problem.

PROBLEMS

3.1 You borrowed $6,000 and have agreed to repay this amount over the next 3 years at 12% interest. Make a table similar to Table 3.1 to show repayment accomplished by:
  a. Year-end payment of interest on the principal with the third payment to include repayment of the principal itself.
  b. Year-end payment of one-third of the principal ($2,000) plus interest on the unpaid balance.
  c. Lump-sum repayment at the end of the third year.
  d. Year-end payments of equal size.
  e. Draw a cash flow diagram for part (d).

3.2 Most finance repayment plans (e.g., home payments, car payments) are similar to which plan above?

3.3 Using the discrete interest table in Appendix F, calculate the following sums.
  a. The future worth of a present sum: $P = \$5,000$, $i = 15\%$, and $n = 7$ years.
  b. The present worth of a uniform series: $A = \$400$, $i = 8\%$, and $n = 12$ years.
  c. The present worth of a future sum: $F = \$500$, $i = 20\%$, and $n = 10$ years.
  d. The future worth of a uniform series: $A = \$400$, $i = 8\%$, and $n = 12$ years.
  e. The present worth of a future sum: $P = \$7,591$, $i = 8\%$, and $n = 12$ years.

3.4 Evaluate the following factors.
   a. The future worth of a present sum: $P = \$10,000$, $i = 2.5\%$, and $n = 20$ years.
   b. The future worth of a uniform series: $A = \$1,000$, $i = 0.75\%$, and $n = 25$ years.
   c. The present worth of a uniform series: $A = \$5,000$, $i = 3.75\%$, and $n = 10$ years.

3.5 Evaluate the following factors using the basic equations in Chapter 3.
   a. $(P/A, 16\%, 10)$
   b. $(A/P, 16\%, 10)$
   c. $(F/A, 16\%, 10)$
   d. $(A/F, 16\%, 10)$
   e. $(P/G, 16\%, 10)$
   f. $(A/G, 16\%, 10)$

3.6 Calculate the following by interpolation from the discrete tables of Appendix F. Then calculate the values using the basic formulas. Compare the results.
   a. Find F given $P = \$10,000$, $i = 1.5\%$, and $n = 10$ years.
   b. Find F given $P = \$10,000$, $i = 11\%$, and $n = 10$ years.
   c. Find F given $P = \$10,000$, $i = 21\%$, and $n = 10$ years.
   d. Find P given $F = \$100,000$, $i = 1.75\%$, and $n = 40$ years.
   e. Find P given $F = \$100,000$, $i = 18\%$, and $n = 40$ years.
   f. Find A given $F = \$100,000$, $i = 0.6\%$, and $n = 80$ years.

3.7 Compare the amount of interest earned by $1,000 over 10 years at 15% simple interest with the amount of interest earned by $1,000 for 10 years at 15% compounded annually.

3.8 Construct cash flow diagrams for the following situations.
   a. Twelve end-of-period payments of $500 each.
   b. Seven end-of-period payments, the first being $700 and each succeeding payment being $25 more than its predecessor.
   c. Five end-of-period payments, the first being $500 and each succeeding one being 15% more than its predecessor.

3.9 Construct cash flow diagrams for the following situations.
   a. A receipt of $1,000 today, plus a receipt of $1,000 5 years from today.
   b. A receipt of 10 beginning-of-period payments of $700 each.
   c. Receipts of $500 at time 0, $200 at time 1, $300 at time 2, $400 at time 3, and $500 at time 4.
   d. Receipt of a series of eight payments, the first being today, of amount $1,200, and each succeeding payment being 4% more than its predecessor.

3.10 What future amount may be withdrawn from an account if the following deposits were made today? Draw cash flow diagrams.
  a. $1,000 deposited today for 5 years at 15% compounded annually.
  b. $1,000 deposited today for 10 years at 15% compounded annually.
  c. $1,000 deposited today for 15 years at 15% compounded annually.
  d. $1,000 deposited today for 20 years at 15% compounded annually.

3.11 What is the present value of the following future amounts using an interest rate of 10%?
  a. $10,000 5 years from now compounded annually.
  b. $10,000 10 years from now compounded annually.
  c. $10,000 20 years from now compounded annually.
  d. $10,000 50 years from now compounded annually.

3.12 What amount will be accumulated in an account over a 5-year period if $10,000 is deposited today at the following interest rates: (a) $i = 5\%$; (b) $i = 10\%$; (c) $i = 20\%$? Assume annual compounding.

3.13 What amount has to be deposited in an account today that pays 10% (compounded annually) in order to have $25,000 in the account at the following future dates:
  a. Five years from today?
  b. Ten years from today?
  c. Twenty years from today?

3.14 If you were to deposit $10,000 today, what uniform annual amount could you withdraw each year over the next 5 years at the following interest rates (compounded annually): (a) $i = 6\%$; (b) $i = 12\%$; (c) $i = 20\%$? Draw a cash flow diagram for part (a).

3.15 If you were to deposit $1,000 at the end of each year for the next 5 years, how much could you withdraw in 5 years at the following interest rates (compounding annually): (a) $i = 8\%$; (b) $i = 15\%$; (c) $i = 25\%$? Draw a cash flow diagram for part (a).

3.16 If you wanted to be able to withdraw $1,000 each year over the next 8 years, how much would you have to deposit today at the following interest rates (compounded annually): (a) $i = 6\%$; (b) $i = 10\%$; (c) $i = 15\%$? Draw a cash flow diagram for part (a).

3.17 If you were to deposit $1,000 today, $1,000 3 years from today, and $1,000 6 years from today, how much could you withdraw 9 years from today if $i = 10\%$ compounded annually? Draw a cash flow diagram.

3.18 If you deposit $1,000 today in an account that pays 10% compounded annually over the next 3 years, and then the interest rate drops to 8% over the following 2 years and increases to 12% over the following 3 years, how much can you withdraw in 8 years' time? Draw a cash flow diagram.

3.19 What amount can you withdraw each month over the next 30 months if you deposit $5,000 today and the nominal annual interest rate is 12%?

3.20 What amount could you withdraw at the end of 5 years if you were to deposit $100 per month starting today into an account that pays a nominal interest rate of 10%? The last deposit is made the same day the withdrawal is made. Draw a cash flow diagram.

3.21 You have borrowed $100 from a friend and promise to repay him $110 in 1 month.
a. What interest rate are you paying over the 1-month period?
b. What is the nominal annual interest rate?
c. What is the effective annual interest rate?

3.22 A graduating engineer of age 25 receives an initial salary of $24,000 per year. Assuming that she receives a 10% raise every year, what will her annual salary be when she is 65 years old?

3.23 A couple concerned for their child's future decide to deposit on each birthday, into an account earning 12% interest, an amount equal to 10 times the child's age: that is, $10 on the first birthday, $20 on the second, and so on. What is the balance in the account just after the child's eighteenth birthday?

3.24 Find the present equivalent at 8% of:
a. A uniform series of 20 end-of-year payments of $100.
b. An arithmetic series of 20 end-of-year payments, the first being $100, the second being $105, the third $110, and each succeeding one being $5 more than its predecessor.
c. A geometric series of 20 end-of-year payments, the first one being $100 and each succeeding one being 5% greater than its predecessor.

3.25 Find the present equivalent at 12% of the situations in Problem 3.24.

3.26 Find the present equivalent at 25% of the situations in Problem 3.24.

3.27 Given the uniform series amount of a future sum $i/[(1 + i)^n - 1]$, develop the equation for the uniform series amount of a present sum.

3.28 Show that when the life (n) approaches infinity, the factor (P/A, i, n) approaches $1/i$.

3.29 Show that as the life (n) approaches infinity, the factor (A/P, i, n) approaches i.

3.30 What effective interest rate per annum corresponds to a nominal rate of 24%:
  a. Compounded annually?
  b. Compounded semiannually?
  c. Compounded monthly?

3.31 Convert the following annual effective rates to monthly rates: (a) 19.56%; (b) 42.58%; (c) 35%.

3.32 Assume that the interest rate charged by various finance companies is 2% per month. What nominal and effective rates per annum does this represent?

3.33 You have the opportunity to finance a loan at a nominal rate of 12% compounded monthly (alternative A) or 13% compounded semiannually (alternative B). Which arrangement would you select?

3.34 Which alternative is preferable: a loan at a nominal interest rate of 15% compounded monthly (alternative A) or a loan with an effective annual rate of 16% (alternative B)?

3.35 You have just purchased a new automobile that has a purchase price of $10,800. Your down payment was $1,600 and you have to make payments of $400 per month for the next 30 months. What effective annual interest rate are you paying?

3.36 You have just returned from your honeymoon and have found an unfurnished apartment. You and your spouse have discussed your finances and decide that you can afford to pay a maximum of $500 per month for furniture and other household effects. You have shopped around and found the best bargains available. Your total purchases total $8,000 and your payments are $350 per month for 30 months. You could have borrowed the money from your banker for 1% per month interest. Did you make a good decision? Show your calculations.

3.37 If you were to deposit $1,000 into a bank account that pays a nominal annual interest rate of 10%, how much would be in the account at the end of 4 years:
  a. Using continuous compounding?
  b. Using daily compounding (250 days per year)?
  c. Using monthly compounding?
  d. Using annual compounding?

3.38 If $3,000 is deposited into an account that pays interest at 10%, what will be accumulated in the fund at the end of 10 years:
  a. Using continuous compounding?
  b. Using daily compounding (250 days per year)?

c. Using monthly compounding?

d. Using annual compounding?

3.39 What is the present worth of a uniform series of cash flows totaling $2,000 per year for 5 years using an interest rate of 10% compounded continuously?

3.40 What initial expenditure is justified for a new piece of equipment that is expected to reduce operating costs by $3,000 per year over each of the next 5 years? Assume that the savings represent a uniform series of continuous cash flows with continuous compounding at 15%.

3.41 Given that $P = \$1,000$, $F = \$2,000$, and $n = 6$ years, find i (compounding annually).

a. Use the interest tables and do a straight-line interpolation.

b. Use your calculator and the basic formulas.

3.42 Given that $P = \$1,000$, $F = \$2,000$, and $n = 10$ years, find i (compounding annually).

3.43 You have borrowed $5,000 from a loan company and promise to pay the company $1,500 at the end of each year over the next 5 years. What interest rate are you paying?

3.44 You have the opportunity to invest $10,000 in a small business with the prospect of receiving $2,800 per year each year for the next 5 years. What annual interest rate (annual compounding) are you earning?

3.45 Easy Ed's Used Cars advertises Ed's Easy 6% financing. Upon inquiry you find that Easy Ed offers the following deal on a candy-apple-red whiplash special.

List price	$1,999.00
Minus down payment	500.00
Balance to finance	$1,499.00
Plus credit investigation fee	50.00
Plus Ed's 6% (of $1,499)	89.94
	$1,638.94

So 12 monthly payments of $(1,638.94 \div 12) = \$136.58$. Alternatively, Ed will sell you the car for $1,750 cash. What is the actual effective annual rate that you would be paying?

3.46 Write a program for your programmable calculator to calculate present equivalents. The program should take the rate i and number of periods n as initial parameters. The data entered will be the end-of-year cash flows. Use Problem 3.13 to check your program.

CHAPTER FOUR

Applications Involving
Interest and Compounding

The purpose of this chapter is to provide practice in applying time-value-of-money concepts through application of the equations derived in Chapter 3. Included in this chapter are examples that illustrate the alternative methods that may be utilized in solving interest problems.

Many major applications involve the analysis of personal and corporate investments. In these investments the effect of income taxation plays a major and often decisive role. Taxation methods, depreciation calculations, and some sections on current taxation legislation are also included in this chapter so that the tax effect can be calculated and the economic evaluation conducted on an after-tax basis.

4.1 APPLICATIONS

The many examples to follow are intended to familiarize the reader with calculations involving the time value of money and to demonstrate the importance of recognizing the significance of interest on investment decisions.

Example 4.1

Problem: If one were to invest $1,000 today at an interest rate of 10%, what equivalent annual sum can be withdrawn each year over the next 5 years?

Solution: $P = \$1,000$, $i = 10\%$, and $n = 5$ years. $A = ?$

$$A = P(A/P, i, n) = \$1,000(A/P, 10\%, 5)$$
$$= \$1,000(0.2638) = \$263.80$$

Example 4.2

Problem: If one were to deposit $263.80 at the end of each year for the next 5 years in an account that pays 10% interest compounded annually, what would be the balance in this account at the end of 5 years?

Solution: $A = \$263.80$, $i = 10\%$, and $n = 5$ years. $F = ?$

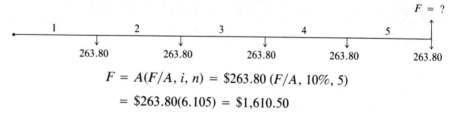

$$F = A(F/A, i, n) = \$263.80\,(F/A, 10\%, 5)$$
$$= \$263.80(6.105) = \$1,610.50$$

Example 4.3

Problem: If one invested $1,610.50 on January 1, 1984, in an investment that pays 10% compounded annually, how much will be in the account on January 1, 1989?

Solution: $P = \$1,610.50$, $i = 10\%$, and $n = 5$ years. $F = ?$

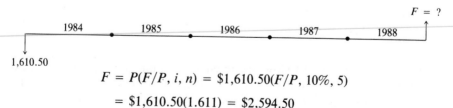

$$F = P(F/P, i, n) = \$1,610.50(F/P, 10\%, 5)$$
$$= \$1,610.50(1.611) = \$2,594.50$$

Example 4.4

Problem: What investment, at 10%, is necessary on January 1, 1979, in order to have a sum on deposit of $2,594.50 on January 1, 1989?

Solution: $F = \$2,594.50$, $i = 10\%$, and $n = 10$ years. $P = ?$

$F = 2,594.50$

79   88

$P = ?$

$$P = F(P/F, i, n) = \$2,594.50 \, (P/F, 10\%, 10)$$

$$= \$2,594.50(0.3856) = \$1,000$$

Comments Regarding Examples 4.1 through 4.4

Once again, the concept of equivalence is demonstrated. $1,000 on January 1, 1979, is equivalent to $263.80 on January 1 each year for the next 5 years, $1,610.50 5 years hence, and $2,594.50 10 years hence, provided that the time value of money remains constant at 10%.

The interest formulas used in this text and in most engineering economy texts assume end-of-period cash flows. For example, an investment made on January 1, 1976, is assumed to be made on December 31, 1975. This concept is also the reason why in converting an annual equivalent sum, A, to a future equivalent sum, F, or vice versa, the last sum A in the series occurs at the same point in time that the future sum, F, occurs.

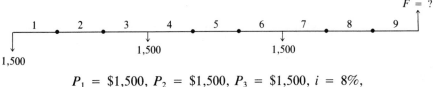

Example 4.5

Problem: If one were to invest $1,500 today, $1,500 3 years hence, and $1,500 6 years hence, what future amount will be accumulated 9 years hence if $i = 8\%$?

Solution: Until one becomes thoroughly familiar with cash flow calculations, the first step in the solution should be to draw a cash flow diagram.

$F = ?$

1   2   3   4   5   6   7   8   9

1,500   1,500

1,500

$$P_1 = \$1,500, \ P_2 = \$1,500, \ P_3 = \$1,500, \ i = 8\%,$$

$$\text{and } n = 9 \text{ years. } F = ?$$

There are several approaches to the solution of this problem.

Method 1:

$$F = P_1(F/P, 8\%, 9) + P_2(F/P, 8\%, 6) + P_3(F/P, 8\%, 3)$$
$$= \$1,500(1.999) + \$1,500(1.587) + \$1,500(1.260)$$
$$= \$2,999 + \$2,381 + \$1,890 = \$7,269$$

Method 2:

$$F = P(A/P, 8\%, 3)(F/A, 8\%, 9)$$
$$= \$1,500(0.3880)(12.488) = \$7,269$$

Method 3:

$$F = [(P_1 + P_2(P/F, 8\%, 3) + P_3(P/F, 8\%, 6)](F/P, 8\%, 9)$$
$$= [\$1,500 + \$1,500(0.7938) + \$1,500(0.6302)](1.999)$$
$$= \$7,268$$

The approach used can often save time-consuming calculations, and time is money. Method 2 is the most desirable approach in this example, particularly if the time span is long. Slight differences can result due to rounding off in the use of the interest table. However, any small error introduced by round-off is not significant from the standpoint of an economy study. We must not lose sight of the fact that economy studies deal with *future estimates* and projections of cash flows based on these estimates. We should exercise extreme care in arriving at these future estimates as the validity of the entire economic analysis is dependent on the validity of the estimates used.

Example 4.6

Problem: How many years would it take to pay back a loan of $5,000 with interest at 10% provided that you can make payments of $1,000 per year?

Solution: $P = \$5,000$, $A = \$1,000$, and $i = 10\%$. $n = ?$

$$(A/P, 10\%, n) = \$1,000/\$5,000 = 0.20000$$

From the tables,

$$(A/P, 10\%, 7) = 0.2054$$

$$(A/P, 10\%, 8) = 0.1875$$

Therefore, seven full payments of $1,000 and one payment of less than $1,000 is required. The eighth payment is equal to

$$[\$5,000 - \$1,000(P/A, 10\%, 7)](F/P, 10\%, 8)$$

$$= [\$5,000 - \$1,000(4.8684)](2.144) = \$282$$

Example 4.7

Problem: Find the present equivalent sum that would have to be invested today to enable one to make withdrawals of $1,000 each at the end of years 16 through 20 inclusive with interest at 10%.

Solution: $A = \$1,000$, $i = 10\%$, and $n = 16, 17, 18, 19$, and 20 years. $P = ?$

$$P = A(F/A, 10\%, 5)(P/F, 10\%, 20)$$

$$= \$1,000(6.105)(0.1487) = \$908$$

or

$$P = \$1,000[(P/F, 10\%, 16) + (P/F, 10\%, 17) + (P/F, 10\%, 18)$$

$$+ (P/F, 10\%, 19) + (P/F, 10\%, 20)]$$

$$= \$1,000(0.2176 + 0.1979 + 0.1799 + 0.1635 + 0.1487)$$

$$= \$908$$

or

$$P = A(P/A, 10\%, 5)(P/F, 10\%, 15)$$

$$= \$1,000(3.7908)(0.2394) = \$908$$

or

$$P = A[(P/A, 10\%, 20) - (P/A, 10\%, 15)]$$

$$= \$1,000(8.5136 - 7.6061) = \$908$$

Example 4.8

Problem: A first-year student anticipates buying an older-model car. He estimates his repair costs to be zero in year 1, $100 at the end of year 2, $200 at the end of year 3, and $300 upon graduation. Find the student's annual equivalent year-end repair cost if interest is equal to 10%.

Solution: $G = \$100$, $i = 10\%$, and $n = 4$ years. $A = ?$

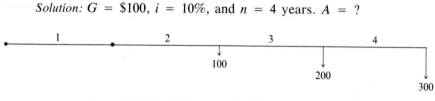

$$A = G(A/G, 10\%, 4) = \$100(1.3812) = \$138$$

Problem: All other costs of operating this vehicle are estimated by the student to be $300 per year. Thus the estimated total operating costs would be $300 in year 1, $400 in year 2, $500 in year 3, and $600 in year 4. Find the student's annual equivalent year-end operating cost if $i = 10\%$.

Solution: $A = \$300$, $G = \$100$, $i = 10\%$, and $n = 4$ years. $A = ?$

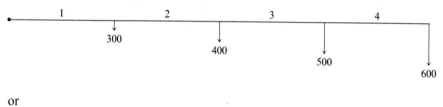

or

plus

$$A = A + G(A/G, 10\%, 4)$$

$$= \$300 + \$100(1.3812) = \$438$$

Problem: The student is also looking at a newer-model truck that needs some work now but will be less expensive for his final years. The estimated costs of operating the truck are $600 in year 1, $500 in year 2, $400 in year

3, and $300 in year 4. Find the annual equivalent year-end cost of operating the truck if $i = 10\%$. Should the student buy the truck?

Solution: $A = \$600$, $G = -\$100$, $i = 10\%$, and $n = 4$ years. $A = ?$

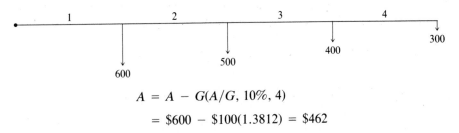

$$A = A - G(A/G, 10\%, 4)$$

$$= \$600 - \$100(1.3812) = \$462$$

The student should not buy the truck because it has a higher annual equivalent cost than the car.

Example 4.9

Problem: A small manufacturing company is planning on buying a certain delivery van which can be purchased for $5,000 cash (plan A) or with a $1,000 down payment now plus 24 monthly payments of $200 each (plan B). The first payment is due 1 month after closing the transaction. Find the rate of return (interest rate) at which the two purchase plans are equivalent.

Solution: $P = \$4,000$, $A = \$200$, and $n = 24$ months. $i = ?$

Note: We need to find the rate of return on the incremental investment.

Period	Plan A	Plan B	B − A
0	$5,000	$1,000	$4,000
1–24	0	200	200

The question to be answered is: At what rate of return is $4,000 today equivalent to $200 per month for 24 months?

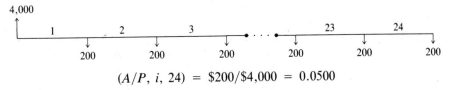

$$(A/P, i, 24) = \$200/\$4,000 = 0.0500$$

Since

$$(A/P, 1\%, 24) = 0.0471 \quad \text{and} \quad (A/P, 2\%, 24) = 0.0529$$

$$i = 1\% + 1\%(0.05 - 0.0471)/(0.0529 - 0.0471) = 1.5\%$$

$$\text{Nominal annual rate} = 12(1.5) = 18\%$$

$$\text{Effective annual rate} = (1 + 0.015)^{12} - 1 = 19.6\%$$

Which plan would you accept? Why? The company should pay cash unless they can invest their money at 19.6% or greater.

Example 4.10

Problem: A woman has just won a lottery game paying $5,000 every 5 years forever. Find the present equivalent of these payments using an interest rate of 10% if the first payment is made today.

Solution: Payment = $5,000, i = 10%, and n = ∞. P = ?

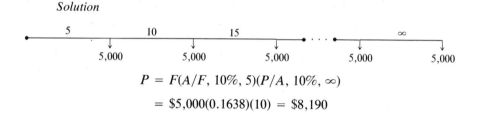

$$P = P(A/P, 10\%, 5)(P/A, 10\%, \infty)$$

$$= \$5,000(0.2638)(10.0) = \$13,190$$

Problem: Recalculate using the same conditions, but assume that the first payment occurs 5 years from now.

Solution

$$P = F(A/F, 10\%, 5)(P/A, 10\%, \infty)$$

$$= \$5,000(0.1638)(10) = \$8,190$$

4.2 CONSUMER FINANCE

It is the business of those who provide goods and services to sell you commodities. That is how they earn a living. Some of the goods they market are products you want and need, but many are not.

If you want value for your money, you must constantly evaluate both the advertising that confronts you and your own motives for buying to determine whether or not your decision is right for you. Before you go shopping, do your homework. Decide what you want to buy, when you want to buy, and how you want to buy. This concept is often referred to as *planned budgeting.*

Some of the more important decisions you will be faced with during your life span are:

1. Should I buy or rent a home?
2. How much insurance should I have (e.g., life, property, medical)?
3. What should my pension portfolio include (e.g., insurance, retirement savings plan)?
4. Should I purchase an automobile?

4.3 RENT OR BUY

The purchase of a home holds a place of high priority with many families. The decision to buy is certainly not entirely financial. However, the fact is that from the viewpoint of getting value for your money, both owning and renting have distinct advantages and disadvantages.

Owning your own home is a good hedge against inflation but do not overlook the additional expenses incurred by owning versus renting. These include paying for repairs, maintenance, landscaping, utilities, property taxes, and others. The statement that "money spent on rent is money down the drain" may be true, but many costs associated with home ownership also fall into this class. In the initial years of home ownership, only a small percentage of the house payment goes to increasing your equity in the property.

All of these factors must be weighed against the undeniable privileges and pleasures of home ownership, such as the sense of privacy and security, and the freedom to alter and improve your surroundings.

Buying a home normally involves the single largest cash outlay that a family is likely to make. Before you make this important decision, carefully consider:

1. Where you want to live
2. How much you want to spend
3. How large a house you need

Example 4.11

Problem: You are interested in purchasing a $70,000 home. You have $10,000 in cash; therefore, you require a $60,000 mortgage over 30 years at 9% per annum. Assume annual payments. Calculate the annual payment.

Solution: $P = \$60,000$, $i = 9\%$, and $n = 30$ years. $A = ?$

60,000

$$A = \$60,000(A/P, 9\%, 30) = \$60,000(0.0973) = \$5,838$$

Problem: Determine the amount of principal and interest paid each year during the first 5 years.

Solution

Year	Balance Remaining Beginning of Period	Interest Payment	Principal Payment	Balance Remaining End of Period
1	$60,000.00	$5,400.00	$440.40	$59,559.60
2	59,559.60	5,360.36	480.04	59,079.56
3	59,079.56	5,317.16	522.24	58,557.32
4	58,557.32	5,270.16	570.24	57,987.08
5	57,987.08	5,218.84	621.56	57,365.52

Problem: Calculate the monthly payment.

Solution: Assume that 9% is the nominal rate and that the period is monthly. $P = \$60,000$, $i = 0.75\%$, and $n = 360$ periods. $A = ?$

$$A = \$60,000(A/P, 0.75\%, 360)$$

$$= \$60,000\{i(1 + i)^n/[(1 + i)^n - 1]\}$$

$$= \$60,000\{(0.0075)(1.0075)^{360}/[(1.0075)^{360} - 1]\}$$

$$= \$482.77$$

Assuming $50 in taxes, the total principal, interest, and taxes $(P - I - T) = \$532.77$.

Problem: What savings would result if you were to rent an apartment for $450 per month instead of buying the home?

Solution: You could rent an apartment for $450 per month and invest the $10,000 in a bond or savings certificate at 9%. Under this option, your position 30 years hence would be:

Value of $10,000 investment:

$$F = \$10,000(F/P, 9\%, 30)$$

$$= \$10,000(13.268)$$

$$= \$132,680$$

Value of monthly savings:

$$F = (\$532.77 - \$450.00)(F/A, 0.75\%, 360)$$

$$= 82.77\{[(1.0075)^{360} - 1]/0.0075\}$$

$$= 82.77(1,830.74)$$

$$= \$151,530.63$$

Total value of savings:

$$F = \$132,680 + \$151,530.63$$

$$= \$284,211$$

Problem: If the house were to increase in value by 5% per year, what would its value be in 30 years?

Solution: The house, 30 years hence, could be an old derelict, or, assuming normal maintenance and appreciation of say 5% per year, could have a value equal to

$$F = \$70,000(F/P, 5\%, 30)$$

$$= \$70,000(4.322)$$

$$= \$302,540$$

The difference, as seen from the calculation, is not very large. Other factors that should enter into the comparison are:

1. Some estimate of annual repairs and maintenance.
2. The monthly saving of $82.77 may not be invested but may be spent on current consumption. If it is used to reduce the amount of money borrowed for car loans or financing of appliances, it is equivalent to investing at a rate higher than 9%.
3. The freedom that comes with home ownership may be a major consideration.

4.4 PERSONAL INVESTMENTS

There are many alternative investments that you may consider for an interim period or as a long-range pension plan. Several of these alternatives are listed below.

1. Place your money under the mattress.
2. Open a bank or credit union savings account.
3. Purchase a term deposit.
4. Buy a government bond.
5. Buy a share of preferred stock.

6. Buy a share of common stock.

7. Speculate in the commodities market.

8. Buy lottery tickets.

9. Go to Las Vegas.

Each of these offers some combination of security and rate of return. In fact, that is how the stock market works; the potential rate of return on an investment usually increases as the risk increases. Evaluation of risk is postponed to later chapters. For the present we shall concentrate on determining the rate of return.

Example 4.12

Problem: If an investment of $1,000 will bring $350 a year for 5 years, what is the rate of return on the investment?

Solution

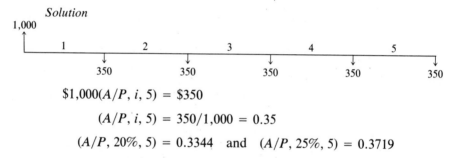

$$\$1,000(A/P, i, 5) = \$350$$

$$(A/P, i, 5) = 350/1,000 = 0.35$$

$$(A/P, 20\%, 5) = 0.3344 \quad \text{and} \quad (A/P, 25\%, 5) = 0.3719$$

Interpolate:

$$20\% + 5\%(0.35 - 0.3344)/(0.3719 - 0.3344) = 22\%$$

Example 4.13

Problem: What is the required amount of an investment that will pay a perpetual annuity of $100 if the rate of return is 10%?

Solution

$$P = \$100(P/A, 10\%, \infty) = \$100(1/i) = \$1,000$$

$P = ?$

Example 4.14

Problem: What is the amount of an endowment that will provide an annuity of $100 for the next 50 years if the rate of return is 10%?

Solution

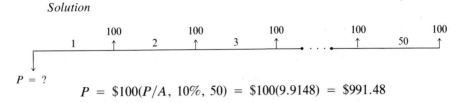

$$P = \$100(P/A,\ 10\%,\ 50) = \$100(9.9148) = \$991.48$$

Assets with long lives (e.g., 50 years plus) can, for all practical purposes, be considered to have an infinite life.

Note:

$$(A/P,\ i\%,\ \infty) = i$$

$$(P/A,\ i\%,\ \infty) = 1/i$$

Example 4.15

Problem: An 8% $1,000 bond matures on July 1, 1992, and interest payments are to be made twice annually. In terms of the money market this means that the $1,000 principal is to be paid at the date of maturity and 4% interest on the principal is to be paid directly to the bondholders (redeeming their coupons) at the end of every 6 months. How much would you be willing to pay for this bond on July 1, 1980, if you want a nominal 10% annual rate of return compounded semiannually?

Solution: $A = \$40$, $i = 5\%$, and $F = \$1,000$. $P = ?$

$$P = \$40(P/A,\ 5\%,\ 24) + \$1,000(P/F,\ 5\%,\ 24)$$

$$= \$40(13.7987) + \$1,000(0.3101)$$

$$= \$551.95 + \$310.10 = \$862.05$$

4.5 TAXATION OF INCOME

An unpleasant fact of life is that the only inevitable things appear to be "death and taxes." Governments require money for their operations and citizens are the source of that money. There are many forms of taxation, such as business taxes, property taxes, excise taxes, license fees, and income taxes. Most of these taxes vary with the locale and character of the investment or undertaking. As such, they are specifically included in the analysis of a particular situation. Income taxes, however, are not specific to a project but rather, affect you directly as an individual or corporation.

Although the particular rules concerning the treatment of some deductions or the percentage of tax charged will vary from year to year as the legislation changes, certain general concepts are usually followed. In this section we present the general concepts, and in Section 4.10 we discuss the specific rules needed to evaluate an investment.

4.5.1 Taxation of Individual Income

A general concept in taxation is that money spent to earn money is a legitimate expense and that taxes are paid only on the surplus. This works well for a company or corporation, where it can be argued that all, or most, activities are directed toward earning revenue and therefore all, or most, expenditures will be deductible expenses. In the case of a person on a fixed salary or hourly wage, very few deductions are allowed. It could be argued theoretically that eating and keeping warm are necessary to earning a living, and thus living expenses should be deductible against income. However, arguments of this type do not impress taxation officials, and for the employed individual taxpayer, there is a very limited range of permissible deductions. For the self-employed person there is more scope for deductions, but also the requirement to keep detailed records to show that expenses were connected to the generation of income.

Marginal Tax Rate

A progressive tax rate is applied to individual incomes. The term *progressive* does not mean modern, up to date, or any of the usual connotations of the word; rather it means that as your taxable income increases, so does the portion of it that the tax collector claims. Table 4.1 illustrates the structure of personal income tax rate schedules.

Example 4.16

Problem: Calculate your tax for a taxable income of $16,500.

Solution: Table 4.1 shows that this taxable income falls between $15,000 and $18,200.

Tax on the first $15,000 of taxable income	$2,330.00
Tax on the next $1,500 is 27%, or	405.00
	$2,735.00

TABLE 4.1 Sample tax rate schedule.

Taxable Income	Tax
Less than $ 2,300	$ 0
2,300	0 + 12% on the next $1,000
3,400	132 + 14% on the next $1,000
4,400	272 + 16% on the next $2,100
6,500	608 + 17% on the next $2,000
8,500	948 + 19% on the next $2,300
10,800	1,385 + 22% on the next $2,100
12,900	1,847 + 23% on the next $2,100
15,000	2,330 + 27% on the next $3,200
18,200	3,194 + 31% on the next $5,300
23,500	4,837 + 35% on the next $5,300
28,800	6,692 + 40% on the next $5,300
34,100	8,812 + 44% on the next $7,400
41,500	12,068 + 50% on the excess

The important factor to notice in this calculation is that while the average tax rate is

$$\frac{\$2,735}{\$16,500} \times 100\% = 16.58\%$$

if you increase your taxable income by one dollar ($1.00), you will increase your tax liability by (27% × 1.00) or 27 cents. The rate of tax charged on the next dollar of income, 27% in this instance, is the *marginal tax rate*. The total tax paid divided by taxable income is called the *average tax rate*.

Since investments generally imply a movement from a current position, it is the marginal tax rate that should be used to evaluate the investment.

4.5.2 Tax Shelters

Occasionally, the government wishes to encourage or reward certain types of behavior. It does this by structuring the tax law in such a way that the desired behavior yields some positive benefit. An excellent example of this is the situation of Registered Retirement Savings Plans (RRSPs). The government would really like people to save for their old age. Therefore, it permits taxpayers, subject to certain limits, to deduct from current income money placed in a retirement account. Of course, you must pay tax on the money when you withdraw it from the account, but in the interim period you have gained in two ways. First, you have deferred the payment of taxes to a later date and thus have been able to use this money to earn a rate of return. Second, because of the marginal tax rate structure, you are removing the amount from income at a time when you are in a high income tax bracket, and adding it to income at a time when you are likely to be in a lower tax bracket.

Example 4.17

Problem: Assume that you are able to invest $1,500 per year over a period of 20 years in a Registered Retirement Savings Plan at an interest rate of 10%. Your current marginal tax rate is 40%. How much would you have in the fund at the end of 20 years?

Solution: $A = \$1,500$, $i = 10\%$, and $n = 20$ years. $F = ?$

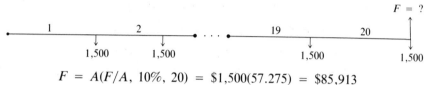

$$F = A(F/A, \ 10\%, \ 20) = \$1,500(57.275) = \$85,913$$

Problem: Assume that a uniform amount is withdrawn each year (annual payments are assumed for ease of calculation) over the following 10 years and are subject to a 20% effective income tax rate. What would the annual payments be?

Solution: $P = \$85,913$, $i = 10\%$, and $n = 10$ years. $A = ?$

85,913

$$A = P(A/P, \ 10\%, \ 10) = \$85,913(0.1628) \quad = \$13,987$$

Income tax payment $= \$13,987(0.20) \quad = \$ \ 2,797$

Net income $= \$13,987 - \$2,797 = \$11,190$

Problem: Assume, instead, that you paid taxes on the money at your marginal tax rate of 40% and invested the money into a regular savings account at 10%. The amount invested each year would be $\$1,500(1 - t) = \$900$. Also, since the interest earned is taxable, the after-tax interest rate is $10\%(1 - t) = 6\%$.

Solution: $A = \$900$, $i = 6\%$, and $n = 20$ years. $F = ?$

$$F = \$900(F/A, 6\%, 20)$$

$$= \$33,107$$

Problem: Withdraw the amount over a 10-year period, assuming that your tax rate is now 20%.

Solution: Your after-tax interest rate is $10\%(1 - 0.2) = 8\%$. $P = \$33,107$, $i = 8\%$, and $n = 10$ years. $A = ?$

$$A = \text{(net income)} = P(A/P, 8\%, 10)$$

$$= \$33,107(0.1490)$$

$$= \$4,933 \text{ per year}$$

Of course, this considerable advantage rests on the assumption that you drop to a lower marginal tax rate after retirement, but for most of us that is a reasonable assumption.

4.6 CORPORATE TAXATION

Corporations pay income tax on annual profits earned. Although there are special tax rates for small businesses and certain specified industries, for most corporations the effective tax rate is between 40 and 50%.

In arriving at taxable income, a corporation is allowed certain deductions. See the portion of the corporate cash flow diagram shown in Figure 4.1. The relationship between before-tax and after-tax cash flows can be derived from a consideration of the following general income statement.

XYZ Corporation
Income Statement
for Year Ending MM DD YY

Operating revenue	OR
Operating costs	OC
Before-tax cash flow	OR − OC
Minus interest on debt	I
Minus depreciation allowance	D
Taxable income	OR − OC − D − I
Income tax (at rate t)	t(OR − OC − D − I)
Net income after tax	(OR − OC − D − I)(1 − t)

After-tax cash flow = before-tax cash flow − income tax

$$\text{ATCF} = \text{OR} - \text{OC} - t(\text{OR} - \text{OC} - D - I)$$

$$= \text{OR}(1 - t) - \text{OC}(1 - t) + tD + tI \tag{4.1}$$

From this statement it can be seen that:

1. To convert operating revenues to an after-tax value, multiply by $(1 - t)$.

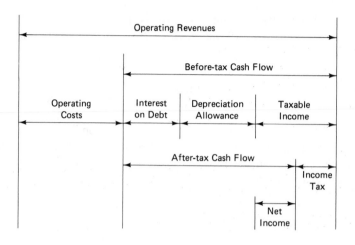

FIGURE 4.1 Cash deductions to arrive at ATCF.

2. To convert operating cost to an after-tax value, multiply by $(1 - t)$.
3. The effect on the after-tax cash flow of depreciation allowance is equivalent to a positive cash flow equal to the depreciation allowance times the tax rate.
4. The effect on the after-tax cash flow of debt interest is equivalent to a positive cash flow equal to the debt interest times the tax rate.

Because interest is generally a function of the capital structure of the corporation, and not specifically attached to a single investment opportunity, it is usually accommodated by the use of a tax-sheltered cost of debt capital $[i_{dt} = i_d(1 - t)]$ (see Section 5.4).

Example 4.18

Problem: Assume that a company has $1,000,000 in operating revenues and $200,000 in operating costs. The depreciation allowance is $100,000 and interest payments on borrowed money are $50,000. If the tax rate is 40%, what will the income tax payment be?

Solution: OR = $1,000,000, OC = $200,000, D = $100,000, interest payments = $50,000, and t = 40%.

$$\text{BTCF} = \text{OR} - \text{OC} = \$1,000,000 - \$200,000 = \$800,000$$

$$\text{TI} = \text{BTCF} - (D + \text{interest payments})$$

$$= \$800,000 - (\$100,000 + \$50,000) = \$650,000$$

$$\text{IT} = t(\text{TI}) = 0.40(\$650,000) = \$260,000$$

4.7 DEPRECIATION

Income and taxes are computed at least on an annual basis. However, many assets, such as buildings, equipment, furniture, vehicles, and so on, have lives much longer than 1 year. To include the total cost of these assets in a 1-year calculation of income does not properly represent the true operations of that year. Therefore, there must be some method of allocating the cost of an asset over its useful life. This method of allocation is called *depreciation*.

Depreciation may be defined as the loss in value of a physical asset with the passage of time. It is generally agreed that all assets, with the possible exception of land, lose value with the passage of time. This loss of value is usually considered in two broad categories, physical loss and functional loss.

Physical loss is use related (e.g., "wear and tear") and time related, which includes such factors as corrosion and action of the elements. Although physical loss would appear to be the primary cause of loss of value, it is, in fact, the lesser of the two types. More important is *functional* loss. This usually results from inadequacy due to growth in demand beyond the capacity of the asset, obsolescence due to changes in technology and customer tastes, and social pressures resulting in pollution control measures and safety regulations.

The aim of the accountant with respect to depreciation is to allocate the asset's original cost less some estimated net salvage value over the estimated useful life of the investment. The accountant is concerned with allocating the asset's original cost over the period during which the asset is being used to produce income.

It is important to realize that the method of allocation for accounting purposes or pricing purposes may vary significantly from the method of allocation for an economy study. Our major concern throughout this book is the economy study, where *depreciation allowance is an allowable deduction (capital cost allowance) in computing taxable income*. Government regulations specify quite clearly the amount of depreciation allowance that can be deducted in any given year. This deduction may not have any direct relationship to the depreciation expense charged on the financial statements.

4.7.1 Straight-Line Depreciation

The most straightforward method of depreciation is to apportion the cost uniformly over the life of the asset. This is called *straight-line* (SL) *depreciation*. Let

P = first cost of the asset

SV = estimated salvage (resale) value

n = useful life

D = annual depreciation allowance

Then

$$\text{Depreciation allowance} = \frac{\text{first cost} - \text{salvage value}}{\text{useful life}}$$

or

$$D = (P - SV)/n \qquad (4.2)$$

The *book value* (amount of unallocated original cost still on the books) at the end of year z is

$$BV = P - \frac{z}{n}(P - SV) \qquad (4.3)$$

Example 4.19

Problem: A $120,000 investment in factory equipment has a salvage value of $20,000 5 years hence. Calculate the annual depreciation and remaining book value for each year.

Solution: By the straight-line method,

$$D = (P - SV)/n$$

$$= (\$120,000 - \$20,000)/5 = \$20,000 \text{ per year}$$

Year	Depreciation Allowance	Accrued Depreciation	Book Value
1	$20,000	$ 20,000	$100,000
2	20,000	40,000	80,000
3	20,000	60,000	60,000
4	20,000	80,000	40,000
5	20,000	100,000	20,000 = SV

4.7.2 Declining-Balance Depreciation

The straight-line method takes a fixed percentage of the original cost less the estimated salvage value each year. *Declining-balance* (DB) *depreciation* takes a fixed percentage of the remaining undepreciated cost (book value) each year. If d is the declining-balance rate, then

$$D = dP(1 - d)^{n-1} \qquad \text{for year } n \qquad (4.4)$$

and

$$BV = P(1 - d)^{n-1} \quad \text{at the beginning of year } n \quad (4.5)$$

Example 4.20

Problem: A \$120,000 investment is made in factory equipment. Assume that a 20% rate applies. Determine the depreciation allowance that may be taken each year over the first 5 years.

Solution: The depreciation allowance in year 1 is

$$(\$120,000)20\% = \$24,000$$

In year 2,

$$D = 20\%(\$120,000 - \$24,000) = \$19,200$$

Year	Depreciation Allowance	Accrued Depreciation	Book Value
1	\$24,000	\$24,000	\$96,000
2	19,200	43,200	76,800
3	15,360	58,560	61,440
4	12,288	70,848	49,152
5	9,830	80,678	39,322

Notice that this method causes the depreciation allowance to be large in the initial years and progressively smaller in the subsequent years.

In the United States prior to 1981, declining-balance rates as high as twice the straight-line rate $(2/n)$ were allowed as an accelerated depreciation method. A rate of $2/n$ is referred to as *double-declining-balance depreciation*.

Example 4.21

Problem: Compare straight-line depreciation over 5 years using a zero salvage value with declining-balance depreciation using a rate of $2/n$. $P = \$100,000$ and $SV = 0$.

Solution

Year	Declining Balance, 40% Rate Depreciation Allowance	Accrued Depreciation	Straight Line, 20% Rate Depreciation Allowance	Accrued Depreciation
1	\$40,000	\$40,000	\$20,000	\$ 20,000
2	24,000	64,000	20,000	40,000
3	14,400	78,400	20,000	60,000
4	8,640	87,040	20,000	80,000
5	5,184	92,224	20,000	100,000

4.7.3 Sum-of-the-Years'-Digits Depreciation

Sum-of-the-years'-digits (SYD) *depreciation* is another accelerated write-off method that was quite popular in the United States prior to 1981. With the SYD method the depreciation rate changes each year. The denominator is equal to the sum of the years' digits and the numerator is equal to the life, n, in year 1 and decreases by one each year thereafter. This rate is applied to the depreciable base $(P - SV)$.

$$D_z = (P - SV)\left(\frac{\text{remaining life in years}}{\text{sum of the years' digits for the total useful life}}\right) \quad (4.6)$$

$$= (P - SV)\left(\frac{n - z + 1}{1 + 2 + 3 + 4 + \cdots n}\right)$$

$$= (P - SV)\left[\frac{n - z + 1}{n(n + 1)/2}\right]$$

and

$$BV_z = (P - SV)\left[\frac{(n - z)(n - z + 1)}{n(n + 1)}\right] + SV \quad (4.7)$$

Example 4.22

Problem: Compute SYD allowance for an asset with $P = \$100,000$, $SV = 0$, and $n = 5$ years.

Solution

Year z	$n - z + 1$	SYD Factor	Depreciation Allowance	Accrued Depreciation
1	5	5/15	$33,333	$ 33,333
2	4	4/15	26,667	60,000
3	3	3/15	20,000	80,000
4	2	2/15	13,333	93,333
5	1	1/15	6,667	100,000
Sum of digits $= 15^a$				

^{a} Sum-of-the-years-digits $= n(n + 1)/2 = 15$.

4.7.4 Usage-Related Depreciation Methods

The methods described previously all depended on time of ownership. An asset was depreciated at an equal rate whether it was used continuously or only once a month. Other methods, which relate to the use of the asset, could also be employed. For example, depreciation could be based on units of production or machine hours or income earned. These methods can be developed as follows.

For units of production, let

u = units produced during year

U = total units asset will produce

Then

$$D = \frac{u}{U}(P - SV)$$

For machine hours, let

m = hours run during year

M = total life of machine in hours

Then

$$D = \frac{m}{M}(P - SV)$$

For income-earning property such as a film, let income for the year = r and total estimated income = R. Then

$$D = \frac{r}{R}(P - SV)$$

4.8 DEPLETION

Depletion represents the consumption of an exhaustible natural resource. Property subject to depletion includes mines, oil and gas wells, timber, and other exhaustible natural deposits. As the resource is consumed or depleted through production, the value of the asset decreases. This loss in value is subject to a depletion allowance, which is similar to a depreciation allowance.

In the United States the depletion allowance may be computed by two methods: the cost method and the percentage method.

4.8.1 Cost Method

Cost depletion of minerals is similar to the units of production method of depreciation. The depletion allowance is based on the amount of the resource available for consumption and the initial cost of the resource. The total number of units (tonnes, barrels, cubic metres, etc.) in the deposit is divided into the cost of the deposit. Assume that an oil reservoir is estimated to contain 100,000 barrels of oil and the investment cost to develop the reserve is $1,000,000. The unit depletion rate is

$$\$1,000,000/100,000 = \$10 \text{ per barrel}$$

The depletion allowance for the year on 5,000 barrels of production would be $50,000.

4.8.2 Percentage Method

The *percentage method of depletion* is based on a percentage rate applied to annual gross income derived from the resource. The applicable percentage in the United States depends on the type of property being depleted and presently ranges from 5 to 22%. If the option is open to compute the depletion allowance under either method, one must use the method that produces the larger deduction, which is usually the percentage method.

In Canada, basically, a taxpayer is entitled to deduct 25% of the "resource profits" for the year to the extent of the "earned depletion base" at year end. However, there are special allowances that apply in certain industries. These allowances may also vary by province and should be checked accordingly.

4.9 CAPITAL TAX FACTORS

When establishing the accounts of a company, the accountant will select that method of depreciation that best represents the actual activities and use of the assets. However, when calculating the income of a company for the purpose of paying income tax, the federal government specifies in the tax legislation the depreciation procedure to be used. The specific procedures and percentages applicable to different assets are in Section 4.10. The present section deals with one method of including the depreciation allowance in an economic evaluation.

The purchase of a capital asset gives rise initially to a large negative cash flow. The depreciation allowance (CCA) on this investment then produces a series of small positive cash flows (tax credits). This is illustrated in Figure 4.2, where d_n is the depreciation allowance percentage at the end of year n.

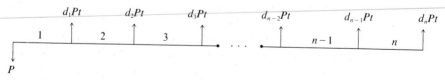

FIGURE 4.2 Tax credits resulting from the purchase of a depreciable asset.

The present equivalent of the series of tax credits is

$$Pt\left[\frac{d_1}{1 + i} + \frac{d_2}{(1 + i)^2} + \frac{d_3}{(1 + i)^3} + \cdots + \frac{d_n}{(1 + i)^n}\right]$$

so the after-tax cost of the asset is

$$P - Pt\left[\frac{d_1}{1 + i} + \frac{d_2}{(1 + i)^2} + \frac{d_3}{(1 + i)^3} + \cdots + \frac{d_n}{(1 + i)^n}\right]$$

$$= P\left\{1 - t\left[\frac{d_1}{1 + i} + \cdots + \frac{d_n}{(1 + i)^n}\right]\right\}$$

The quantity in braces is referred to as a *capital tax factor* (CTF). The value of the CTF will vary depending on the applicable tax legislation, but its use in economy studies is uniform.

Generally, the sum of the tax credits has a simple summation. For example, the CTF for straight-line depreciation to zero salvage value is

$$\text{CTF} = 1 - \frac{t}{n}\left[\frac{1}{1 + i} + \frac{1}{(1 + i)^2} + \frac{1}{(1 + i)^3} + \cdots + \frac{1}{(1 + i)^n}\right]$$

$$= 1 - \frac{t}{n}\left[\frac{(1 + i)^n - 1}{i(1 + i)^n}\right]$$

and since the series sum is the uniform series factor,

$$\text{CTF} = 1 - \frac{t}{n}(P/A, i, n) \tag{4.8}$$

The CTF for the tax credits from declining-balance depreciation of percentage amount d is

$$\text{CTF} = 1 - td\left[\frac{1}{1 + i} + \frac{1 - d}{(1 + i)^2} + \frac{(1 - d)^2}{(1 + i)^3} + \cdots + \frac{(1 - d)^n}{(1 + i)^{n+1}} + \cdots\right]$$

$$= 1 - \frac{td}{i + d} \tag{4.9}$$

Tables for commonly used capital tax factors are given in Appendix F.

Example 4.23

Problem: A patent that has been purchased for $175,000 has a remaining life of 13 years and zero salvage value. It is estimated that the patent will generate operating revenues of $35,000 per year and operating costs of

$8,000 per year. Find the after-tax present value of the investment using straight-line depreciation, an after-tax rate of return of 12%, and a tax rate of 46%.

Solution: The before-tax cash flow diagram is as follows:

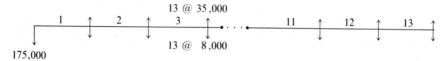

Convert these data into an after-tax cash flow diagram using the relations from equation 4.1:

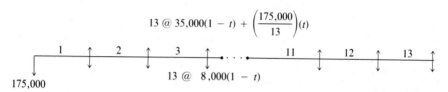

However, the depreciation tax credits ($175,000/13)(t) may be converted to a present sum by the capital tax factor. Therefore, our cash flows can be represented as follows:

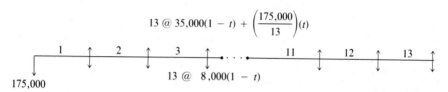

The equation for the present worth then follows from the diagram.

$$PE = (OR - OC)(1 - t)(P/A, i, n) - P(\text{CTF})$$

$$= (\$35,000 - \$8,000)(0.54)(6.4236) - \$175,000(\text{CTF})$$

$$\text{CTF} = 1 - (t/n)(P/A, i, n)$$

$$= 1 - (0.46/13)(6.4236)$$

$$= 0.7727$$

$$PE = (\$27,000)(0.54)(6.4236) - \$175,000(0.7727)$$

$$= -\$41,566$$

We probably would not use the capital tax factor (CTF) with the straight-line method because it does not reduce the calculation time in this instance. However, as will be illustrated, the CTF does simplify calculations with most depreciation methods.

The capital tax factor converts an initial investment, P, to an after-tax value by subtracting from the amount (P) the present equivalent of the depreciation tax credits that result from the investment.

In situations where the capital tax factor can be used, it usually provides a considerable reduction in the calculations required. However, because the CTF considers all future depreciation allowance, down to a book value of zero, in those situations where the asset is disposed of before the full depreciation term it is necessary to adjust the salvage value as follows,

Net salvage value = receipts from disposal + tax effect of disposal

+ adjustment of tax credits already taken by the CTF

The calculations necessary to adjust the tax credits already taken by the CTF usually do not warrant its use in solving problems where the asset is disposed of before it has reached a zero book value, except for the declining-balance method, which is fully discussed in section 4.12.

Fortunately, the general situation in industry is that the asset is not disposed of until after it is fully depreciated (zero book value), so the CTF can normally be used. However, instances of early disposal do occur, for example, when doing economic life (Chapter 9) and replacement (Chapter 10) calculations.

4.10 CURRENT TAX LEGISLATION IN CANADA

Tax laws are subject to continuous review. Minor changes are made in these laws every year and major revisions occur periodically. The following discussion is intended to give the reader an understanding of how the tax laws affect investment decisions. However, Canadian tax law covers many volumes the size of this book. Therefore, with respect to any specific economy study, check with your accountant or tax specialist to ensure that you have included any special tax considerations.

4.10.1 Capital Cost Allowance

Canadian income tax law permits corporations to depreciate most capital assets by the declining-balance method at a rate specified in the tax legislation. (A *capital asset* is one with a useful life of more than 1 year.) With very few exceptions, the taxpayer is given no option as to the depreciation method or the depreciation rate to be used. Since the depreciation allowance for income tax purposes represents an allowable deduction in computing taxable income, it is appropriately referred to as *capital cost allowance* (CCA) in Canada.

The legislation is also very specific about how the capital cost allowance is to be calculated. Basically, the prescribed method is what is considered asset class accounting. That is, all assets of a single class are grouped together into a single ledger account. When additional assets of that class are acquired, their cost is added to the account; when assets are disposed of, the proceeds are deducted from the account. The capital cost allowance for any year is the account total at the end of the year times the capital cost allowance rate for the class. The capital cost allowance rate that applies to each class is the maximum rate. The company can apply any rate between zero and the maximum. The maximum capital cost allowance that a company would take in any one year would just be sufficient to reduce the taxable income to zero. In addition, in the first year of ownership of a depreciable asset, only one-half of the maximum rate can be applied.

4.10.2 The Half-Year Rule

In the year most depreciable assets are acquired the maximum capital cost allowance that can be claimed is one-half of the maximum capital cost allowance rate with respect to assets acquired after November 12, 1981. However, for most classes of assets, disposal of assets during the year are first netted against acquisitions made in the same year. Consequently, the effect of the half-year rule is mitigated when there are major disposals of fixed assets that can be netted against additions.

There are exceptions to this netting rule. For example, for assets with a normal capital cost allowance rate of 50% (note classes 24, 27, 29, 34), the maximum capital cost allowance in the year of acquisition is 25%. However, any claim forgone in one year may be claimed in full in future years (e.g., if no CCA is taken in year 1, 75% may be claimed in year 2).

In all examples and problems to follow where the assets were acquired after November 12, 1981, we will apply the half-year rule. The method of calculating capital cost allowance is outlined in the following example.

Example 4.24

Problem: A firm has six vehicles; the make, age, and current book value of each is as follows:

Description	Book Value
1982 Chevy van	$12,400
1975 Ford station wagon	3,910
1981 Chrysler sedan	9,770
1980 Ford Bronco	8,760
1983 Dodge half-ton	13,580
Total book value	$57,580

What depreciation deduction (CCA) is permitted at the end of the current year (year 1)?

Solution: According to the tax legislation, automotive equipment is a class 10 asset and can be depreciated using a CCA rate of 30%. Therefore, all the vehicles would be grouped into a single CCA schedule as follows:

(1)	(2)	(3)	(4)	(5)	(6)	(7)	(8)	(9)
Class No.	Undepreciated Capital Cost at Beginning of Year	Cost of Additions during Year	Proceeds from Disposals during Year	Adjust-ments Pursuant to Subsection 13(7.1)[a]	Undepreciated Capital Cost (Col. 2 plus 3, less Cols. 4 and 5)	Rate (%)	Capital Cost Allowance	Undepreciated Capital Cost at End of Year (Col. 6 less Col 8)
10	$57,580	0	0		$57,580	30	$17,274	$40,306

[a] Column 5 is where one includes any government assistance or investment tax credits received.

The year 1 CCA = $17,274.

Problem: In year 2 the company sells the Ford Wagon for $3,000. What CCA is permitted at the end of the year?

Solution: When the firm disposes of an asset, the proceeds from the disposal are subtracted from the current book value total. The sale of the wagon for $3,000 is treated in the schedule as follows:

(1)	(2)	(3)	(4)	(5)	(6)	(7)	(8)	(9)
Class No.	Undepreciated Capital Cost at Beginning of Year	Cost of Additions during Year	Proceeds from Disposals during Year	Adjust-ments Pursuant to Subsection 13(7.1)	Undepreciated Capital Cost (Col. 2 plus 3, less Cols. 4 and 5)	Rate (%)	Capital Cost Allowance	Undepreciated Capital Cost at End of Year (Col. 6 less Col. 8)
10	$40,306	0	$3,000	—	$37,306	30	$11,192	$26,114

Notice that the book value of the wagon was = $3,910(1 - 0.30) = $2,737. So the sale resulted in a recapture of ($3,000 - $2,737) = $263 of capital cost allowance. However, recaptures and losses are not explicitly calculated; only the class total is adjusted. The result is that the recapture/loss is subtracted/added back into income at the same CCA rate as it was taken out. The year 2 CCA = $11,192.

Problem: In year 3 the company buys a Toyota Van for $14,000 and a Ford Thunderbird for $23,000. What CCA is permitted at the end of year 3?

Solution: When assets are acquired, only one-half of the normal CCA is permitted in the first year. Therefore, with the purchase of the van and the Thunderbird for a total capital cost of $37,000 in effect, only $37,000/2 = $18,500 is added to the depreciation schedule this year. The remaining $18,500 is added next year.

(1)	(2)	(3)	(4)	(5)	(6)	(7)	(8)	(9)
Class No.	Undepreciated Capital Cost at Beginning of Year	Cost of Additions during Year	Proceeds from Disposals during Year	Adjustments Pursuant to Subsection 13(7.1)	Undepreciated Capital Cost (Col. 2 plus 3, less Cols. 4 and 5)	Rate (%)	Capital Cost Allowance	Undepreciated Capital Cost at End of Year (Col. 6 less Col. 8)
10	$26,114	$18,500	0	0	$44,614	30	$13,384	$31,230

The CCA in year 3 = $13,384.

Problem: In year 4 there are no transactions. What CCA is permitted at the end of year 4?

Solution: Although there are no transactions during the year, there is still the $18,500 to add from the previous year due to the half-year rule.

(1)	(2)	(3)	(4)	(5)	(6)	(7)	(8)	(9)
Class No.	Undepreciated Capital Cost at Beginning of Year	Cost of Additions during Year	Proceeds from Disposals during Year	Adjustments Pursuant to Subsection 13(7.1)	Undepreciated Capital Cost (Col. 2 plus 3, less Cols. 4 and 5)	Rate (%)	Capital Cost Allowance	Undepreciated Capital Cost at End of Year (Col. 6 less Col. 8)
10	$31,230	$18,500	—	—	$49,730	30	$14,919	$34,811

The CCA in year 4 = $14,919.

Problem: In year 5, all the vehicles except the Thunderbird are sold for a total of $30,000. What CCA is permitted at the end of year 5?

Solution: Subtracting the $30,000 from the sale gives the following schedule:

(1)	(2)	(3)	(4)	(5)	(6)	(7)	(8)	(9)
Class No.	Undepreciated Capital Cost at Beginning of Year	Cost of Additions during Year	Proceeds from Disposals during Year	Adjustments Pursuant to Subsection 13(7.1)	Undepreciated Capital Cost (Col. 2 plus 3, less Cols. 4 and 5)	Rate (%)	Capital Cost Allowance	Undepreciated Capital Cost at End of Year (Col. 6 less Col. 8)
10	$34,811	—	$30,000	—	$4,811	30	$1,443	$3,368

The CCA in year 5 = $1,443.

Observe that at this point the firm has a 2-year-old Thunderbird which should have a book value of

$$BV = \text{original cost} - \text{accumulated capital cost allowance } (D)$$

$$D_1 = (\$23,000)0.15 = \$3,450$$

$$D_2 = (\$19,550)0.30 = \$5,865$$

$$BV = \$23,000 - \$9,315$$

$$= \$13,685$$

or

$$BV = \$23,000(1 - 0.15)(1 - 0.30) = \$13,685$$

However, the amount on which the CCA is calculated is \$4,811. This illustrates the difference between the asset class account totals and actual book values. The account totals reflect the transactions for the asset class and are a result of assets that are no longer possessed. Generally, as long as the firm possesses a single asset in any one class, in this case one vehicle, the account remains open and the CCA is calculated annually. If the last

TABLE 4.2 A sample table listing assets by class and the corresponding declining balance rate normally applied within each class[a].

	Rate (%)	Class
Automobiles	30	10
Bridges	4	1
Buses	30	10
Contractors movable equipment	30	10
Contractors: heavy equipment (excavating, compacting)	50	22
Cutting part of a machine	100	12
Dies	100	12
Distributing equipment for gas production and distribution	6	2
Distributing equipment for power	6	2
Distributing equipment for water	6	2
Manufacturing and processing machinery and equipment	50	29
Most other machinery and equipment not listed under class 29	20	8
Jigs	100	12
Logging equipment	30	10
Mining equipment and machinery	30	10
Oil pipelines	6	2
Oil storage tanks	10	6
Oil equipment for use above ground	30	10
Power plants: electric	6	2
Pulp and paper mills	10	5
Refrigeration equipment	20	8
Roads	4	1
Tangible capital assets not specifically listed	20	8
Telegraph and telephone equipment	8	17
Tools under \$100	100	12
Tractors	30	10
Building: brick, stone cement frame, log, stucco, corrugated iron	10	6
Land: nondepreciable		

[a] Rates are subject to change and special rates are periodically initiated by the government.

asset is sold, the account is closed and the remaining balance (loss or recapture) is added to that year's income.

4.10.3 Asset Classes

The tax legislation lists 37 different classes that cover all conceivable capital assets. Table 4.2 lists typical capital assets that are encountered in economy studies.

4.11 CANADIAN TAX RATES

The basic federal corporation tax rate is 46%. This amount is reduced by a provincial tax credit of 10%, but then the provinces and territories impose their own corporate tax. Their rates range from 10 to 16%. In addition, the federal government sometimes imposes a surtax on top of the federal tax. The calculation of the effective income tax rate is basically as follows:

Basic federal tax rate	46%
Minus provincial tax credit	10%
Net amount	36%
Plus 5% federal surtax (5% × 36%)	1.8%
Federal taxes	37.8%
Plus provincial tax (depends on province)	13%
Corporate tax rate	50.8%

There are also special deductions for manufacturing and processing industries which reduce the effective tax rate to about 46%. A small business is entitled to a small business deduction in basic federal tax. This results in small businesses being taxed at an effective rate of about 25%.

4.12 CAPITAL TAX FACTORS

The basic cash flow pattern for an asset depreciated according to the declining-balance method at rate d with the one-half-year rule is illustrated in Figure 4.3.

The present equivalent of the CCA tax credits is

$$PE = \frac{(d/2)Pt}{1 + i} + \frac{(1 - d/2)dPt}{(1 + i)^2} + \frac{(1 - d/2)(1 - d)dPt}{(1 + i)^3}$$
$$+ \frac{(1 - d/2)(1 - d)^2 dPt}{(1 + i)^4} + \cdots$$
$$PE = P\left[\left(\frac{td}{i + d}\right)\left(\frac{1 + i/2}{1 + i}\right)\right]$$

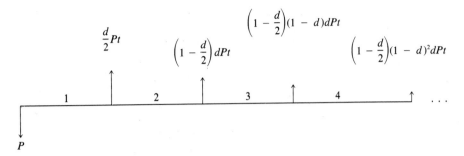

FIGURE 4.3 Cash flow pattern of tax credits due to CCA.

Therefore, the after-tax cost of an asset is

$$P\left[1 - \left(\frac{td}{i + d}\right)\left(\frac{1 + i/2}{1 + i}\right)\right]$$

The value in brackets is referred to as the capital tax factor.

Since the CCA is a fraction of the declining balance, the book value of an asset never reaches zero. Also, the asset class method of calculation provides that the recapture or loss on disposal is spread over the subsequent years at the CCA rate. The effect of this procedure is to produce a cash flow pattern such as illustrated in Figure 4.4.

The present equivalent at the end of year n due to tax adjustments is

$$\begin{aligned}
PE_n &= \frac{SV(td)}{1 + i} + \frac{SV(i - d)td}{(1 + i)^2} + \frac{SV(i - d)td}{(1 + i)^3} + \cdots \\
&= SV\left(\frac{td}{i + d}\right)
\end{aligned}$$

The salvage value, adjusted for tax effects, is then

$$SV = SV\left(1 - \frac{td}{i + d}\right)$$

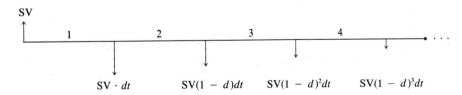

FIGURE 4.4 Tax effects of salvage.

The term $1 - [td/(i + d)]$ is, in fact, the capital tax factor for declining balance when there is no half-year rule.

Example 4.25

Problem: A capital expenditure of $125,000 is made for equipment (CCA class 8, 20% declining balance). The investment is estimated to generate the following before-tax cash flows for 5 years:

Year	Amount
1	$30,000
2	45,000
3	60,000
4	75,000
5	90,000

At the end of 5 years the equipment is disposed of for $20,000. The tax rate is 46% and the after-tax minimum acceptable rate of return is 12%. Find the present equivalent value of this investment.

Solution: This problem will be solved in two ways: first, using the capital tax factors, and second, using a tabular format. The CTF simplifies calculations, but the tabular format is more illustrative.

1. *Solution using CTFs:*
The before-tax cash flows (values in thousands) are

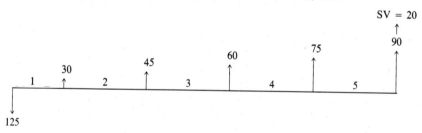

The after-tax cash flows are

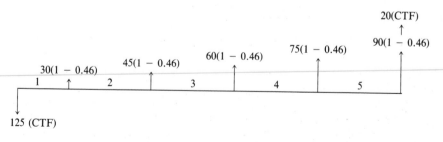

$\text{CTF}(P) = 0.7279$ and $\text{CTF}(\text{SV}) = 0.7125$.

$$PE = \$30(1 - 0.46)(P/A, 12\%, 5) + \$15(1 - 0.46)(P/G, 12\%, 5)$$
$$- \$125 \, CTF(P) + \$20(CTF(SV))(P/F, 12\%, 5)$$
$$= + \$30(0.54)(3.6048) + \$15(0.54)(6.3970) \$ - 125(0.7279)$$
$$+ \$20(0.7125)(0.5674)$$
$$= + \$58.3978 + \$51.8157 - \$90.9875 + \$8.0855$$
$$= \$27,300$$

Two items to notice in the calculation. First, with the declining-balance method, theoretically the book value never reaches zero. Second, the CCA (depreciation) continues after disposal of the asset.

2. *Solution using tabular format* (thousands of dollars):

Item	Year	BTCF	CCA	Tax-able Income	In-come Tax (46%)	ATCF	(P/F, 12%, n)	Present Equiv-alent
Equipment	0	$ − 125				$ − 125	1.0	$ − 125
Earnings	1	30	$12.5	$17.5	8.05	21.95	0.8929	19.60
Earnings	2	45	22.5	22.5	10.35	34.65	0.7972	27.62
Earnings	3	60	18.0	42	19.32	40.68	0.7118	28.96
Earnings	4	75	14.4	60.6	27.88	47.12	0.6355	29.95
Earnings	5	90	11.52	78.48	36.10	53.90	0.5674	30.58
Salvage	5	20				20	0.5674	11.35

At time of sale the book value was $46,080. Thus, after the sale, there was a remaining book value of ($46,080 − $20,000) = $26,080 to be taken in CCA over future years.

6	5.216	−2.40	2.4	0.5066	1.22	
7	4.173	−1.92	1.92	0.4524	0.87	
8	3.338	−1.54	1.54	0.4039	0.62	
9	2.671	−1.23	1.23	0.3606	0.44	
10	2.137	−0.98	0.98	0.3220	0.32	
11	1.709	−0.79	0.79	0.2875	0.23	
12	1.367	−0.63	0.63	0.2567	0.16	
13	1.094	−0.50	0.50	0.2292	0.12	
14	0.875	−0.40	0.40	0.2046	0.08	
15	0.700	−0.32	0.32	0.1827	0.06	
16	0.560	−0.26	0.26	0.1631	0.04	
	⋮			Total PE	= 27.22	

If we were to continue taking CCA on the remaining book value, we would arrive at a total PE = $27,300.

4.13 DISPOSAL OF CAPITAL ASSETS IN CANADA

The general rule on the disposal of a capital asset is to deduct the proceeds of the sale from the undepreciated capital cost (book value) of the capital assets in a particular class.

After-tax cash flows for economy study purposes generally deal with individual assets, not with groups, so it is necessary to determine the net salvage value of a capital asset on disposal. (The exception to the rule occurs when using the declining balance method with capital tax factors.) If an asset is disposed of for more or less than its undepreciated capital cost (book value), the tax implications of this difference must be considered. Three possible situations can occur and are discussed below.

4.13.1 Capital Gains Tax in Canada

A capital gain occurs when an asset is sold for more than its initial purchase price. A capital gain can occur without recaptured capital cost allowance only if the asset sold is a nondepreciable asset such as land or if the asset is sold in the first year of ownership before any capital cost allowance has been taken. Only one half of a capital gain is subject to tax. In a corporation with a constant tax rate, this means that the capital gains tax rate is 50% of the effective tax rate.

Example 4.26

Problem: A piece of land was purchased 3 years ago for $100,000 and sold today for $120,000. The book value today of this land is $100,000 (capital cost allowance cannot be taken on land). The effective tax rate (t) is 40%. What are the capital gains tax and the net salvage value?

Solution

$$\text{Capital gain} = \$120,000 - \$100,000 = \$20,000$$
$$\text{Capital gains tax rate} = t/2 = 20\%$$
$$\text{Capital gains tax} = 0.20(\$20,000) = \$4,000$$
$$\text{Net salvage value} = \text{selling price} - \text{capital gains tax}$$
$$= \$120,000 - \$4,000 = \$116,000$$

4.13.2 Recapture or a Terminal Loss on Disposal

If an individual asset is sold for more than its book value but for no more than its original cost, recapture of capital cost allowance occurs. Thus

$$\text{Recaptured CCA} = \text{selling price}$$

$$- \text{book value if selling price} < \text{original cost}$$

If an individual asset is sold for less than its book value a terminal loss on disposal occurs. Thus

$$\text{Loss on disposal} = \text{book value} - \text{selling price}$$

From an accounting viewpoint, with the pooling of assets in each class the "pool" is credited/penalized with any recapture/terminal loss that may occur on the sale of an individual asset. Therefore, the disposal of an individual asset at more or less than book value is not normally a concern to the accountant. In accounting terms with pooled assets, recapture of capital cost allowance happens only when a negative balance occurs within a specific class ledger at the end of a taxation year. A terminal loss occurs only if all the assets in a given class have been disposed of but a balance of undepreciated capital cost remains on the books. However, in conducting an economic analysis the disposal of an individual asset resulting in recapture or a terminal loss may have a significant impact on the feasibility of the investment. The normal situation when using the declining balance method is to apply capital tax factors. Under these conditions, the purchase price (P) is multiplied by the CTF, and the salvage value (SV) less any capital gain is multiplied by the CTF (see example 4.25). These calculations consider any tax implications due to recapture or a loss on disposal. If a net salvage value is required when using the declining balance method it can be calculated as follows.

As discussed previously under capital tax factors, the present equivalent of the tax adjustments for the declining balance method is equal to

$$PE = td/(i + d) \text{ (the recapture or terminal loss on disposal)}$$

Consider the following example.

Example 4.27

Problem: A capital asset was purchased 3 years ago for $10,000 and is sold today for $7,000. The asset has been depreciated for capital cost allowance purposes using a 20% declining-balance rate. The effective tax rate is 46% and $i = 15\%$. What is the net salvage value from the sale?

Solution

$$\begin{aligned}
\text{Book value (BV)} &= \text{original cost} - \text{cumulative depreciation} \\
&= \$10,000(1 - d/2)(1 - d)(1 - d) \\
&= \$10,000(0.9)(0.8)(0.8) \\
&= \$5,760
\end{aligned}$$

$$\begin{aligned}
\text{Recaptured CCA} &= \text{selling price} - \text{book value (BV)} \\
&= \$7,000 - \$5,760 \\
&= \$1,240
\end{aligned}$$

The present equivalent of additional taxes in future years due to recapture:

$$PE = \$1,240(td)/(i + d)$$

$$= \$1,240(0.46 \times 0.20)/(0.15 + 0.20)$$

$$= \$1,240(0.2629)$$

$$= \$326$$

Net salvage value = selling price − (present equivalent of future additional taxes)

$$= \$7,000 - \$326$$

$$= \$6,674$$

Problem: If the asset was sold for $1,000, what is the net salvage value from the sale?

Solution

Loss on disposal = book value − selling price

$$= \$5,760 - \$1,000$$

$$= \$4,760$$

Present equivalent of tax savings in future years due to the loss on disposal:

$$PE = \$4,760(0.2629)$$

$$= \$1,251$$

Net salvage value = selling price + (present equivalent of future tax savings)

$$= \$1,000 + \$1,251$$

$$= \$2,251$$

There is the occasional (rare) situation where one sells a depreciable asset for more than the original cost. In this case, to arrive at the net salvage value requires calculation of both recapture and capital gain.

Example 4.28

Problem: A piece of equipment was purchased 2 years ago for $100,000 and sold today for $120,000. The machinery was depreciated using the declining-balance method and a CCA rate of 20%. The effective tax rate is 40% and MARR = 12%. What is the net salvage value?

Solution

$$\text{Capital gain} = \$120,000 - \$100,000 = \$20,000$$

$$\text{Capital gains tax } 0.20(\$20,000) = \$4,000$$

$$\text{Book value (BV)} = \$100,000(1 - d/2)(1 - d)$$

$$\text{Book value} = \$100,000(1 - 0.1)(1 - 0.2)$$

$$= \$72,000$$

$$\text{Recaptured CCA} = \$100,000 - \$72,000$$

$$= \$28,000$$

$$\text{PE of tax on recapture} = \$28,000(td)/(i + d)$$

$$= \$28,000(0.4)(0.2)/(0.32)$$

$$= \$7,000$$

$$\text{Net salvage} = \$120,000 - \$4,000 - \$7,000$$

$$= \$109,000$$

4.14 FAST-WRITE-OFF TAX PROVISIONS

The government uses the tax act to encourage certain types of investment. This is especially obvious with respect to manufacturing and process machinery, energy-efficient equipment such as solar heating systems, waste-heat recovery systems, and water and air pollution control property. The investment in these capital assets can be depreciated to zero in 3 years using a CCA rate of 50% and applying the half year rule. (25% in year 1, 50% in year 2, and 25% in year 3). The pattern of tax credits is illustrated in Figure 4.5. Because the deduction is short and straightforward, it is more convenient to work with the actual capital cost allowances than to calculate a capital tax factor.

FIGURE 4.5 Tax credits from Fast-Write-Off tax provision.

Example 4.29

Problem: Global Industries is considering the purchase of a piece of manufacturing equipment worth \$100,000. The estimated salvage value in 5 years is zero. i (after-tax) $= 10\%$ and $t = 46\%$.

Estimated annual operating revenues $50,000
Estimated annual operating costs   20,000

The equipment can be depreciated using the fast-write-off method to calculate depreciation allowance (CCA).

Determine the present equivalent after-tax value of this investment.

Before-tax cash flow diagram (values in thousands of dollars):

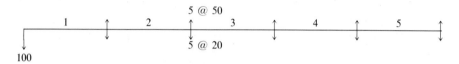

After-tax cash flow diagram:

$$D_1 = (\$100,000)0.25 = \$25,000$$

$$D_2 = (\$100,000)0.50 = \$50,000$$

$$D_3 = (\$100,000)0.25 = \$25,000$$

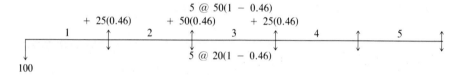

Solution: $i = 10\%$, and $t = 46\%$.

$$PE = + (\$50,000 - \$20,000)(1 - 0.46)(P/A, 10\%, 5) - \$100,000$$
$$+ \$25,000(0.46)(P/F, 10\%, 1) + \$50,000(0.46)(P/F, 10\%, 2)$$
$$+ \$25,000(0.46)(P/F, 10\%, 3)$$
$$= \$61,410 - \$100,000 + \$38,103$$
$$= \$-487$$

The investment pays slightly less than 10%.

4.15 STRAIGHT-LINE WRITE-OFF

Straight-line write-off is permitted for leasehold improvements, patents, franchises, concessions, and licenses. These assets are generally related to a specific period (e.g., life of the lease), so the annual depreciation allowance

(CCA) is obtained by apportioning the cost of the asset over the remaining life. Therefore, use equation 4.2 with zero salvage value to calculate the annual depreciation allowance. These assets are not subject to the half-year rule.

Example 4.30

Problem: A company is interested in purchasing the patent rights to a specialized piece of equipment. The patent rights are good for the next 6 years and can be purchased for $36,000. The patent is expected to generate revenues of $20,000 per year. Operating costs are expected to be $8,000 per year. $t = 40\%$ and i (after-tax) = 10%. Determine the present equivalent after-tax value of the investment.

Solution: The before-tax cash flow diagram is as follows:

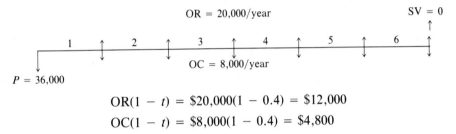

$$OR(1 - t) = \$20,000(1 - 0.4) = \$12,000$$
$$OC(1 - t) = \$8,000(1 - 0.4) = \$4,800$$

To calculate tax savings due to depreciation allowance (CCA) each year:

$$D = (\$36,000 - 0)/6 = \$6,000$$
$$\text{Tax savings each year} = (\$6,000)t = \$2,400$$

The after-tax cash flow diagram is as follows:

$$OR (1 - t) = \$12,000/\text{year}$$
$$\text{Tax savings} = \underline{\quad 2,400/\text{year}}$$
$$\text{Total} = \$14,400/\text{year}$$

6 @ 14,400

| | | 1 | ↑ | 2 | ↑ | 3 | ↑ | 4 | ↑ | 5 | ↑ | 6 | ↑ |

$P \doteq 36,000$ 6 @ 4,800

$$PE = (\$14,400 - \$4,800)(P/A, 10\%, 6) - \$36,000$$
$$= + \$9,600(4.3553) - \$36,000$$
$$= \$5,811$$

The positive present equivalent means that the investment pays more than 10% after taxes.

Example 4.31

Problem: A patent was purchased by a company 5 years ago for $200,000. At the time of purchase the patent rights were good for 10 years. The book value of the patent account was $500,000 prior to the purchase. Assume that the average life of patents in this account is 10 years for capital cost allowance purposes. Determine the net salvage value to be used in an economy study if the patent is sold today for $150,000. $i = 15\%$ and $t = 46\%$.

Solution

$$D = \$200,000/10 = \$20,000/\text{year}$$

$$\text{Book value (BV)} = \$200,000 - (\$20,000)5$$

$$= \$100,000$$

$$\text{Recaptured CCA} = \$150,000 - \$100,000 = \$50,000$$

Find the present equivalent of additional taxes in future years due to recaptured CCA. The recapture is spread over a period equal to the patent life.

$$\text{Spread } \$50,000 \text{ over 10 years} = \$5,000/\text{year}$$

Additional taxes each year:

$$(\$5,000)t = (\$5,000)0.46 = \$2,300/\text{year}$$

$$PE = \$2,300(P/A, 15\%, 10)$$

$$= \$2,300(5.0188) = \$11,543$$

$$\text{Net salvage value (NSV)} = \$150,000 - \$11,543 = \$138,457$$

Problem: Find the net salvage value if the patent is sold for $100,000.

Solution: Selling price = $100,000, BV = $100,000, and NSV = $100,000.

Problem: Find the net salvage value if the patent is sold for $60,000.

Solution

$$\text{Terminal loss on disposal} = \$100,000 - \$60,000$$

$$= \$40,000$$

Find the present equivalent of tax savings due to $40,000 loss on disposal.

$$\text{Spread } \$40,000 \text{ over } 10 \text{ years} = \$4,000 \text{ per year}$$
$$\text{Tax savings each year} = (\$4,000)0.46$$
$$= \$1,840$$

Present equivalent of tax savings:

$$PE = \$1,840(P/A, 15\%, 10) = \$9,235$$
$$\text{Net salvage value (NSV)} = \$60,000 + \$9,235$$
$$= \$69,235$$

4.16 SUMMARY

In this chapter we have presented several examples of applications involving interest and compounding. In addition, taxation has been introduced, both at the individual and corporate level. Income tax is an important component associated with most economic studies. Depreciation allowance (CCA) represents a deduction allowed by the government to arrive at taxable income. This deduction represents an incentive to capital investment and the amount may vary depending on government policy. In considering the depreciation allowance using the declining-balance method, the capital tax factor saves tedious calculations that can result in solution errors.

The net salvage to be used in an economy study will usually differ from the selling price of the asset. To arrive at the net salvage value, any capital gain, recaptured capital cost allowance, or loss on disposal must be considered.

PROBLEMS

4.1 Solve for the following equivalents using an interest rate of 15%.
   a. An amount on January 1, 1995, equivalent to an annual amount of $100 from January 1, 1986, to January 1, 1995, inclusive.
   b. What present amount must be invested to realize $1,000 at the end of year 5 and $1,000 at the end of year 10?
   c. An amount today equivalent to a uniform annual payment of $100 for the next 100 years.
   d. If $5,000 is deposited today, what uniform annual amount can be withdrawn so as to completely deplete the investment in 20 years?
   e. What present investment is needed to secure a perpetual income of $5,000 per year?

4.2 You have just won a lottery paying $1,000 a year for life (use $n = \infty$). You have also been offered $11,000 today in lieu of the prize. Assuming an annual interest rate of 10%, would you accept the annual prize or the $11,000?
a. Assume that the first payment is today.
b. Assume that the first payment is 1 year from today.

4.3 The bank of your choice has just offered a new savings account whereby one can deposit any uniform annual amount for 10 years at 8% interest. What would the balance of the account be at the end of year 10 if $1,500 were deposited each year starting today and the last deposit were made 9 years from today?

4.4 The maintenance costs of a student's motorcycle are estimated to be $50 in year 1, $100 in year 2, $150 in year 3, $200 in year 4, and $250 in year 5. With $i = 10\%$, find the annual and present equivalents of these costs.

4.5 Because of the cyclical pattern of a flood-fighting business, the net cash flows for the company may be $1,500, $2,000, $2,500, $3,000, $3,500, $3,000, $2,500, $2,000, and $1,500 for the next 9 years. With an interest rate of 8%, find the annual equivalent of these cash flows.

4.6 A company owns a fleet of trucks and operates its own maintenance shop. A certain type of truck, normally used for 5 years, has a first cost of $4,000 and a salvage value of $1,000. Maintenance costs are $100 the first year and increase by $200 each year. Assuming that the before-tax rate of return required is 20%, find the annual equivalent cost of owning and maintaining the truck.

4.7 Your company is trying to fix the economic life of its delivery trucks. The cost of running and maintaining one of these trucks is $3,000 in the first year and is expected to rise by $500 per year thereafter. A truck costs $8,000 new: its secondhand value at the end of each year is

Year	1	2	3	4	5	6
Value ($)	5,600	3,920	2,745	1,920	1,345	940

The company's before-tax rate of return (BTRR) is 15%. Compare the annual equivalent costs of replacing the trucks after:
a. Three years.
b. Four years.

4.8 Assume that you have agreed to repay a loan of $5,000 with payments of $1,200 a year for 5 years. What is the actual interest rate you are paying?

4.9 Assume that you have agreed to repay a loan of $12,000 over the next 4 years at $3,000 per annum. You agreed with the loan company to prepay $2,000 and therefore received only $10,000. What is the actual interest rate you are paying?

4.10 A car may be purchased for $5,500 cash or $1,500 down payment and $100 a month for 60 months. Find the rate of return at which these two purchase plans are equivalent. Which plan would you choose? Interpolate between 1 and 2% to determine the monthly rate.

4.11 You are considering the purchase of a new boat. Your options are to pay $10,000 cash or $2,000 as a down payment and $330 per month for 30 months. If you can invest your money (net after taxes) for 12%, which plan should you choose?

4.12 In planning for your retirement you have decided you would like to be able to provide $1,000 per month for a 10-year period. The first withdrawal will occur 20 years from today.
  a. What amount will you need to deposit in the bank today at an interest rate of 12% to provide for these withdrawals?
  b. What amount could you deposit annually for the next 20 years to receive the same benefits?

4.13 If you were to invest $500 today, $500 2 years from today, $500 4 years from today, and $500 6 years from today, what future amount will be accumulated at the end of year 10 if the interest rate is 12%? Try to minimize your calculations.

4.14 Using an effective annual interest rate of 10%, find the present equivalent of the following cash deposits into a bank account:
  a. $1,000 now, $1,000 3 years from now and every 3 years thereafter, for a total of 10 deposits.
  b. $2,000 5 years from today and every 5 years thereafter, for a total of 20 deposits.

4.15 Using an interest rate of 10%, find the annual equivalent cost for the following investment. The first cost equals $50,000, the estimated economic life is 10 years, and the estimated salvage value is $10,000. Periodic overhauls of equipment are expected to cost $1,000 and will occur every 2 years starting at the end of year 4. The last overhaul is in year 10. Maintenance costs will be $5,000 in year 1 and will increase by $500 per year until the end of year 5 and then level off at $7,000 per year for the remaining years. Draw a cash flow diagram.

4.16 You purchased a home 5 years ago (60 months) for $100,000. The down payment was $20,000 and the mortgage was for 25 years at a 12% nominal annual interest rate.

a. What are your monthly payments?

b. To pay off the mortgage, how much money do you need?

4.17 You have purchased a home with a first mortgage of $50,000 at 10% over 25 years and a second mortgage of $20,000 at 15% over 10 years.

a. What is your monthly payment on each mortgage?

b. How much cash will you need to pay off the second mortgage 36 months from now?

4.18 What semiannual deposit in a bank account will be necessary to accumulate a fund of $10,000 in 10 years assuming that you make the deposits at the beginning and midpoint of each year and that the final fund is to be accumulated 6 months after the last payment? Interest is at 8% per annum compounded semiannually.

4.19 An $80,000 machine can be purchased for $7,500 per month for 12 months or in 12 equal monthly payments with interest at 12% per annum. Under which set of conditions should the machine be purchased?

4.20 You presently owe $10,000 on a loan that will be paid off over the next 20 years at 8% per annum. You can refinance this loan at 6% interest per annum. How much can you afford to pay today for refinance charges, legal fees, penalties, and so on, without losing any money by refinancing? Assume annual payments.

4.21 You have your house mortgage with your bank and you would like to consolidate your other outstanding debts with a single loan from the bank. They are:

1. Car payments of $300 per month for the next 24 months. Principal owing = $6,200.
2. Furniture payments of $250 per month for the next 30 months. Principal owing = $7,000.
3. Boat payments of $400 per month for the next 20 months. Principal owing = $7,400.

Your banker has agreed to loan you the $20,600 for 24 months at 1% per month. Should you make the bank loan?

4.22 Tom and Shirley recently purchased their first home. Their desire for all the modern conveniences and appreciation of fine engineering led them to the showrooms of the Ace Appliance Company. There they found a suitable refrigerator, stove, washer, and dryer. After careful negotiation the salesperson agreed to part with these valuable items for a total sum of $1,978.79. However, if they were short of ready cash, he could let them have the appliances for only $60 a month for the next 5 years (nothing down and the first payment due 1 month from today). Their credit is very good, and they can borrow money from the bank at a nominal 12% annual interest rate, compounded monthly. Should Tom and Shirley borrow the money from the bank or take the time-payment plan? Why?

4.23 A corporation sells a bond issue of $10,000,000 to a trust firm for $9,000,000. The bonds are $1,000 denomination, pay 10% interest annually, and mature in 20 years. Annual expenses to take care of this transaction amount to $50,000. An initial expense of $500,000 is involved in legal and accounting fees to close the transaction. Find the cost of debt capital to the corporation expressed as an interest rate (e.g., show all cash flows on a cash flow diagram and calculate i).

4.24 A $1,000, 10% bond that matures in 6 years is offered for sale for $900. If you could invest your money for 12% elsewhere, would you be interested in buying this bond? Draw a cash flow diagram.

4.25 The city in which you live has a severe cash flow shortage and has offered the taxpayer the opportunity to pay property taxes in advance for the next 5 years. Your property taxes next year will be $1,000 and are expected to increase by 5% each year over the next 5 years. You can pay a lump sum of $4,500 now in lieu of annual property taxes over the next 5 years. You estimate that you can invest your money elsewhere for 8% after income taxes. Should you pay your property taxes in advance?

4.26 You expect the sales of a small company you are interested in buying to be $100,000 in year 1 and to grow by 10% per year over the next 5 years. You would now like to determine the present value of these projected sales figures using a discount rate of 20%. Draw a cash flow diagram.

4.27 Maintenance costs for a company are expected to be $10,000 in year 1 and increase by $1,000 per year over the next 3 years and then increase by 5% per year for a period of 4 years. Use a discount rate of 10% and determine the present equivalent maintenance cost over the 8-year period. Draw a cash flow diagram.

4.28 Your annual salary today is $25,000. If inflation increases by 6% per year and you are able to maintain salary increases each year equivalent to the increase in inflation, what will your annual salary be:
a. Five years from now?
b. Ten years from now?
c. Twenty-five years from now?

4.29 Assume that you are able to invest $5,500 per year over a period of 20 years in a Registered Retirement Savings Plan at an interest rate of 10%.

a. How much would you have in the fund at the end of 20 years?

b. What net after-tax annual equivalent amount can you withdraw over the following 20-year period? Assume that you are subject to a 30% effective income tax rate.

4.30 The taxable income generated by a company is $100,000. Assume that their effective tax rate is 35%.

a. What will their income tax payments be?

b. What is their net income for the year?

4.31 A company generates $500,000 in operating revenues (OR) and incurs operating costs (OC) of $200,000. The depreciation allowance (CCA) equals $100,000 and interest payments on debt capital are zero.

a. If the tax rate is 46%, what will the income tax payment be?

b. What is the net income?

4.32 Assume that the following cash flows and deductions occur for a company that has an effective income tax rate of 50%:

Operating revenues (OR)	$1,000,000
Operating costs (OC)	600,000
Depreciation allowance (D)	150,000
Interest payments on borrowed money (debt capital)	100,000

a. What will the income tax payment be?

b. What will the net income be?

4.33 Assume that your taxable income after deductions is $10,500. Use Table 4.1 and determine your income tax payment, the marginal tax rate, and the average tax rate.

4.34 Depreciation has different meanings to different people.

a. Define depreciation for income tax purposes.

b. Define depreciation for accounting purposes.

4.35 What is the difference between functional and physical depreciation?

4.36 What impact does depreciation have on the income statement?

4.37 A new machine has a first cost of $60,000. Capital cost allowance can be calculated using the declining balance method and a CCA rate of 20%.

a. What is the CCA in year 2?

b. Compute the book value at the end of 6 years.

e. Assume a before tax cash flow (BTCF) of $10,000 in year 3 and calculate the income tax payment to be made. Use a tax rate of 40%.

4.38 A new machine has a first cost of $100,000. The life of this machine is estimated to be 10 years. The salvage value in 10 years is estimated to be $20,000. Use the fast-write-off method to calculate CCA. $t = 50\%$, $i = 15\%$.
  a. What is the CCA in year 2?
  b. What is the book value at the end of year 2?
  c. What is the net salvage value in year 10?
  d. Calculate the income tax payment in year 3 if the before tax cash flow (BTCF) is $25,000. Use a tax rate of 50%. Interest payments on debt capital are zero.

4.39 A saleswoman has just purchased a new car for $12,000 which she expects to have zero salvage value in 3 years. She has the opportunity to depreciate this car over 3 years using either the fast write-off method or the declining balance method and a CCA rate of 30%. Which method should she use provided that she has sufficient income each year that she has to pay income tax? With either method she can claim the full CCA each year. Draw a cash flow diagram.

4.40 A new heating and ventilating system is under installation in an office building for $300,000. What is the annual depreciation charge and the book value each year for the first 3 years using the declining-balance method (20% rate)?

4.41 Using the declining-balance method and a capital cost allowance rate of 30% determine:
  a. The CCA for each of the first 3 years for a new automobile with a first cost of $12,000.
  b. The income tax payment in year 2 if BTCF = $20,000, interest on debt capital = 0%, and $t = 50\%$.

4.42 A new milling machine was just purchased for $100,000. The salvage value in 5 years is estimated to be $30,000. Determine in tabular format the assets CCA, book value, and accumulated CCA each year.
  a. If the machine is eligible for the fast-write-off tax provision.
  b. If the asset is depreciated using the declining-balance method and a CCA rate of 20%.

4.43 A company is considering the purchase of some new equipment for $10,000. The equipment is estimated to have a useful life of 5 years and an estimated salvage value in 5 years of zero. Using the fast-write-off tax provision, determine:
  a. The CCA, book value, and accumulated CCA each year over the 3-year recovery period.

b. The income tax payment assuming that the BTCF generated by the equipment in year 2 is $12,000, the interest on debt capital is zero, and $t = 40\%$.

c. The net income in year 2 if the interest on debt capital is $2,000.

4.44 A power company has just installed a new transformer worth $50,000. Using the declining-balance method and a CCA rate of 6% determine

a. The CCA for year 5.

b. The book value at the end of year 5.

c. The net income for year 5 assuming that BTCF = $40,000, interest on debt = 0, and $t = 46\%$.

4.45 A capital expenditure of $100,000 is expected to generate revenues of $60,000 per year over the next 5 years. Assume operating costs of $12,000 per year and a selling price in 5 years of zero. Use an income tax rate of 40%, an after-tax discount rate of 20%, and assume 100% equity capital.

a. Generate a table showing all cash flows and determine the present equivalent value for this investment. Use the declining-balance method and a CCA rate of 30%

b. Repeat part (a) using the fast-write-off tax provision.

4.46 Assume that a corporation purchased a piece of land 3 years ago for $50,000 and sold it today for $300,000. The effective income tax rate is 40%. Assume that capital gains from other sources in the company do not affect this sale.

a. Calculate the capital gains tax.

b. Calculate the net salvage value.

4.47 A piece of factory equipment was purchased for $300,000 3 years ago and sold just after the third year for $150,000. The effective tax rate is 46%, $i = 15\%$. Use declining-balance and a CCA rate of 20%. Assume that gains or losses from other sources in the company do not affect the sale.

a. What is the net salvage value from the sale?

b. If the equipment is sold for $350,000, what is the net salvage value from the sale?

c. If the equipment is sold for $230,000, what is the net salvage value from the sale?

4.48 Find the present equivalent after-tax cost of the following investment. $P = \$100,000$, $i = 20\%$, $n = 5$ years, $t = 46\%$, and SV = $30,000. Use CTF with the declining-balance method and a CCA of 30%.

4.49 A company just purchased a piece of equipment which has an estimated useful life of 6 years. $t = 46\%$.

Estimated annual operating revenues	$40,000
Estimated annual operating costs	20,000
Initial cost	60,000
Estimated salvage value in 6 years	0

Use the declining-balance method and a CCA rate of 30% to determine
a. The present equivalent before-tax value of the investment $i = 20\%$. Draw a cash flow diagram.
b. The present equivalent after-tax cash flow value for the investment $i = 12\%$. Draw an after-tax cash flow diagram. Do not use the CTF.
c. Work part (b) using the CTF.

4.50 A company just purchased a piece of equipment which has an estimated useful life of 4 years. $t = 46\%$.

Estimated annual operating revenues	$55,000
Estimated annual operating costs	25,000
Initial cost	80,000
Estimated salvage value in year 4	0

Use the fast-write-off tax provision to determine
a. The present equivalent before-tax value of the investment $i = 25\%$. Draw a cash flow diagram.
b. The present equivalent after-tax value of the investment $i = 15\%$. Draw a cash flow diagram.

CHAPTER FIVE
A Firm's Capital

A firm's balance sheet illustrates that for all assets, such as cash, inventory, and fixed assets, there is a corresponding claim, either by the firm's creditors, or by its owners. Money is provided to a firm under certain conditions and with the expectation of earning a profit. Thus it is incumbent on the firm to manage its assets in a manner that will provide this expected return. In this chapter we consider the sources of capital available to the firm, the conditions under which it is available, and the monetary costs associated with its availability.

5.1 CAPITAL

Capital in its broadest sense means the total resources of a person or organization. In business, the word "capital" takes on various meanings when used in different contexts. For example:

1. *Capital expenditure* is the expenditure of money on a long-term asset (building, equipment) that is used by the organization. This is opposed to an operating expenditure, which represents the purchase of such

items as materials, labor, and utilities that are consumed within a short period (usually, a maximum of 1 year).

2. *Share capital or equity capital* is the total amount of money invested in the firm by its owners or shareholders.

3. *Debt capital* is the total amount of money loaned to a firm by its creditors.

4. *Capital employed* generally means the sum of fixed assets plus working capital.

5. *Capital structure* is the mixture of the types of debt and equity that are listed on the right-hand side of the balance sheet, but excluding short-term debt.

Capital (money) is provided to an organization by lenders and investors, under certain conditions and with certain expectations. It is the job of management to conduct the firm's business in such a way that the expectations of shareholders and creditors are fulfilled. In this chapter we consider the sources of capital available to a firm, the conditions under which it is available, and the monetary costs associated with its availability. The two basic sources are through loans and share purchases, and the corresponding amounts are called debt capital and equity capital.

5.2 TYPES OF FUNDS

There are two basic types of investment capital available: debt capital and equity capital. *Debt capital* is an obligation incurred by borrowing. *Equity capital* is the owners' (investors') share of the assets. Four features are used to distinguish these sources of capital: (1) maturity, (2) claim on income, (3) claim on assets, and (4) a right to a voice in management.

5.2.1 Maturity

Debt capital must be repaid at the time specified in the agreement between the company and its creditors. Thus debt capital is said to mature at the specified time. Equity capital has no date of maturity. The equity holders basically recover their investment through share appreciation and dividend payout. If the company is a public company (listed on the stock market) the shareholders have the option to sell their shares at any time. In small companies not listed on the market, an owner who wishes to sell an equity interest will have to find another buyer or if possible liquidate the company. However, in many instances, buy–sell agreements between owners place major restrictions on either alternative.

5.3 DEBT CAPITAL

The distinguishing features of *debt capital* are that it must be repaid at a specific time, that a specified amount must be paid for its use (interest), that its claim on the income and assets of the firm take priority over other specified claims, and that there are specified remedies the lender has in the event that the conditions are not met.

5.3.1 Terms of Repayment

At the time of borrowing, the lender and the borrower enter into an agreement as to the time and manner in which the loan will be repaid. Debt funds are usually classified as short term, intermediate term, or long term, depending on when repayment is due.

Short-term funds can have terms up to 1 year but generally are due within the next 30 to 90 days or are callable. To be *callable* means that the money is due when the lender asks for it. Most short-term bank loans are of this type and are referred to as *demand loans*. That is, the loan is due and payable when the bank demands payment. Individuals, via their savings and checking accounts, lend money to banks on this basis, so accounts that carry the right of immediate withdrawal are called *demand deposits*.

Trade credit is a major source of short-term funds. A firm with an established credit rating receives its supplies and materials, is invoiced for them, and 30 to 60 days later pays the bill. Thus the firm has the use of the money for that period. To encourage prompt payment, suppliers will often offer a discount, such as 1% if paid within 30 days. When a supplier is uncertain about a firm, or the firm has a bad credit rating or reputation, the supplier will probably deliver the goods only on a C.O.D. (cash on delivery) basis.

The main sources of short-term loans are commercial banks and finance companies. They offer money, usually demand loans, to cover fluctuations in working capital and to finance inventory. Large, well-established firms also have access to the money market, where they can borrow directly from investors by issuing their own promissory notes, which are called commercial paper.

Intermediate-term loans are those due within the next 1 to 10 years, whereas short-term loans are concerned with day-to-day fluctuations and repayment comes from working capital. Intermediate-term loans are usually used to finance a change in assets, such as the purchase of new equipment, and repayment comes from the money generated by the new equipment.

A distinguishing feature of this form of credit is that there is a repayment schedule, and the loan is repaid in regular monthly, quarterly, or yearly amounts over the term of the loan. The main sources of intermediate-term funds are commercial banks, finance and insurance companies, and government agencies.

Long-term debt is technically debt for which the term is 10 years or more. In fact, long-term debt is usually a stable and continuing part of the capital structure of the firm. When repayment of the debt principal comes due, the normal course is to borrow more long-term money for this repayment. Large firms raise long-term debt by issuing bonds and selling them on the capital markets. For small firms, the long-term debt is usually the money provided by the owners or shareholders in the form of shareholder loans.

5.3.2 Payment for Debt

Money is loaned in the expectation of earning still more money. The firm or person receiving the money must pay for the use of the money. This payment, which is usually related to both the amount involved and the time period involved is called *interest*. The most common forms, simple interest and compound interest, are explained in detail in Chapter 3.

The amounts and timings of interest payments are generally specified in the loan agreement. They can be a fixed dollar amount, a percentage amount, or an amount tied to some external variable. For example, it is common for demand loans to be tied to the official bank prime rate, and the interest rate is usually on the order of "prime plus 2%."

The making of interest payments is a contractual obligation undertaken by the firm. It is not dependent on the fortunes of the firm or on whether the firm is profitable. If the firm fails to make an interest payment on the date specified, the loan is then in default and the creditor can take action as specified in the loan agreement or prescribed by law. The type of action taken generally varies from the seizing of assets to forcing the company to declare bankruptcy.

5.3.3 Priority of Creditors

It is unlikely that one would loan money to a firm that is likely to go bankrupt. However, the eventuality must be considered, and some form of protection must exist. In a case of bankruptcy there is generally not enough money left to satisfy all the claims. Therefore, the priority of claims becomes important. Priority implies that if, for example, a firm went bankrupt and after liquidation of the assets there was $1,000,000 left, then if there were only two creditors, each with a claim of $800,000 and neither with any priority, they would share the $1,000,000, each receiving $500,000 and each sustaining a $300,000 loss. If, on the other hand, one of the creditors had a prior claim, he or she would receive $800,000 and the other creditor would receive only $200,000. Priority conditions can be included in the terms and conditions of a loan, and all creditors' claims take precedence over claims of the shareholders.

5.4 EQUITY CAPITAL

Equity capital is the money invested in a firm by owners and shareholders. Its distinguishing features are its claim on the income of the company and its right to a voice in management. Contrary to debt capital, with equity capital there is no guaranteed return and no obligation of repayment.

Firms generally raise equity capital by issuing shares. For example, a new firm wishing to raise $1,000,000 in equity capital might offer for sale 100,000 shares at $10 each. The purchasers of these shares would then become the owners, or shareholders, of the firm. The holder of one share of stock would own one hundred thousandth of the company. The firm would not receive the full $10 from each share, because there would be costs associated with the sale, such as printing, commissions, and underwriting costs.

5.4.1 Resale of Shares

Unlike debt, which carries an obligation of repayment, equity capital has no date of maturity. If an investor wishes to recover the investment, this is done by selling the share to someone else. The shares of large public companies are frequently sold or traded, so formal stock markets, such as the New York Stock Exchange, exist to manage these transactions. A shareholder in a smaller public company wishing to sell shares must locate a buyer. Private companies often have restrictions on the sale such as that the shares must first be offered to existing shareholders.

There is no fixed price for the buying or selling of shares. The original cost of the share is not relevant. A share represents a portion of ownership of a company, and if the market feels that the future prospects of that company are good, or bad, the price of the share will reflect that feeling.

5.4.2 Shareholders and Management

The shareholders are the owners of a company, but they do not manage the company; that is the job of the management. Shareholders do direct the affairs of a firm through the board of directors. The directors are elected annually by the shareholders, and it is the directors' job to set policy, to oversee the activities of the firm, and to hire and fire managers.

Formally, creditors have no direct say in management. As long as the interest and principal payments are made when due, the conduct of the firm is the concern of the managers, directors, and shareholders. In reality, creditors often exert a considerable influence on management and it is not uncommon to have a representative of the bank on the board of directors.

5.4.3 Shareholders, Dividends, and Retained Earnings

At the bottom of the income statement, after the managers have been paid, the creditors have received their interest, and the tax authorities their share, there remains the net income after taxes, or profit. This is the property of the shareholders. The board of directors will decide what portion to keep in the company and reinvest, referred to as retained earnings, and what portion to pay out to the shareholders in the form of dividends.

In a successful, well-run company, the claim of the shareholders can be quite substantial. The creditor is guaranteed interest payments regardless of the fortunes of the company, which in bad times means that often there is little left for the shareholders. However, when the company is doing very well, the creditor still receives only the interest specified; all the surplus goes to the shareholders.

The shareholders benefit according to the amount of profit, regardless of whether it is retained or paid out as dividends. If it is paid out as dividends, the shareholder receives the benefit directly. If the earnings are retained, the value of the company increases and the shareholder benefits indirectly through share appreciation; that is, the selling price of the share increases.

5.5 ACCEPTABLE RATE OF RETURN

A company acquires capital by borrowing, by selling equity, and by plowing back monies from retained earnings, sale of assets, and depreciation accounting. This capital has a cost, and for the economic operation of the firm, it should be used in such a manner that the return exceeds the cost.

In subsequent chapters, methods of evaluating investments will be introduced. These methods are designed to measure the rate of return that an investment will yield, or to measure whether the investment return exceeds some minimum acceptable level. One measure of the minimum acceptable rate of return is the cost of capital to the company. This is the rate used when the capital available exceeds the opportunities for investment (*capital-surplus position*). When the supply of investment opportunites with returns in excess of the firm's cost of capital exceeds the funds available for investment, the company is in a *capital-rationing position* and the minimum acceptable rate of return is arrived at by laddering the opportunities in order of descending rates of return and moving down the list until the funds are exhausted. The rate of return on the first project rejected is the minimum acceptable rate of return and is termed the *cutoff rate*. These concepts are investigated further in the following sections.

5.6 COST OF CAPITAL

A company's supply of capital comes from different sources and each source may have a different cost. Furthermore, different sources are treated in a different manner in income tax accounting, and under certain circumstances the tax effect will change the cost to the firm. Also, the cost from one source is not independent of the cost from other sources, as the investors and creditors tend to look carefully at a firm's capital mix.

The question of the best capital mix, and of the interaction of costs of different classes of capital, are important subjects of finance but are beyond the scope of this text. Herein we shall assume that the firm's mix is comparable to the industry averages and we shall assume the cost of each source to be independent of the cost from other sources.*

5.7 COST OF DEBT CAPITAL

Capital borrowed by a company is generally contracted on a long-term basis. This debt capital is known as *funded debt, fixed liability,* or *long-term debt* and is secured by mortgages or bonds having long maturity dates.

The contract between the company and the lender will specify certain costs and dates of interest payments. However, the rate specified on the bond or mortgage is not usually the true cost of debt since there are usually also brokerage and underwriting expenses, sales commissions, and price reductions incurred in the sale or placement of the loan. The true estimated cost of debt capital should include all costs.

Example 5.1

Problem: An 8%, $1,000 bond series is offered to the public for $950 to assure its sale by a broker, who, after commissions, turns over $915 to the firm. If the interest is paid annually and the date of maturity is 10 years, what is the cost of debt?

Solution: Ignoring, for the present, the effect of income tax, the cost can be estimated by solving for i in the equation

$$PE = \$80(P/A, i, 10) + \$1,000(P/F, i, 10) - \$915$$

* The exception is discussed in connection with leasing in Chapter 14, where the effect of changing the capital structure is explicitly treated in the formulation.

At 8%:

$$PE = \$85$$

At 10%:

$$PE = \$37.90$$

$$i = 8\% + 2\%(85)/(85 + 37.90) = 9.4\%$$

However, expenses and interest are tax deductible. If the company is a profit-making concern, these expenses will result in a reduction of tax paid, and thus the actual cost of the debt capital is lowered. Specifically, if the tax rate is t:

1. The company receives a credit of $\$35t$ ($\$35$ = brokerage expenses) at year 0.
2. The annual payments are reduced by an amount equal to $\$80(1 - t)$ ($\$80$ = annual interest payment).
3. The company sold the bonds initially for $\$950$. Thus $\$50$ of the $\$1,000$ payment in year 10 can be treated as an expense. The after-tax cost is $\$50 (1 - t)$.

The estimated after-tax cost of capital is

$$\$950 - \$35(1 - t) = \$80(1 - t)(P/A, i, 10)$$
$$+ [(\$950 + \$50(1 - t)](P/F, i, 10)$$

If $t = 48\%$:

$$\$950 - \$18.20 = \$80(1 - 0.48)(P/A, i, 10) + (\$950 + \$26)(P/F, i, 10)$$
$$\$931.80 = \$41.60(P/A, i, 10) + (\$976)(P/F, i, 10)$$
$$PE = \$41.60(P/A, i, 10) + (\$976)(P/F, i, 10) - \$931.80$$

Try 4%:

$$PE = \$65$$

Try 5%:

$$PE = \$ - 11.40$$

$$i = 4\% + 1.0\%(65/76.4) = 4.85\%$$

The effect of income tax is to reduce the cost of debt capital to the company. An approximation commonly used in economy studies is

$$i_{dt} = i_d(1 - t) \tag{5.1}$$

where i_{dt} = after-tax cost of debt capital
i_d = before-tax cost of debt capital

Usually, the timing and amount of the extra charges, commissions, and so on, are not known; therefore, the error introduced by this approximation is small.

5.8 ESTIMATING THE COST OF EQUITY CAPITAL

In evaluating the cost of equity capital we are concerned with approximating the rate of return required by the investor. There are two approaches that may be used to measure this cost: (1) approach the analysis from the viewpoint of a prospective purchaser in search of good investments to maximize a stock portfolio or (2) approach the analysis from the viewpoint of the company considering a stock issue to the public. The company has a responsibility to the existing shareholders to obtain a fair value for its shares. If the stock is overpriced, the shares will be difficult, if not impossible, to market. If the stock is underpriced, dilution results with respect to the existing shares. Either approach to the cost of equity evaluation should produce the same results.

The value of a share of stock to the investor can be viewed as the present value of the expected future stream of income received by the investor. This future stream of income is represented by dividends if the share is retained indefinitely, or by dividends plus a final liquidating dividend, that is, the market value of the share at a final disposal date.

Therefore, in effect the cost of equity capital may be defined as the market rate of discount, i, that equates the present value of all expected future dividends per share with the current market price of the stock.

This approach, although theoretically appealing, contains major practical difficulties. First, it requires an estimate of the future dividends, which is, in effect, an estimate of the future growth and dividend policy of the firm.

Second, it requires the determination of the present price of a share of stock. Since this can fluctuate from day to day, an analysis done on one day is not necessarily correct on the next day.

Despite these difficulties, the market rate of discount is the most sound method of estimating the cost of equity capital. Therefore, one makes the best forecast, and most reasonable assumptions, and calculates accordingly.

5.8.1 Models for the Cost of Equity Capital

Let

P_0 = estimate of the current market price of a share of stock
D_t = dividend expected at time t
i = interest rate that makes the stream of future dividends equivalent to the market price of the stock

If this year's dividend has just been paid, we have the following cash flows:

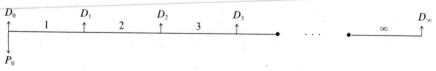

Then

$$P_0 = \frac{D_1}{1 + i} + \frac{D_2}{(1 + i)^2} + \frac{D_3}{(1 + i)^3} + \cdots + \frac{D_\infty}{(1 + i)^\infty}$$

$$P_0 = D_1(P/F, i, 1) + D_2(P/F, i, 2) + \cdots + D_\infty(P/F, i, \infty)$$

or

$$P_0 = \sum_{n=1}^{\infty} \frac{D_n}{(1 + i)^n} \tag{5.2}$$

If this year's dividend is about to be paid, we have the following cash flows:

Then

$$P_0 = D_0 + \frac{D_1}{1 + i} + \frac{D_2}{(1 + i)^2} + \frac{D_3}{(1 + i)^3} + \cdots + \frac{D_\infty}{(1 + i)^\infty}$$

$$P_0 = D_0 + D_1(P/F, i, 1) + D_2(P/F, i, 2) + D_3(P/F, i, 3) + \cdots$$

$$+ D_\infty(P/F, i, \infty)$$

or

$$P_0 = \sum_{n=0}^{\infty} \frac{D_n}{(1 + i)^n} \tag{5.3}$$

If the share is sold in year t for an amount P_t, the equations are

$$P_0 = \sum_{n=1}^{t} \frac{D_n}{(1 + i)^n} + \frac{P_t}{(1 + i)^t} \tag{5.4}$$

or

$$P_0 = \sum_{n=0}^{t} \frac{D_n}{(1 + i)^n} + \frac{P_t}{(1 + i)^t} \tag{5.5}$$

The upper range of the summation will change to $t - 1$ if the share is sold before the dividend date. To avoid errors it is best to construct the cash flow diagram for the particular situation under analysis.

Example 5.2

Problem: The company sells a stock for $40 and is expected to pay a $1.75 dividend at the end of the year. Its market price after paying the dividend is expected to be $43 per share. What is the expected return on this investment?

Solution

$$1.75 + 43$$

$$40$$

$$P_0 = P_t/(1 + i)^t + \sum_{n=1}^{t} D_n/(1 + i)^n$$

$$\$40 = \$43(P/F, i, 1) + \$1.75(P/F, i, 1)$$

$$(P/F, i, 1) = 40/44.75 = 0.8938$$

$$i = 12\%$$

Problem: Assume that the stock is retained for 5 years. Dividends are expected to increase in value by $0.20 per year and the stock is expected to sell for $66 at the end of 5 years.

Solution

$$66 + 2.55$$

1.75		1.95		2.15		2.35	
1	2	3	4	5

$$40$$

$$P_0 = P_5/(1 + i)^5 + \sum_{n=1}^{5} D_n/(1 + i)^n$$

$G = \$0.20$, $D = \$1.75$, $P_0 = \$40$, and $P_5 = \$66$.

$$PE = 0 = \$-40 + \$1.75(P/A, i, 5) + \$0.20(P/G, i, 5) + \$66(P/F, i, 5)$$

Try 15%:

$$PE = \$-40 + \$5.87 + \$1.16 + \$32.82 = \$-0.15$$

$$i = 15\%$$

Geometric Growth

If dividends per share are growing (geometrically) by a constant rate per year and if the growth rate, k, is less than the valuation rate, i, equation 3.19 becomes applicable.

$$D_1(1 + k)^n - 2D_1(1 + k)^{n - 1}$$

$$D_1(1 + k)^n$$

$$D_1(1 + k)^2$$

$$D_1(1 + k)$$

$$D_1$$

1   2   3   ...   n - 1   n

$$P_0 = D_1\{[1 - (1 + k)^n/(1 + i)^n]/(i - k)\} \qquad (5.6)$$

If $n \to \infty$ and i is greater than k, the expression

$$(1 + k)^n/(1 + i)^n \to 0$$

and

$$P_0 = D_1/(i - k) \qquad (5.7)$$

$$i = D_1/P_0 + k \qquad (5.8)$$

For many companies the assumption that dividends are expected to grow geometrically forever is reasonable. However, the model can be adjusted to handle any expected growth rate.

If reinvested funds (retained earnings) earn the same rate of return as present capital earns, the dividend payout ratio (the ratio of dividends over earnings per share) should not have any bearing on the cost of equity capital. Therefore, in evaluating the cost of equity capital, we may realistically assume that all earnings are paid out as dividends. However, if we so desire, we may model earnings per share (E/S) rather than dividends per share. The end result should be identical.

Example 5.3

Problem: Assume that a company's stock presently sells for $100 and will pay a dividend of $5 per share in year 1 with an expected growth rate in dividends of 5% per year. What is the expected rate of return on this investment?

Solution

$$i = D_1/P_0 + k = 5/100 + 5\% = 10\%$$

Problem: Assume that the expected dividend in year 1 is $5 and that it increases by 8% over the first 5 years and then levels off to a 5% growth rate. What would a prospective investor be willing to pay for this stock if she requires 10% before taxes on any stock that she purchases?

Solution: The dividends represent a geometric progression.

$$P_0 = \sum_{n=1}^{5} \frac{D_1(1.08)^{n-1}}{(1+i)^n} + \sum_{n=6}^{\infty} \frac{D_5(1.05)^{n-5}}{(1+i)^n}$$

The present equivalent of the dividends for the first 5 years equals

$$P_0 = D_1\{[1 - (1.08/1.10)^5]/(0.10 - 0.08)\}$$

$$= \$5(4.3831) = \$21.92$$

The present equivalent of the cash flow stream of dividends after 5 years equals

$$D_6 = \$5(1.08)^4(1.05) = \$7.14$$

$$P_5 = D_6/(i - k)$$

$$= \$7.14/(0.10 - 0.05)$$

$$= \$142.85$$

$$P_0 = \$142.85(P/F, 10\%, 5) = \$88.66$$

The purchaser should be willing to pay

$$\$21.92 + \$88.66 = \$110.58$$

5.9 WEIGHTED COST OF CAPITAL

A firm will have a certain mix of debt and equity capital. This is referred to as the *capital structure* of the firm. Although there are no fixed rates as to what the proportions should be, banks, investors, and lending institutions tend to be cautious and are governed by averages. Table 5.1 lists some representative average debt/equity ratios. A company whose ratio is considerably higher than the industry's average will probably encounter difficulty obtaining additional capital. The values in Table 5.1 are intended as representative only. Debt ratios have been slowly moving upward over the past 15 years. However, some reversal in this trend is likely over the next few years due to the significant increase in bankruptcies throughout North America in 1981–1985.

For a company with a specific capital structure, the cost of capital is estimated by calculating the weighted-average after-tax cost of externally

TABLE 5.1 Representative debt ratios for specific industries in North America.

Industry	Division	Debt Ratio, $r = \dfrac{Debt}{Debt + Equity}$
Construction	Building contractors	60
	Highway and bridge	20
Utilities	Telephone	55
	Power	70
	Gas distribution	60
Transportation	Air transport	45
	Railways	45
	Truck transport	60
	Pipelines	65
Wholesale trade	Drugs	20
	Farm machinery	15
	Industrial machinery	15
	Metal products	25
	Lumber and building products	20
Retail trade	Food stores	20
	Department stores	15
	Auto accessories and parts	30
	Motor vehicle dealers	35
	Hardware stores	15
	Furniture stores	20
	Drugstores	15

generated funds as in Table 5.2. Thus the company should use 10% as the weighted-average cost of capital.

TABLE 5.2 Estimating the weighted-average after-tax cost of capital.

Type of Funds	Amount	Percentage of Total		After-Tax Cost		Weighted Cost
Bonds (10%)	$ 2,500,000	25%	×	4.8%[a]	=	1.2%
Shares	4,000,000	40%	×	12%	=	4.8%
Retained earnings	3,500,000	35%	×	12%	=	4.2%
	$10,000,000	100%				10.2%[b]

[a] $t = 52\%$ and $i_{dt} = i_d(1 - t) = (10\%)(0.48) = 4.8\%$.
[b] 10.2% or 10% is the weighted-average cost of capital.

For economy studies, this breakdown often introduces a level of sophistication that is not warranted by the data. The more common situation is where the following are available:

1. An estimate of the debt ratio

2. An estimate of the cost of debt capital
3. An estimate of the return expected by the shareholders

Defining

i_d = before-tax cost of debt capital

i_e = rate of return, required or prospective, on equity capital

t = marginal taxation rate on incremental income

r = debt ratio = debt capital/(debt + equity capital)

i_a = weighted-average after-tax cost of capital

As earlier outlined, the tax-sheltered cost of debt capital is

$$i_{dt} = i_d(1 - t)$$

So the weighted-average after-tax cost of composite (debt + equity) capital is

$$i_a = ri_d(1 - t) + (1 - r)i_e \tag{5.9}$$

5.10 CUTOFF COST OF CAPITAL

The weighted-average cost of capital is derived by considering the sources of capital. To ensure long-run survival, a firm must be receiving a return on the capital employed which exceeds its cost. Thus the weighted-average cost is the minimum rate of return that a firm should consider.

Often, a firm has more opportunities than it has capital, and all these opportunities offer returns in excess of the minimum. In this situation, one alternative is to raise more capital and take advantage of the opportunities. However, this is not always possible, due to market forces and the desires of the directors and shareholders. The other alternative is to allocate the existing capital to the opportunities in such a way as to maximize the firm's profit. This is known as the *capital-rationing situation*.

The minimum acceptable rate of return in the capital-rationing situation is arrived at in the following manner:

1. First, rates of return are calculated for all existing and potential opportunities using an incremental analysis.
2. Second, the opportunities are listed in order of descending rate of return, the opportunity with the highest rate of return being placed at the top of the list.

3. Finally, one moves down the list, accepting opportunities until the funds are exhausted.

The rate of return of the last opportunity accepted becomes the new minimum acceptable rate of return. A rate of return derived in this manner is known as the *cutoff rate of return*. This concept is developed further in Section 14.6.

5.11 SOCIAL DISCOUNT RATE

Whereas private organizations attempt to maximize profit, the objective of governments is to maximize social welfare. Thus, for public projects, although the calculation of cash flows and the use of the interest arithmetic is the same as for private firms, the choice of the correct rate of return to use is much more difficult.

A complete analysis of this topic is beyond the scope of this text, although some discussion is contained in Chapter 11. At this point it is sufficient to note that the minimum acceptable rate of return for public projects is referred to as the *social discount rate*.

5.12 EFFECT OF CAPITAL STRUCTURE ON THE COST OF CAPITAL

Debt capital, because of its characteristics of repayment, interest, and its claim on assets, is generally available at a cost lower than that of equity capital. When this is combined with the tax-sheltering effect, the after-tax cost of debt capital is generally lower than equity capital. Thus it would appear from equation 5.9 that an organization could substantially lower its cost of capital by increasing the percentage of debt in its capital structure. This, however, does not tend to be the case, for the following reasons.

Leverage is the term used to refer to the ratio of debt to equity within an organization's capital structure. The term is taken from physics and is analogous to where the fulcrum is placed in a first-class lever. A company with a high debt ratio, r, is referred to as a highly leveraged company. Real estate transactions tend to be highly leveraged, so we shall use one as an example.

Suppose that a building is purchased for $10,000,000. The financing of the purchase is with $1,500,000 in equity and $8,500,000 of debt. If the market is buoyant and prices are rising, the building may be sold 1 year later for 10% more, that is, for $11,000,000. In this case, the equity holders will realize a $1,000,000 gain on their $1,500,000 investment, a 67% increase. Conversely, the market may fall by 10%, in which case the building would be sold for $9,000,000. In this situation the investors have sustained a 67% loss. The leverage, high debt ratio, magnified a 10% market shift into a 67% equity shift.

Speculators and developers tend to favor highly leveraged transactions. It permits them to realize large gains from small investments. Companies and general lenders and investors tend to shy away from this type of investment. Although they may be impressed by the large potential gains, they remain very aware of the downside risks. Thus a company trying to change its debt ratio encounters considerable resistance from management, lenders, and investors.

One way in which the financial community judges a company is by monitoring certain financial ratios. This is discussed in Chapter 2. Lenders are especially concerned with the debt ratio and the times-interest-earned ratio. If a company's debt ratio is high relative to the industry average, the company will find it increasingly more difficult to raise further debt capital. Lenders will insist either that additional equity capital accompany the additional debt capital, or will charge a premium on the debt capital. Considering these changes with equation 5.9,

$$i_a = ri_d(1 - t) + (1 - r)i_e$$

In case (1), r stays constant, so i_a stays constant. In case (2), as r increases, i_d also increases, and the resulting effect is little change in i_a.

A change in the debt ratio also has an effect on the equity holders. As the leverage increases, so does the possible variability of the dividends. There is a relationship between risk and expected return; investors expect to pay less for the prospect of variable or uncertain income. The suggested net effect of this expectation is that as the debt ratio increases, so does i_e.

This discussion is not to be interpreted to mean that the cost of capital is fixed and that a company is powerless to change it. Instead, it is intended to illustrate that there are strong market forces that work to keep the cost of capital within certain limits and that the process is much more complex than that which is implied by equation 5.9. The significant increase in bankruptcies in the period 1981–1985 is indicative of the catastrophic effect that can result from companies using too much debt in their capital structure.

5.13 SUMMARY

It is important to note that estimating the cost of capital is far from a precise science. The cost of debt capital can be estimated with a reasonable degree of accuracy when this cost is tied to the cost of a new bond issue. However, borrowing from the bank is usually tied to the prime rate, which can experience large fluctuations over short time spans (1 to 2 years). Under these conditions, the cost of debt capital becomes somewhat more difficult to calculate.

The usual approach to determining the cost of equity capital is to equate the projected future earnings stream of a share to its present market value. This value becomes difficult to evaluate with a reasonable degree of accuracy. Fortunately, within most companies the cost of capital does not need to be quantified to a large degree of accuracy. This is especially true when the demand far exceeds the supply of capital, which is normally the case. Under these conditions the minimum attractive rate of return will be somewhat higher than the weighted-average cost of capital.

PROBLEMS

5.1 Define (a) debt capital; (b) equity capital; (c) capital structure.

5.2 Determine the after-tax cost of capital for a firm with the following capital structure: cost of debt capital = 10%, cost of equity capital = 15%, r = 25%, and t = 40%.

5.3 Why should we calculate the after-tax cost of capital using equation (1) rather than equation (2)?

$$i_a = ri_d(1 - t) + (1 - r)i_e \tag{1}$$

$$i_c = ri_d + (1 - r)i_e \tag{2}$$

5.4 Assume that a company borrows money from the bank with repayment over a 5-year period at prime plus 2%. In view of the fluctuations in the prime rate over this period of time, how would you estimate the cost of this debt capital to the firm?

5.5 List two major difficulties in estimating the cost of equity capital to a firm.

5.6 How would you arrive at the cutoff rate of return in a capital-rationing situation?

5.7 Determine the weighted-average after-tax cost of capital for a firm with the following capital structure: t = 46%, i_d = 10%, i_e = 15%.

Type of Funds	Amount	Percentage of Total
Bonds	$3,000,000	30
Shares	5,000,000	50
Retained earnings	2,000,000	20

5.8 You are considering the purchase of a small apartment block for $500,000 which you expect to sell within 1 year at a good profit. Two alternatives of financing this investment are available to you.

1. Borrow $400,000 from the bank at 12% (a highly leveraged investment).
2. Finance the investment out of your savings, which are invested elsewhere at 15%.

a. If you borrowed the money from the bank and sell the property for $600,000 in 1 year, what is the before-tax profit, in dollars and as a percentage on your $100,000 investment?
b. If you borrowed the money from the bank and were forced to sell the property in 1 year for $400,000, what is the before-tax loss, in dollars and as a percentage, on your $100,000 investment?
c. Answer parts (a) and (b) if your decision was to finance the total investment from savings.

5.9 A company sells a 10%, $100 bond issue. The plan is to sell 5,000 bonds which will be redeemed at the end of 20 years for $100 each. Interest coupons worth $5 can be redeemed by each bondholder at the end of every 6 months. The cost of administering these interest payments is estimated to be $0.25 per bond every 6 months. It is predicted that the company will initially receive $90 net per bond after allowance for underpricing and broker's fees. Compute the before-tax effective interest rate on debt capital.

5.10 Assuming an income tax rate of 46%, what is the after-tax cost of debt capital in Problem 5.9?

5.11 The following conditions exist for the Agro Company, a large farm machinery company: $i_d = 10\%$, $i_e = 15\%$, $r = 15\%$, and $t = 46\%$.
a. Determine the after-tax cost of capital.
b. Using an r (debt ratio) value of 30%, determine the company's after-tax cost of capital.
c. Why doesn't the Agro Company increase its debt ratio above its present value? Discuss.

5.12 A company's stock presently sells for $50 per share and will pay a dividend of $3 per share in year 1. The expected growth rate in dividends is 4% per year. What is the expected rate of return on the investment?

5.13 Assume that you are an analyst for the Cangrow Co. and you have been asked to develop a share price for a new issue of stock the company is considering. Your projections indicate that the company should pay a dividend of $4 per share in year 1 and that dividends should increase by 8% per year for the first 10 years and level off at 5% per year from there on. What price should the company place on the stock issue if an investor wants a minimum of 12% on Cangrow stock?

5.14 A stock sells for $30 per share. This stock is expected to pay a dividend of $2 per share in year 1. Projections indicate that the dividend will increase by $0.10 per year from year 1 on. If you expect to receive 12% before taxes on any investment you make, would you buy this stock? $[(P/G, 12\%, \infty) = 69.44]$

5.15 A stock pays a dividend in year 1 of $3 per share. Projections indicate that the dividend will increase by 5% per year. The stock can be sold for $50 in 5 years (assume all cash flows to be year-end cash flows). What would you be willing to pay for this stock if you expect to receive 15% before taxes on any investments you make?

5.16 Find the cost of new equity capital for a company faced with the following conditions.

1. Present market price per share = $40.
2. Present earnings per share (E/S) is $2 and earnings are expected to grow by 5% per year.
3. Dividend payout = 100%.
4. Flotation costs, which include such items as brokerage fees, legal fees, and issue costs, are estimated at 15%.

5.17 A new company is financed with $1,000,000 in equity capital plus a bank loan of $400,000 to be paid back uniformly over a 5-year period ($80,000 per year). The company has 100 shareholders and each owns 200 shares. The company plans to pay a dividend each year of $5 per share. Projected revenues in year 2 are $500,000. Operating costs in year 2 are estimated to be $100,000. Depreciation allowance can be taken on the total investment using the declining-balance method and a 20% rate. (The full 20% allowance applies in year 1.) $i_d = 10\%$ and $t = 25\%$.

a. What rate of return did the equity holder receive on his or her investment in year 2 assuming that the equity investment remaining at the beginning of year 2 is $800,000?
b. How much will the company have available for reinvestment from operations in year 2?

5.18 The following estimates apply to an investment alternative:

	First Cost	Salvage Value (Year 5)
1. Building	$400,000	$200,000
2. Land	100,000	250,000
3. Working capital	150,000	150,000
4. Equipment	500,000	300,000

5. Annual operating costs = $200,000
   Annual operating revenues = $450,000
   Major overhaul in year 3 = $75,000
   $t = 40\%, r = 0$
6. Use the declining balance method. The depreciation rate on the building is 5% and the depreciation rate on the equipment is 30%. (The full 5% rate on the building and 30% rate on equipment apply in year 1.)

a. Assume that taxable income in year 2 is $100,000. What is the net income for the year?
b. If the company paid a dividend of $1 per share based on 10,000 shares, what total funds will be available for reinvestment from operations during year 2? Use the data given above in items 1 to 6.

CHAPTER SIX

Basis for Comparison of Investment Alternatives

The purpose of performing cash flow calculations throughout this text is comparison. The classic comparison is the one between the high-first-cost, low-operating-cost facility; and the low-first-cost, high-operating-cost facility. This is the problem that continuously confronts the analyst, and the one we address. The interest formulas were introduced in Chapter 3. In Chapter 5 we discussed how to arrive at an interest rate that suitably reflects the position (opportunity cost, social welfare cost, cost of capital) of the organization. Now the interest formulas and the interest rate are combined to do equivalence calculations using the more common methods.

6.1 EQUIVALENCE

Sometimes equivalence calculations are performed on a before-tax basis. At other times it is the after-tax situation that is analyzed. The following abbreviations will be used in this text:

BTRR: the minimum acceptable before-tax rate of return required

MARR: the minimum acceptable after-tax rate of return required (i_a)

The equivalence concept states that a company or person whose minimum acceptable rate of return is 15%:

1. Would be indifferent to a choice between $100 today and $115 1 year from today
2. Would choose $100 today over $114 1 year from today
3. Would choose $116 1 year from today over $100 today

These decisions are made by using the interest formula and MARR/BTRR to convert the sums to the same point in time. For example, they could be converted to today's sums and then:

1. The choice is between $100 and 115(P/F, 15\%, 1)$ = $100.00.
2. The choice is between $100 and 114(P/F, 15\%, 1)$ = $99.13.
3. The choice is between $100 and 116(P/F, 15\%, 1)$ = $100.87.

The key to the decision situation is the acceptance by company or individual that the conversion is valid. For example, that $114 due 1 year from today is equivalent to $99.13 today. If the concept of equivalence is accepted, the decision becomes straightforward.

It is not the interest calculations that usually raise questions about equivalence. Interest, and its arithmetic, has been with us for centuries, and it is easy to accept that $99.13 invested for 1 year at 15% will result in $114. However, our purpose initially is not to invest but to make a choice between alternatives and to realize that this choice does not change regardless of the point in time used as a *comparison date*. The comparison date can be today or at some other point in time. As long as one accepts that sums of money can be transformed to equivalent sums at a different date, comparison is possible.

Obviously, the basis to this acceptance must be the minimum acceptable rate of return (MARR for private corporations, BTRR for governments and some public corporations). If this MARR/BTRR has been arrived at by the methods of Chapter 5 and accurately reflects the opportunities and capabilities of the organization, the idea of equivalence can be accepted and the concept is valid.

The MARR/BTRR is therefore the key, and it must be recognized that it is meaningful only for a particular organization at a particular time. A decision made at one time, say when the MARR/BTRR requirement is low, may be reversed when the MARR/BTRR requirement is high. The MARR/BTRR will be different for different companies; therefore, when they are faced with similar situations, different companies make different decisions. The company that is short of funds but long on opportunities will accept only those investments promising a very high rate of return, whereas one company's rejected projects may be ideal for another company with a different set of circumstances.

6.2 METHODS OF COMPARISON

The basis of comparison is to bring all the cash flows to the same point in time. This point could be the present, the future, or some intermediate point, as long as it is the same for all alternatives.

Convention, and historical use have dictated that two time patterns are the most common. One is to convert all the cash flows to time zero using MARR/BTRR. The sum resulting from this calculation is called the *present value* or *present worth* or *present equivalent*.

The second method is to convert the cash flows using MARR/BTRR into equal annual amounts over the life of the project. This is referred to as the *annual cost* or *annual equivalent sum* or *uniform annual cost method*.

A third method is to solve for the interest rate that equates the positive and negative cash flows (the rate that makes the present value equal to zero). This value is referred to as the project's *rate of return* or, more correctly, *internal rate of return*.

These are not the only methods whereby comparison can be effected, but they are the most common methods. Also, as will be explained in subsequent sections, some of these methods have developed a specific meaning in the areas of financial decision making.

6.3 THE ACCEPT/REJECT DECISION

The basic decision that any project must face is the *accept/reject decision*. Do we do this project at all?

For this decision, any of the three classical methods are directly applicable. The decision rules are:

1. If the present equivalent sum is positive, or zero, using MARR/BTRR, accept the project.
2. If the annual equivalent sum is positive or zero, using MARR/BTRR, accept the project.
3. If the calculated rate of return is greater than or equal to MARR/BTRR, accept the project.

Example 6.1

Problem: The Greentree Forest Company is considering starting a new sawmill. Their market forecasts indicate that the proposed mill can expect a gross revenue of $2,000,000 per year for the next 30 years. The construction cost of the mill is estimated to be $7,000,000, and the annual operating costs are estimated to be $1,200,000 per year. The expected salvage value at the end of 30 years is zero. Since the proposed mill is to be located in an area of high unemployment, the government has agreed not to charge any income tax for the first 30 years. If the company's BTRR = 8%, what should the managers do?

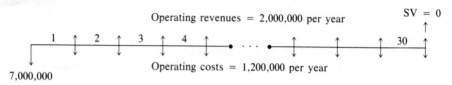

Solution

Present Value

$$PE = (\$2,000,000 - \$1,200,000)(P/A, 8\%, 30) - \$7,000,000$$

$$= \$2,006,240$$

Annual Equivalent

$$AE = \$800,000 - \$7,000,000(A/P, 8\%, 30)$$

$$= \$178,400$$

Rate of Return

$$PE = 0 = \$-7,000,000 + \$800,000(P/A, i, 30)$$

$$(P/A, i, 30) = 7/0.8 = 8.75$$

$$i = 11\%$$

In this situation the decision rules imply:

Present value: PE > 0; therefore, accept.
Annual equivalent: AE > 0; therefore, accept.
Rate of return: 11% > 8%; therefore, accept.

Problem: If the Company's BTRR = 15%, what should the managers do?

Solution

Present Value

$$PE = \$800,000(P/A, 15\%, 30) - \$7,000,000$$

$$= \$-1,747,200$$

Annual Equivalent

$$AE = \$800,000 - \$7,000,000(A/P, 15\%, 30)$$

$$= \$-266,100$$

Rate of Return

$$i = 11\%$$

In this situation the reverse is the case.

Present value: PE < 0; therefore, reject.
Annual equivalent: AE < 0; therefore, reject.
Rate of return: 11% < 15%; therefore, reject.

Note that the calculated rate of return did not change in the two situations, and in fact will not change, regardless of changes in the company's rate-of-return requirement. The calculated rate of return depends only on the project's cash flows and is independent of the company conditions. The present equivalent and annual equivalent methods, however, depend on both the cash flows and the company's MARR/BTRR.

Example 6.2

Problem: Evaluate the following investment on a before-tax basis:

$P = \$50,000 = $ first cost
$SV = \$10,000 = $ estimated salvage value
$n = 10$ years $= $ estimated useful life of the investment
$OR = \$37,000 = $ annual operating revenues
$OC = \$25,000 = $ annual operating costs
$BTRR = 20\%$ (before-tax rate of return)

Solution: Using the present equivalent approach, we have

$BTCF = $ annual operating revenues (OR) $-$ annual operating costs (OC)

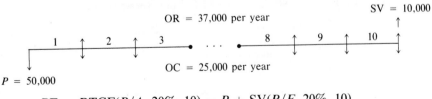

$$PE = BTCF(P/A, 20\%, 10) - P + SV(P/F, 20\%, 10)$$

$$= \$12,000(4.1925) - \$50,000 + \$10,000(0.1615) = \$1,925$$

The positive net cash flow of $1,925 indicates that the investment pays in excess of the 20% before-tax rate of return (BTRR) required and is therefore an acceptable investment.

Using the annual equivalent approach, we have

$$AE = \text{net annual value}$$

$$= BTCF - P(A/P, i, n) + SV(A/F, i, n)$$

$$= \$12,000 - \$50,000(A/P, 20\%, 10) + \$10,000(A/F, 20\%, 10)$$

$$= \$12,000 - \$50,000(0.2385) + \$10,000(0.0385)$$

$$= \$12,000 - \$11,925 + \$385 = \$460$$

The annual positive cash flow of $460 indicates that revenues are more than sufficient to meet all costs, including a 20% before-tax rate return on the investment.

Note: A present equivalent value of $1,925 is identical to an annual equivalent value of $460 when $i = 20\%$:

$$\$1,925(A/P, 20\%, 10) = \$460$$

Example 6.3

Problem: Evaluate the following opportunity by the rate-of-return method. The company's BTRR = 15%. Use the following information:

Initial investment, P	$100,000
Estimated useful life, n	5 years
Estimated salvage value, SV	$20,000
Estimated annual revenues, OR	$50,000
Estimated annual costs, OC	$25,000

Solution: The prospective rate of return on the investment is that rate of return for which

$$PE(\text{all positive cash flows}) - PE(\text{all negative cash flows}) = 0$$

or

$$AE(\text{all positive cash flows}) - AE(\text{all negative cash flows}) = 0$$

$$BTCF = OR - OC = \$50,000 - \$25,000 = \$25,000$$

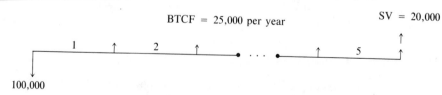

$$PE = 0 = \$25,000(P/A, i\%, 5) - \$100,000 + \$20,000(P/F, i, 5)$$

Try 12%:

$$PE = \$25,000(3.6048) - \$100,000 + \$20,000(0.5674)$$

$$= \$1,468$$

A positive value of $1,468 indicates that cash income exceeds cash outflow at a BTRR of 12%; therefore, we should try a higher rate of return. The next table value is 15%.

Try 15%:

$$PE = \$25,000(3.3522) - \$100,000 + \$20,000(0.4972)$$

$$= \$-6,251$$

The prospective rate of return is between 12 and 15%. For all practical purposes a straight-line interpolation between these values is acceptable.

$$i = 12\% + 3\%[1,468/(1,468 + 6,251)] = 12.6\%$$

Since $i <$ BTRR, the company should reject the proposal.

Note: The cash flow curve in Figure 6.1 is not linear. Therefore, it

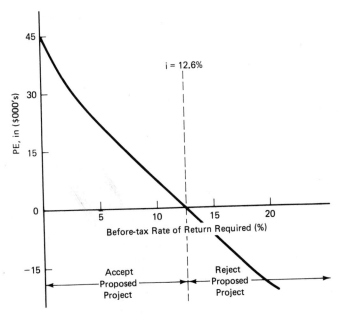

FIGURE 6.1 Present equivalent cash flows for Example 6.3.

is important to do a linear interpolation between the two closest table values; otherwise, a significant error may result.

6.4 COMPARISON BETWEEN TWO ALTERNATIVES

In the comparison of alternatives the usual situation represents a choice between high first cost and low operating costs versus low first cost and high operating costs. These comparisons can be handled by any one of the three basic methods: (1) present equivalent, (2) annual equivalent, or (3) rate of return.

However, there are two methods of approaching the comparison. One is by calculating the present equivalent or the annual equivalent, as in the accept/reject decision. The second is by evaluating the difference between the alternatives on the basis of present equivalent, annual equivalent, or rate of return. The rate-of-return method can be used only when the difference (or incremental investment) is considered. The rate of return should not be calculated on each individual project as a basis for comparison. The reason for this is illustrated in Example 6.4.

Example 6.4

Problem: Choose between the following two alternatives.

	Machine A	Machine B
First cost	$50,000	$10,000
BTCF	$15,000	$5,000
Net salvage value	$8,000	$2,000
Analysis period	10 years	10 years
BTRR	15%	15%

Solution: Applying the present equivalent approach:

Machine A

$PE = \$15,000(P/A, 15\%, 10) - \$50,000 + \$8,000(P/F, 15\%, 10)$

$\quad = \$15,000(5.0188) - \$50,000 + \$8,000(0.2472)$

$\quad = \$27,260$

Machine B

$PE = \$5,000(P/A, 15\%, 10) - \$10,000 + \$2,000(P/F, 15\%, 10)$

$\quad = \$5,000(5.0188) - \$10,000 + \$2,000(0.2472)$

$\quad = \$15,588$

Using the PE criteria, select machine A, with the higher positive present equivalent value.

Applying the annual equivalent approach:

Machine A

$$AE = \$15,000 - \$50,000(A/P, 15\%, 10) + \$8,000(A/F, 15\%, 10)$$

$$= \$15,000 - \$50,000(0.1993) + \$8,000(0.0493)$$

$$= \$5,429$$

Machine B

$$AE = \$5,000 - \$10,000(A/P, 15\%, 10) + \$2,000(A/F, 15\%, 10)$$

$$= \$5,000 - \$10,000(0.1993) + \$2,000(0.0493)$$

$$= \$3,106$$

Using the AE criteria, select machine A, with the higher positive annual equivalent value.

Applying the rate-of-return (ROR) approach to this problem produces the following, apparently contradictory result.

Machine A

$$PE = 0 = \$15,000(P/A, i, 10) - \$50,000 + \$8,000(P/F, i, 10)$$

Try 30%:

$$PE = \$-3,048$$

Try 25%:

$$PE = \$4,417$$

$$i = 25\% + 5\%[4,417/(4,417 + 3,048)]$$

$$= 28\%$$

Machine B

$$PE = 0 = \$-10,000 + \$5,000(P/A, i, 10) + \$2,000(P/F, i, 10)$$

Try 50%:

$$PE = \$-138$$

Try 40%:

$$PE = \$2,137$$
$$i = 40\% + 10\%[2,137/(2,137 + 138)]$$
$$= 49.4\% \quad \text{say, } 50\%$$

If the project with the highest rate of return was selected, machine B would be the choice. This is the reverse of the choice using the PE and AE method.

The reconciliation of this paradox, and the explanation of why machine A is the correct choice, lies in an understanding of the derivation and meaning of the BTRR/MARR rate. If it is derived by the methods of Chapter 5, it represents either the organization's minimum attractive rate of return or the cutoff rate of return. In either case the derivation of the rate is directed at maximizing the organization's profits for all of its invested capital. Assume that there is $50,000 to invest and that the choice must be made between machines A and B.

For machine A: The rate of return on the $50,000 invested in machine A is 28%.

For machine B: The rate of return on the $10,000 invested in machine B is 50%. The remaining $40,000 will be invested elsewhere at the firm's BTRR of 15%, so the rate of return on the $40,000 is 15% and the average rate of return $= (10/50)50\% + (40/50)15\% = 22\%$.

22% < 28%: therefore, invest the $50,000 in machine A.

An alternative method of assessment is as follows. Machine B will generate $5,000 per year for 10 years and then be sold for $2,000 for an annual equivalent over 10 years of

$$\$5,000 + \$2,000(A/F, 15\%, 10) = \$5,099$$

The remaining $40,000 can be assumed to be invested at the organization's BTRR (or MARR if after taxes) which is

$$\$40,000(A/P, 15\%, 10) = \$7,972$$

for a total annual equivalent of $13,071. Machine A, on the other hand, will generate

$$\$15,000 + \$8,000(A/F, 15\%, 10) = \$15,394$$

Thus, although machine B has the higher rate of return, machine A is the correct choice. Machine A maximizes profits.

Example 6.5

Problem: Let us look now at the difference between alternatives. Applying the rate-of-return approach does not cause any problems if the analysis is based on the difference between the two alternatives. That is, if one subtracts the project with the smaller first cost from the one with the higher first cost, it creates a third project. Consider proposals A and B from Example 6.4, and construct a proposal of A minus B.

		$A - B$
First cost = $50,000 - $10,000		$40,000
Annual BTCF = $15,000 - $5,000		$10,000
Salvage value = $8,000 - $2,000		$6,000
Analysis period (life)		10 years

This project can be treated as an accept/reject proposal. If proposal A − B is rejected, the extra benefits do not justify the extra cost, and the lower cost alternative, machine B, should be selected. If A − B is accepted, the higher-cost alternative, project A, is the correct decision.

Apply the three methods to proposal A − B.

1. Present equivalent:

   $$PE = \$10,000(P/A, 15\%, 10) - \$40,000 + \$6,000(P/F, 15\%, 10)$$

   $$= \$10,000(5.0188) - \$40,000 + \$6,000(0.2472)$$

   $$= \$11,673$$

   The present equivalent method says accept proposal A − B; therefore, choose machine A.

2. Annual equivalent:

   $$AE = \$10,000 - \$40,000(A/P, 15\%, 10) + \$6,000(A/F, 15\%, 10)$$

   $$= \$10,000 - \$40,000(0.1993) + \$6,000(0.0493)$$

   $$= \$2,324$$

   The annual equivalent method also accepts proposal A − B; therefore, choose machine A.

3. Rate of return:

$$PE = 0 = \$10,000(P/A, i, 10) - \$40,000 + \$6,000(P/F, i, 10)$$

Try 20%:

$$PE = \$2,894$$

Try 25%:

$$PE = \$-3,651$$

$$i = 20\% + 5\%[2,894/(2,894 + 3,651)]$$

$$= 22.2\%$$

Since 22.2% is greater than the BTRR of 15% required, the rate-of-return method also accepts proposal A − B and consequently, machine A.

When the analysis is directed at the differences between projects, all three methods will give the same result. Our objective is to maximize profits, not rate of return. To accomplish this objective we must evaluate each increment of investment when using the rate-of-return approach.

6.5 COST COMPARISON BETWEEN TWO ALTERNATIVES

In many comparison situations only the cost components are given. Often, the competing alternatives provide the same service, and the revenue, or benefit, side of the equation is the same for each alternative.

Example 6.6

Problem: We want to make a cost comparison of two alternatives.

	Machine A	Machine B
First cost	$12,000	$30,000
Annual operating cost	$10,000	$5,000
Salvage	$2,000	$6,000
Analysis period (life)	10 years	10 years
BTRR	20%	20%

Solution: Notice that since it is a cost comparison, costs are treated as positive numbers. Therefore, the selection criterion is directed toward the lowest present equivalent cost or annual equivalent cost (PEC or AEC).

Machine A

$$PEC = OC(P/A, 20\%, 10) + P - SV(P/F, 20\%, 10)$$
$$= \$10,000(P/A, 20\%, 10) + \$12,000 - \$2,000(P/F, 20\%, 10)$$
$$= \$10,000(4.1925) + \$12,000 - \$2,000(0.1615)$$
$$= \$53,602$$

Machine B

$$PEC = \$5,000(P/A, 20\%, 10) + \$30,000 - \$6,000(P/F, 20\%, 10)$$
$$= \$5,000(4.1925) + \$30,000 - \$6,000(0.1615)$$
$$= \$49,994$$

Select machine B, with the lower present equivalent cost.
Applying the AEC approach:

Machine A

$$AEC = OC + P(A/P, 20\%, 10) - SV(A/F, 20\%, 10)$$
$$= \$10,000 + \$12,000(0.2385) - \$2,000(0.0385)$$
$$= \$12,785$$

Machine B

$$AEC = \$5,000 + \$30,000(0.2385) - \$6,000(0.0385)$$
$$= \$11,924$$

Select machine B, with the lower annual equivalent cost.

Example 6.7

Problem: We want to analyze the difference between the proposals.

First cost = $-30,000 - ($-12,000)	$-18,000
Annual operating cost = $-5,000 - ($-10,000)	$5,000
Salvage value = $6,000 - $2,000	$4,000
Estimated life	10 years
BTRR	20%

Confusion could be introduced into this calculation because of the previous sign change (i.e., costs given positive signs). Therefore, it is best to revert to the earlier convention and treat receipts as positive cash flows and costs as negative cash flows.

Proposal B

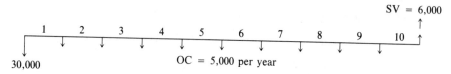

Proposal A

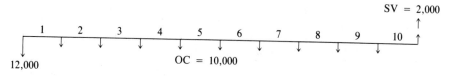

Proposal B — A

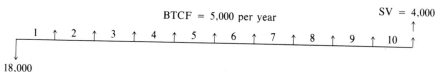

$$PE = \$5,000(P/A, 20\%, 10) - \$18,000 + \$4,000(P/F, 20\%, 10)$$

$$= \$3,608$$

This value could also have been developed from the previous calculations.

$$PEC_B - PEC_A = \$-49,994 - \$-53,602$$

$$= \$3,608$$

Even though the data given were only costs, when differences are considered the cost savings become positive cash flows and the conventional criterion apply.

Applying the annual equivalent calculation:

$$AE = AEC_B - AEC_A$$

$$= \$-49,994(A/P, 20\%, 10) - (\$-53,602)(A/P, 20\%, 10)$$

$$= \$-11,924 + \$12,785$$

$$= \$861$$

Since AE (and PE) give a positive result, proposal B — A is accepted and the higher-cost alternative, proposal B, is chosen.

6.6 COMPARISON OF ALTERNATIVES HAVING DIFFERENT ECONOMIC LIVES

To be valid, an analysis must be made between comparable alternatives. It is not correct, for example, to compare directly a roofing system with a 5-year life with a roofing system that has a 20-year life. The economy study must determine the period for which the service or facility is needed, and then ensure that the alternatives will provide service for that period. This usually involves either the forecasting of replacement facilities or the estimation of salvage values.

A common assumption is that of *like-for-like replacement*. That is, the facility will be replaced by a like facility with the same cash flow pattern. If the study period is a multiple of the economic lives of the alternatives, the annual equivalents can be compared directly.

Example 6.8

Problem: We now compare two alternatives having different economic lives.

	Proposal A	*Proposal B*
First cost	$25,000	$50,000
Life	5 years	10 years
Salvage value	$5,000	$10,000
Operating costs	$30,000	$20,000
Before-tax rate of return	25%	25%

Solution: Assuming like-for-like replacement and using the annual equivalent cost approach:

Proposal A

$$AEC = OC + P(A/P, 25\%, 5) - SV(A/F, 25\%, 5)$$

$$= \$30,000 + \$25,000(0.3719) - \$5,000(0.1219)$$

$$= \$38,688$$

Proposal B

$$AEC = OC + P(A/P, 25\%, 10) - SV(A/F, 25\%, 10)$$

$$= \$20,000 + \$50,000(0.2801) - \$10,000(0.0301)$$

$$= \$33,704$$

If we assume that proposal A can be replaced with another facility at year 5 which has the identical cash-flow pattern, then accept proposal B with the lower annual equivalent cost.

The result of spreading the cash flows to a uniform annual amount during the second 5-year period gives an identical answer to the result of spreading the cash flows to a uniform annual amount over the first 5-year period. Thus with proposal A it is only necessary to evaluate the cash flows over a 5-year period *provided that the cash flows for replacement assets repeat the costs that have been forecast for the original assets.*

Using the present equivalent cost approach, it is again necessary to assume the existence of a replacement for A so that the comparisons are made for equal periods.

Proposal A

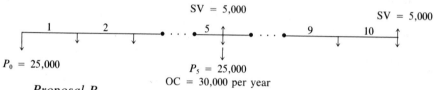

Proposal B

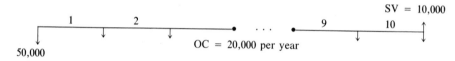

Proposal A

$$\text{PEC} = P_0 + P_5(P/F, 25\%, 5) + \text{OC}(P/A, 25\%, 10)$$
$$- \text{SV}(P/F, 25\%, 5) - \text{SV}(P/F, 25\%, 10)$$
$$= \$25{,}000 + \$25{,}000(0.3277) + \$30{,}000(3.5705)$$
$$- \$5{,}000(0.3277) - \$5{,}000(0.1074)$$
$$= \$138{,}132$$

Proposal B

$$\text{PEC} = P + \text{OC}(P/A, 25\%, 10) - \text{SV}(P/F, 25\%, 10)$$
$$= \$50{,}000 + \$20{,}000(3.5705) - \$10{,}000(0.1074)$$
$$= \$120{,}336$$

Choose proposal B, with the lower present equivalent cost. The decision, as we would expect, is identical to the result using the annual equivalent approach. When comparison of alternative proposals using the present equivalent approach is desired, we must be very careful to compare the proposals on an equal-life basis. A study period can be selected that is the least common multiple of the lives of the various assets involved. Sometimes it is convenient to choose a period for analysis that is shorter than the expected service period. This procedure requires a value being placed on all assets at the end of the service period selected. Under certain conditions the evaluation may be made assuming a perpetual period of service. The present equivalent of a perpetual period of service is often referred to as *capitalized costs,* a misnomer but nevertheless, an existing one.

6.7 COMPARISON OF PROJECTS THAT DO NOT MEET THE MARR/BTRR REQUIREMENT

Often the decision maker is faced with a choice between two projects, neither of which is acceptable by the standard decision criteria: for example, the choice may be between providing subsidized transit fares for employees or building a parking lot. Although one or the other is necessary, both represent a drain on the company's treasury. Exactly the same methods of analysis are used in this situation as discussed previously, but care must be exercised in the interpretation of the results.

Example 6.9

Problem: Choose between two alternatives with negative returns.

	Proposal C	*Proposal D*
First cost	$20,000	$6,000
BTCF	$5,000	$500
Salvage value	0	0
Analysis period (life)	5 years	5 years
MARR	15%	15%

Solution: Using the present equivalent approach:

Proposal C

$$PE = \$5,000(P/A, 15\%, 5) - \$20,000$$
$$= \$5,000(3.3522) - \$20,000$$
$$= \$-3,240$$

Proposal D

$$PE = \$500(P/A, 15\%, 5) - \$6,000$$
$$= \$-4,324$$

From the present equivalent calculations it is apparent that neither proposal C nor proposal D meets the BTRR criterion. However, if a choice must be made, project C, with the higher present equivalent value (least negative value), should be chosen.

The annual equivalent approach gives similar results and the same conclusion.

$$AE = \$-20,000(A/P, 15\%, 5) + \$5,000 = \$-966$$

$$AE = \$-6,000(A/P, 15\%, 5) + \$500 = \$-1,290$$

As in the preceding example, an analysis of the differences between proposals is most revealing.

First cost = \$20,000 − \$6,000 = \$14,000
BTCF = \$5,000 − \$500   = \$ 4,500
Salvage value   = 0
Estimated economic life   = 5 years

$$PE = \$-14,000 + \$4,500(P/A, 15\%, 5) = \$1,084$$

Accept C − D and choose C.

Using the rate-of-return method:

$$PE = \$-14,000 + \$4,500(P/A, i, 5)$$

$$(P/A, i, 5) = 14,000/4,500 = 3.111$$

For $i = 15\%$:

$$(P/A, 15\%, 5) = 3.3522$$

For $i = 20\%$:

$$(P/A, 20\%, 5) = 2.9906$$

$$i = 15\% + 5\%[(3.3522 - 3.111)/(3.3522 - 2.9906)] = 18.3\%$$

$18.3\% > 15\%$; therefore, accept C − D and choose C.

Proposal C − D has a present equivalent of \$1,084, indicating that it is an acceptable proposal. In fact, as the rate-of-return calculation shows, it gives an 18.3% rate of return on the \$14,000 investment. Thus, although one would reject proposal C if it were possible, given the choice between C and D, C is the better choice.

6.8 MEANINGS FOR THE METHODS

One of the difficulties that a decision maker has when using or explaining the decision-making basis for the methods explained above is the problem of attaching an intuitive meaning to the calculations. For example, a calculation of the present equivalent value for a new machine tool is illustrated below.

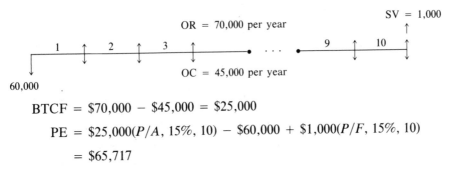

$$BTCF = \$70,000 - \$45,000 = \$25,000$$

$$PE = \$25,000(P/A, 15\%, 10) - \$60,000 + \$1,000(P/F, 15\%, 10)$$

$$= \$65,717$$

This calculation is presented to the management as the principal justification for purchasing the machine. The manager who is not familiar with present equivalent calculations may question the significance of the value $65,717. This value represents the present equivalent of the cash flow stream using a BTRR of 15%, and the fact that it is positive indicates that the investment should be accepted. The value has no other significance. The same philosophy applies to an annual equivalent calculation.

On the other hand, if the calculations presented were

$$PE = 0 = \$25,000(P/A, i, 10) - \$60,000 + \$1,000 (P/F, i, 10)$$

Try $i = 40\%$:

$$PE = \$375$$

$$i \approx 40\%$$

and the report stated: "The machine tool will require a $60,000 investment and will provide a 40% before-tax rate of return."

The manager has little difficulty with this terminology. Investment and rate of return are terms with which managers are familiar. They are terms used by bankers, stockbrokers, trade associations, and on the financial pages of newspapers. However, the rate of return calculation is more difficult to make and as was shown in Example 6.4, is liable to misapplication and misinterpretation if not properly applied. Further, it is possible for rate-of-return calculations to give multiple, and thus ambiguous, answers.

For the accept/reject example just cited there are two possible methods of presentation that should be acceptable. One is to calculate and present the rate-of-return values, together with the organization's BTRR value, and point out that the estimated rate of return exceeds the BTRR. A second method may be to suppress the figure $65,717, stating only that the present equivalent is positive and that a positive present equivalent indicates that the project more than satisfies the organization's BTRR, and thus should be accepted. For the accept/reject decision it is only necessary to know whether or not the present equivalent is positive or negative. The actual numbers, although necessary for ranking decisions, are not necessary for the accept/reject decision.

This is not intended to imply that present, future, and annual equivalents do not have a meaning that can be understood in regular jargon. They do, and these will be discussed in the following sections. However, as will be seen, these meanings are often restricted to a specific field or industry. Furthermore, sometimes the use of a meaning from another field is inappropriate or confusing. Thus if, as in the preceding example, the situation is one of accept/reject, it may be best to describe the function of the present equivalent calculation in the accept/reject context rather than trying to introduce explanations from, for example, the valuation field.

There are, however, places where the equivalents have a specific and well-understood meaning. Calculations of valuations, annuities, capital recovery factors, sinking funds, and revenue requirements are examples of just such areas. These are described in the following sections.

6.9 VALUATIONS

The value of many assets and properties is dependent on their ability to generate income. The earning value of a property is the present worth of probable future net earnings estimated on the basis of recent and projected expenses and earnings and the business outlook.

The most straightforward example of an income-generating asset is a bond. If it is a government bond, there is little risk attached and the income amounts and timings are stated on the bond.

Example 6.10

Problem: We want to determine the value of a bond.

In January 1976 the government issued 20-year, 10% bonds. The bonds have a maturity value of $1,000 in January 1, 1996. They pay interest of $50 every 6 months (January 1 and June 1) until maturity.

Knowing the market rate for this security or a similar-quality security, the market value of this bond can be calculated as of any date.

Solution: To determine the June 1982 market value of the bond, it is first necessary to determine the remaining payments. Assume that the market rate is 15% (effective annual rate).

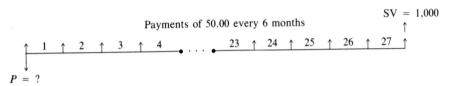

$P = ?$

The problem of semiannual payments must be dealt with. This can be done either by converting the interest rate (15%) to a semiannual equivalent, or by moving the payments to an end-of-year payment

$$15\% \text{ per annum} = (1 + i)^2 - 1$$

$$i = \text{semiannual rate}$$

$$0.15 = (1 + i)^2 - 1$$

$$i = 0.0724 \quad \text{or} \quad 7.24\%$$

$$P = \$50 + \$50(P/A, 7.24\%, 27) + \$1,000(P/F, 7.24\%, 27)$$

$$= \$50 + \$50(11.7188) + \$1,000(0.1515)$$

$$= \$50 + \$585.94 + \$151.50$$

$$= \$787.44$$

Converting to an annual cash flow:

$$\$50(F/P, 7.24\%, 1) + \$50 = \$103.62$$

Payments of 103.62 annually   SV = 1,000

June 1 Jan. 1

$P = ?$

$$P = (P/F, 7.24\%, 1)[\$103.62 + \$103.62(P/A, 15\%, 13)$$

$$+ \$1,000(P/F, 15\%, 13)]$$

$$= \$787.61$$

The difference between the two methods of calculation is due to round-off errors.

The present value, $787, is what one would pay for the bond in order to receive a 15% return on the investment.

Example 6.11

Problem: We now determine the value of an apartment building. A tax-free organization with a BTRR value of 20% is considering purchasing a 21-suite apartment building, and after careful examination they estimated the cash flows to be as follows. What should the company be willing to pay for the building? Operating revenues = $80,000, operating cost = $30,000, and estimated salvage value in 20 years = $100,000.

Solution

$$PE = \$50,000(P/A, 20\%, 20) + \$100,000(P/F, 20\%, 20)$$

$$= \$246,090$$

To this organization, the value of the building is close to $250,000, and they would be unlikely to pay more than that for its purchase.

Another organization, say one with a BTRR of 15%, would perform the following calculation:

$$PE = \$50,000(P/A, 15\%, 20) + \$100,000(P/F, 15\%, 20)$$

$$= \$319,075$$

and evaluate the building at close to $320,000.

Obviously, the interest rate used in the valuation calculations is crucial. For expropriation hearings, the court will set a rate, but for private valuations, the appraiser must estimate a suitable rate. Organizations such as those in Example 6.11 have their own discount rates, and assets to them have differing values. Suppose that the building in Example 6.11 sold for $275,000 to a third party. The first organization would think that the party paid too much, whereas the second organization would be pleased to buy the building at this price.

6.10 ANNUITY

An *annuity* is a sum of money paid regularly (usually monthly or annually) to a person for a fixed number of years. The primary use of annuities is in situations such as retirement planning, where a person who has a large sum of money wants to receive it in regular payments over some time period. Often, life insurance companies will combine the annuity concept with the insurance concept and will sell policies that guarantee the payment of a fixed sum until the death of the recipient.

Example 6.12

Problem: On retirement at age 65 from the Wombat Wigit Company, Irving Longlive received a lump-sum payment of $130,000. In addition, Irving had savings amounting to $70,000. He came from a family known for its longevity and so estimated that he would live to age 90. The Friendly Loan and Savings Company offered Irving a 20% annuity.

Solution: The result was an annual income to age 90 of

$$A = (\$130,000 + \$70,000)(A/P, 20\%, 25)$$

$$= \$200,000(0.2021) = \$40,420$$

If Irving were to die before age 90, the company would pay the outstanding principal to his estate. For example, if he died at age 76:

$$\text{Remaining payments} = 90 - 76 = 14 \text{ years}$$

$$\text{Remaining principal} = \$40,420(P/A, 20\%, 14)$$

$$= \$40,420(4.6106) = \$186,360$$

6.11 CAPITAL RECOVERY FACTORS

On public works projects the economic analyst is often concerned with allocating the initial cost over the lifetime of the project. This is dealt with in more detail in Chapter 11, but a short introduction is suitable here.

Example 6.13

Problem: We want to determine the annual cost for a specific highway improvement project. It is estimated that subgrade reconstruction will cost $1,200,000 per kilometer and have a life of 20 years. The surfacing will cost $300,000 per kilometer and will have a life of 10 years. To convert these values to annual costs, highway engineers multiply these estimates by the appropriate capital recovery factors (which are, in fact, A/P factors).

Solution

$$\text{Annual cost} = \$1,200,000(A/P, i, 20) + \$300,000(A/P, i, 10)$$

The implicit assumption is that the road will be resurfaced at the end of year 10 for the same cost as today. This figure is then compared with the annual benefits resulting from reduced travel time, reduced accidents, increased vehicle life, and so on.

6.12 SINKING FUNDS

A *sinking fund* is a fund of money that is set aside and invested to earn interest so that it can be used to "sink" (i.e., pay off) a debt or obligation. A good example of a sinking-fund application is a strip-mining company that has a statutory obligation to reclaim and restore the site of its operations after the mine is exhausted. It is possible that the regulatory agencies will not have faith in the company's promises that they will restore the land and therefore require the company to establish a sinking fund with fixed annual payments to ensure that when reclamation time arrives, the money is available. As one would expect, this is simply an application of the future-worth concept. Some texts refer to the uniform series future-worth factor $(F/A, i, n)$ as the sinking-fund factor.

6.13 REVENUE REQUIREMENTS

When dealing with an income-generating asset in a profit-making corporation, the most meaningful calculation is often the *annual equivalent revenue requirements,* or just *revenue requirements,* calculation. Often, a company would like to know how much revenue is required from a new product or service, to justify the incremental cost of supplying the product or service. This revenue requirement can then be compared with the estimated revenues from market projections to determine project feasibility. Consider the before-tax cash flow diagram shown below.

AER = annual equivalent revenue requirements
AEOC = annual equivalent operating costs
AEOR = annual equivalent operating revenues
AED = annual equivalent depreciation allowance

On a before-tax basis:

$$PE = (AEOR - AEOC)(P/A, i, n) - P + SV(P/F, i, n)$$
$$AE = (AEOR - AEOC) - P(A/P, i, n) + SV(A/F, i, n)$$

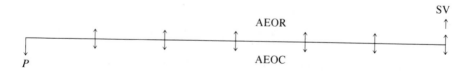

This before-tax cash flow diagram can be converted into an after-tax cash flow diagram by including the tax saving due to depreciation allowance, the tax effect on operating costs and operating revenues, and the tax effect on the salvage value.

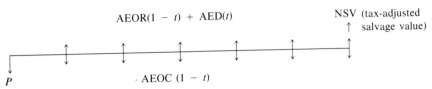

$AEOR(1 - t) + AED(t)$

NSV (tax-adjusted
↑ salvage value)

P

$AEOC (1 - t)$

Note: The depreciation allowance has the same sign as the revenues.

Therefore, when revenues are known to arrive at the after-tax cash flows as a basis for rate-of-return calculations or comparison of alternatives on an after-tax basis, we have the following equations:

For annual equivalent:

$$AE = (AEOR - AEOC)(1 - t) - P(A/P, i_a, n)$$
$$+ NSV(A/F, i_a, n) + t(AED) \qquad (6.1)$$

For present equivalent, let

PER = present equivalent revenue requirements
PEOR = present equivalent operating revenue
PEOC = present equivalent operating costs
PED = present equivalent of depreciation allowance

$$PE = (PEOR - PEOC)(1 - t) - P + NSV(P/F, i_a, n) + t(PED) \quad (6.2)$$

When the determination of the rate of return is our objective, the equations above are set equal to zero.

The question now arises as how to handle a problem when revenues are unknown or difficult to estimate. For example, to determine the revenue generated by an individual machine that performs one of many operations in producing a product is difficult and often impractical. In addition, revenues are difficult to estimate due to market fluctuations. Under these conditions we would like to know what annual revenues must be generated by an individual investment to meet all the costs involved (e.g., operating costs, the after-tax cash flow requirement, and the income tax requirement). Operating costs, depreciation allowance, and taxes can usually be estimated with a reasonable degree of accuracy. With these data and knowing the company's MARR, we can proceed by setting equation 6.1 or 6.2 equal to zero (the accept criteria), substitute revenue requirements for revenues (e.g., AER = AEOR), and rearrange the values to give us the following equations:

$$AER(1 - t) = AEOC(1 - t) + P(A/P, i_a, n) - NSV(A/F, i_a, n) - t(AED)$$

Dividing through by $(1 - t)$ yields

$$AER = AEOC + [P(A/P, i_a, n) - NSV(A/F, i_a, n) - t(AED)]/(1 - t)$$
$$(6.3)$$

And for the present equivalent of revenue requirements:

$$PER = PEOC + [P - NSV(P/F, i_a, n) - t(PED)]/(1 - t) \quad (6.4)$$

The revenue requirements are probably the easiest to explain of all the measures. The AER is the amount of money that the asset must generate annually to cover operating costs, income tax, and provide a recovery of, and rate of return on, invested capital.

For additional clarification of this concept, let us return to the simplified funds flow diagram.

From Figure 6.2 we can see that

$$AER = \text{operating costs} + \text{after-tax cash flow requirement} + \text{income tax}$$

1. AEOC = annual equivalent operating cost
2. $ATCF = P(A/P, i_a, n) - NSV(A/F, i_a, n)$
3. Income tax $(IT) = t(\text{taxable income})$

Assuming 100% equity capital:

$$\text{Taxable income} = IT + \text{net income}$$
$$IT = t(\text{taxable income})$$
$$= t(IT + \text{net income})$$
$$IT(1 - t) = t(\text{net income})$$
$$IT = t/(1 - t)(\text{net income})$$
$$\text{Net income} = ATCF - AED$$
$$IT = t/(1 - t)[P(A/P, i_a, n) - NSV(A/F, i_a, n) - AED]$$

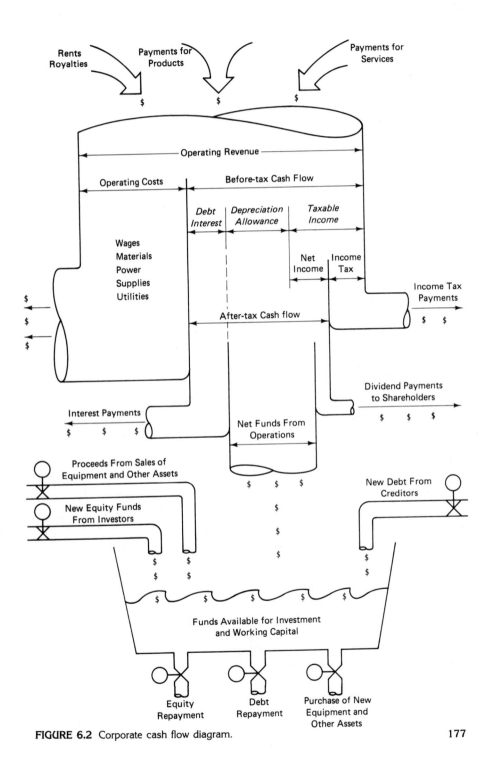

FIGURE 6.2 Corporate cash flow diagram.

Therefore,

$$AER = AEOC + P(A/P, i_a, n) - NSV(A/F, i_a, n)$$
$$+ t/(1 - t)[P(A/P, i_a, n) - NSV(A/F, i_a, n) - AED]$$
$$= AEOC + [P(A/P, i_a, n) - NSV(A/F, i_a, n) - t(AED)]/(1 - t)$$

Bringing all cash flows to a present equivalent, we have

$$PER = PEOC + [P - NSV(P/F, i_a, n) - t(PED)]/(1 - t)$$

From Figure 6.2 we see that operating revenues represent the revenue that must be generated to offset all obligations.

Total annual revenue requirement (AER)

= operating costs + repayment of the original investment
+ interest payments to the debt holder + return to the equity holder (net income) + income tax payment

Example 6.14

Problem: Global Industries is considering the purchase of a lathe:

$$P = \$20,000 = \text{installed cost}$$
$$SV = \$5,000 = \text{estimated salvage value}$$
$$n = 5 \text{ years} = \text{useful economic life}$$
$$OC = \$2,000 = \text{annual operating costs}$$

$r = 0$, $t = 50\%$, and MARR $= 12\%$. The depreciation allowance (CCA) for tax purposes is based on the fast-write-off method. Find the annual revenue requirements. The recaptured CCA is spread over 2 years when using the fast write-off method and a zero book value.

Recaptured CCA = $5,000 (BV = 0)

Tax effects each year over year 6 & year 7 are:

($5,000/2)$t$ = $1,250/year

NSV = $5,000 - $1,250$(P/A, 12\%, 2)$ = $2,887

Solution

$$AED = [\$5,000(P/F, 12\%, 1) + \$10,000(P/F, 12\%, 2)$$

$$+ \$5,000(P/F, 12\%, 3)](A/P, 12\%, 5)$$

$$= \$4,437$$

$$AER = AEOC + [P(A/P, i_a, n) - NSV(A/F, i_a, n) - t(AED)]/(1 - t)$$

$$= \$2,000 + [\$20,000(0.2774) - \$2,887(0.1574) - 0.5(\$4,437)]/0.5$$

$$= \$2,000 + (\$5,548 - \$454 - \$2,219)/0.5$$

$$= \$7,750$$

Thus, to be an acceptable investment, the lathe must be capable of generating annual revenues of $7,750.

Example 6.15

Problem: The information is identical to Example 6.14 except that operating costs increase by $200 per year.

Solution

$$AEOC = \$2,000 + \$200(A/G, 12\%, n) = \$2,355$$

$$AER = AEOC + [P(A/P, 12\%, 5) - NSV(A/F, 12\%, 5)$$

$$- t(AED)]/(1 - t)$$

$$AER = \$2,355 + [\$20,000(0.2774) - \$2,887(0.1574)$$

$$- 0.5(\$4,437)]/0.5$$

$$= \$8,105$$

Example 6.16

Problem: The information is identical to Example 6.14 except that annual operating revenues are given as $7,750. Find the rate of return.

Solution

$$PE = 0 = (AEOR - AEOC)(P/A, i_a, 5)(1 - t) - P + NSV(P/F, i_a, 5)$$

$$+ t(AED)(P/A, i_a, 5)$$

Try 12%:

$$PE = \$5,750(0.5)(P/A, 12\%, 5) - \$20,000 + \$2,887(P/F, 12\%, 5)$$
$$+ 0.5(\$4,437)(P/A, 12\%, 5)$$
$$= \$10,364 - \$20,000 + \$1,638 + \$7,998 = 0$$
$$i = 12\%$$

6.14 SIMPLIFYING EQUATIONS FOR AFTER-TAX CALCULATIONS

In Chapter 4 we derived the capital tax factors for calculating the tax effect of capital cost allowance. When the capital tax factor is used with the declining-balance method, equations 6.1 to 6.4 are changed as follows:

$$AE = (AEOR - AEOC)(1 - t) - P(A/P, i_a, n)CTF$$
$$+ SV(A/F, i, n)CTF \tag{6.5}$$

$$PE = (PEOR - PEOC)(1 - t) - P(CTF)$$
$$+ SV(P/F, i_a, n)CTF \tag{6.6}$$

$$AER = AEOC + [P(A/P, i_a, n)CTF$$
$$- SV(A/F, i_a, n)CTF]/(1 - t) \tag{6.7}$$

$$PER = PEOC + [P(CTF)$$
$$- SV(P/F, i_a, n)CTF]/(1 - t) \tag{6.8}$$

Example 6.17

Problem: The JC Corporation is considering the purchase of a new piece of equipment. $P = \$30,000$, $SV = \$6,000$, $n = 5$ years, $OC = \$25,000$, $r = 0$, $t = 45\%$. Capital cost allowance (CCA) is based on the declining-balance method, CCA rate $= 20\%$, and MARR $= 15\%$. What is the revenue requirement to justify this investment?

Solution: When the half-year rule applies, the CTF for the initial investment (P) is different from the CTF for the salvage value (SV).

$$CTF(P) = 1 - [(td)/(i + d)][(1 + i/2)/(1 + i)] = 0.7596$$

$$CTF(SV) = 1 - (td)/(i + d) = 1 - (0.45)(0.20)/0.35 = 0.7429$$

$$AER = AEOC + [P(A/P, 15\%, 5)CTF$$
$$- SV(A/F, 15\%, 5)CTF]/(1 - t)$$
$$= \$25,000 + [\$30,000(0.2983)0.7596$$
$$- \$6,000(0.1483)0.7429]/0.55$$
$$= \$36,158$$

Problem: The information is as given above, except that the selling price of the equipment is estimated to be $40,000 5 years hence.

Solution

Capital gain = $40,000 − $30,000 = $10,000

Capital gains tax = ($10,000)0.225 = $2,250

Net salvage on capital gain = $10,000 − $2,250 = $7,750

$$AER = \$25,000 + [\$30,000(0.2983)0.7596 - \$30,000(0.1483)0.7429$$
$$- \$7,750(0.1483)]/0.55$$
$$= \$29,260$$

Note: The CTF does not apply to the capital gain.

Problem: The information is identical to the original data except that annual operating revenues are estimated to be $36,158. Find the rate of return (i_a) on the investment.

Solution

$$PE = 0 = (PEOR - PEOC)(1 - t)(P/A, i_a, 5) - P(CTF)$$
$$+ SV(P/F, i_a, 5)CTF$$

Try 15%:

$$PE = \$11,158(0.55)(3.3522) - \$30,000(0.7596)$$
$$+ \$6,000(0.4972)(0.7429)$$
$$= \$20,572 - \$22,788 + \$2,216 = 0$$
$$i_a = 15\%$$

Example 6.18

Problem: Determine the revenue requirements for the following investment.

The Double E Company is considering the installation of a new plant to manufacture circuit boards. The following estimates apply to this investment decision.

	P	SV (10 years)
1. Building	$800,000	$500,000
2. Land	200,000	300,000
3. Working capital	350,000	350,000
4. Equipment	600,000	200,000
5. Annual operating costs	700,000	
6. A major overhaul in year 5	100,000	
7. $t = 40\%$, $r = 20\%$, $i_d = 10\%$, and $i_e = 17.3\%$.		

Use the declining-balance method. The CCA rate on the building $= 5\%$ and the CCA rate on the equipment $= 20\%$.

Solution

$$i_a = ri_d(1 - t) + (1 - r)i_e$$

$$= (0.2)(10\%)(0.6) + (0.8)(17.3\%) = 15\%$$

$$\text{AEOC} = \$700,000 + \$100,000(P/F, 15\%, 5)(A/P, 15\%, 10)$$

$$= \$709,907$$

Building:

$$\text{CTF}(P) = 1 - [(td)/(i + d)][(1 + i/2)/(1 + i)]$$

$$= 1 - [(0.4)(0.05/0.20)(1.075/1.15)]$$

$$= 0.9065$$

$$\text{CTF}(SV) = 1 - (td)/(i + d)$$

$$= 1 - [(0.40)(0.05)/(0.15 + 0.05) = 0.90$$

Equipment: $\text{CTF}(P) = 0.7863$ and $\text{CTF}(SV) = 0.7714$.

Capital gain on land $= \$300,000 - \$200,000 = \$100,000$

Capital gains tax $= \$100,000(0.20) = \$20,000$

Net salvage on land $= \$300,000 - \$20,000 = \$280,000$

	P		SV
Building =			
$800,000(0.9065)	$ 725,200	$500,000(0.90)	$ 450,000
Land	200,000		280,000
Working capital	350,000		350,000
Equipment =			
$600,000(0.7863) =	471,780	$200,000(0.7714) =	154,280
	$1,746,980		$1,234,280

$$\text{AER} = \$709,907 + [\$1,746,980(A/P, 15\%, 10)$$
$$- \$1,234,280(A/F, 15\%, 10)]/0.60$$
$$= \$1,188,750$$

Problem: Use the declining-balance method for the building and a CCA rate of 5%; use the fast-write-off method for equipment.

Solution: For the equipment:

$$\text{Recapture} = SV - BV = \$200,000 - 0 = \$200,000$$

Spread the $200,000 over 2 years when using the fast-write-off method and determine the present equivalent of tax consideration.

$$\text{PE} = \$100,000(P/A, 15\%, 2)0.4$$
$$= \$65,028$$
$$\text{NSV} = \$200,000 - \$65,028 = \$134,972$$
$$\text{PED} = \$150,000(P/F, 15\%, 1) + \$300,000(P/F, 15\%, 2)$$
$$+ \$150,000(P/F, 15\%, 3) = \$455,905$$

	P	SV	PED
Building	$ 725,200	$ 450,000	
Land	200,000	280,000	
Working capital	350,000	350,000	
	$1,275,200	$1,080,000	
Equipment	600,000	134,972	$455,905
	$1,875,200	$1,214,972	

$$\text{AER} = \$709,907 + [\$1,875,200(0.1993) - \$1,214,972(0.0493)$$
$$- 0.40(\$455,905)(0.1993)]/0.60$$
$$= \$1,172,381$$

6.15 USING A TABULAR FORMAT TO DISPLAY RESULTS

If explicit display of the year-by-year cash flows is desirable, a tabular format can be used. This format has the distinct advantage of showing the actual cash flows that will occur each year and thereby is helpful in determining the effect of the investment on the cash requirements. An investment may meet the MARR requirement set by the company but may cause a severe cash flow drain on company funds during certain years and as a result be, in fact, an undesirable investment.

However, caution must be exercised when calculating the rate of return using the after-tax cash flows listed in the table. When the debt interest is included in the table, the rate of return calculated represents the composite cost of capital (i_c) and not the tax-sheltered cost of capital (i_a).

Example 6.19

Problem: We want to calculate the rate of return. The HM Corporation purchased a new forklift. $P = \$40,000$, SV $= 0$, $n = 5$ years, OC $= \$30,000$ per year, OR $= \$50,100$ per year, $r = 0$, and $t = 50\%$. Use the fast-write-off method to calculate CCA.

Solution:

Year	BTCF	AED	TI	IT	ATCF
0	$-40,000				$-40,000
1	20,100	$10,000	$10,100	$ 5,050	15,050
2	20,100	20,000	100	50	20,050
3	20,100	10,000	10,100	5,050	15,050
4	20,100	0	20,100	10,050	10,050
5	20,100	0	20,100	10,050	10,050

$$PE = 0 = \$-40,000 + \$15,050(P/F, i, 1) + \$20,050(P/F, i, 2)$$
$$+ \$15,050(P/F, i, 3) + \$10,050(P/F, i, 4)$$
$$+ \$10,050(P/F, i, 5)$$

Try 25%:

$$PE = \$-40,000 + \$15,050(0.8000) + \$20,050(0.6400)$$
$$+ \$15,050(0.5120) + \$10,050(0.4096)$$
$$+ \$10,050(0.3277)$$
$$= \$-13$$
$$i_a = i_e = i_c = 25\%$$

Using equation 6.2:

$$PED = \$10,000(P/F, i, 1) + \$20,000(P/F, i, 2)$$
$$+ 10,000(P/F, i, 3) = \$25,920$$
$$BTCF = OR - OC = \$50,100 - \$30,000 = \$20,100 \text{ per year}$$
$$PE = \$20,100(P/A, 25\%, 5)0.5 - \$40,000 + 0.5(\$25,920)$$
$$= \$-13$$
$$i_a = 25\%$$

Example 6.20

Problem: The data are identical to Example 6.19 except that financing is by \$10,000 debt capital at 10% and \$30,000 equity capital.

Solution

$$r = \$10,000/\$40,000 = 25\%$$
$$D_1 = \$10,000, \ D_2 = \$20,000, \ D_3 = \$10,000$$

Therefore, assume that the book value of the debt reduces as follows:

Year	Debt Principal	Principal Repayment	Debt Interest
0	$10,000	$ 0	
1	7,500	2,500	$1,000
2	2,500	5,000	750
3	0	2,500	250
4	0	0	0
5	0	0	0

Year	BTCF	AED	Debt Interest	TI	IT	ATCF
0	$ – 40,000					$ – 40,000
1	20,100	$10,000	$1,000	$ 9,100	$4,550	15,550
2	20,100	20,000	750	– 650	– 325	20,425
3	20,100	10,000	250	9,850	4,925	15,175
4	20,100	0	0	20,100	10,050	10,050
5	20,100	0	0	20,100	10,050	10,050

Try 25%:

$$PE = \$-40,000 + \$15,550(P/F, i, 1) + \$20,425(P/F, i, 2)$$
$$+ \$15,175(P/F, i, 3) + \$10,050(P/F, i, 4)$$
$$+ \$10,050(P/F, i, 5)$$
$$= \$+691$$

Try 30%:

$$PE = \$-2,821$$
$$i_c = 25\% + 5\%(691/3,512) = 26.0\%.$$

The tax-sheltered cost of capital can now readily be calculated

$$i_a = i_c - ri_d t = 26.0\% - 0.25(10\%)(0.5) = 24.8\%$$

We suggest that equations 6.1 to 6.8 be used rather than the tabular format unless explicit display of the year-by-year cash flows is desirable.

6.16 DECISION MAKING INVOLVING MULTIPLE ALTERNATIVES

To this point the discussions have been concerned with evaluating a single investment or choosing between two alternative investments. We are often faced with making a choice among a number of investments. Selecting a single investment from among a number of alternatives should not present any additional difficulties, as we can evaluate multiple alternatives in pairs. However, this analysis must be done in a systematic manner, keeping in mind two important criteria:

1. Our prime objective from an economic viewpoint is to maximize profits.
2. It is the profit on the incremental (additional) investment, when evaluating alternatives, that is critical to our investment decision.

6.16.1 Types of Investment Proposals

There are two types of investment proposals of major interest: independent proposals and dependent proposals.

Independent Proposals

When the acceptance of a proposal from a group of proposals has no effect on the selection of any of the other proposals in the group, the proposal is said to be *independent*. Very few investment decisions within a firm are truly independent but for all practical purposes many are considered as such. For example, the selection of a machine to perform a specific function and the decision to buy or lease cars for sales personnel may be treated as independent decisions.

Dependent Proposals

The type of *dependent* proposals of major concern to our discussion are classified as *mutually exclusive proposals*. If the proposals contained in the group of proposals under consideration are related so that the acceptance of one proposal from the group excludes the acceptance of any other proposal in the group, the proposals are said to be mutually exclusive (e.g., only one proposal can be accepted from any mutually exclusive group of proposals). For example, if the firm is considering the alternatives of buying building A, or building B to house their manufacturing operation, these alternatives are mutually exclusive. The selection of one excludes the other.

Another classification of dependent proposals are referred to as *contingent proposals* because their acceptance is dependent on the acceptance of some other proposal. For example, the erection of a building is dependent on being able to purchase the land.

6.16.2 Comparing Mutually Exclusive Alternatives

In evaluating mutually exclusive alternatives, if MARR/BTRR is known and revenues are identical for all alternatives, the alternatives can be compared using the annual equivalent revenue requirements, present equivalent revenue requirements, or future equivalent revenue requirements approaches. If MARR is not known and/or if the sensitivity of the investment choice to various MARR/BTRR values is desired, the rate-of-return approach should be used.

The Do-Nothing Alternative

The *do-nothing alternative* is a feasible alternative to be considered in the analysis of a group of mutually exclusive alternatives if the firm is not committed to investment in any of the projects being considered. For example, assume that a company has the opportunity to accept a contract to supply a specific component to a car manufacturer. However, the acceptance of the contract requires the purchase of some additional milling equipment. On evaluation of the equipment alternatives available, the company decides that they cannot make an acceptable profit. Therefore, their decision is not to accept the contract, or to "do nothing" with respect to this particular investment alternative.

Example 6.21

Problem: Assume that to meet production demand for a large contract, the firm requires five welders of type A or B or two semiautomatic welders, alternative C, or one fully automatic welder, alternative D. The investment does not have to be accepted. Estimates pertaining to these investment alternatives are as follows:

| | 0 | Alternative | | | |
		5 Units of A	5 Units of B	2 Units of C	1 Unit of D
Initial investment	0	$25,000	$40,000	$140,000	$175,000
Life	0	5 years	5 years	5 years	5 years
Salvage value	0	$5,000	$5,000	$5,000	$5,000
Operating costs	0	$100,000	$95,000	$60,000	$42,000
Operating revenue	0	$110,000	$110,000	$110,000	$110,000
Before-tax cash flow	0	$10,000	$15,000	$50,000	$68,000

MARR $= 15\%$ and $t = 46\%$. Use the straight-line method (equation 4.2) and assume 100% equity capital. Use the annual equivalent approach as a method of selecting the preferred alternative.

Solution

$AE = 0$ (do nothing alternative)

Alternative A

$AE = (AEOR - AEOC)(1 - t) - P(A/P, 15\%, 5) + NSV(A/F, 15\%, 5)$
$\quad + t(AED)$

$\quad = (\$10,000)(0.54) - \$25,000(0.2983) + \$5,000(0.1483) + 0.46(\$4,000)$

$\quad = \$5,400 - \$7,458 + \$742 + \$1,840 = \$524$

Alternative B

$AE = (\$15,000)0.54 - \$40,000(0.2983) + \$5,000(0.1483) + 0.46(\$7,000)$

$\quad = \$8,100 - \$11,933 + \$742 + \$3,220$

$\quad = \$129$

Alternative C

$$AE = (\$50,000)(0.54) - \$140,000(0.2983) + \$5,000(0.1483)$$

$$+ \ 0.46(\$27,000)$$

$$= \$27,000 - \$41,762 + \$742 + \$12,420$$

$$= \$-1,600$$

Alternative D

$$AE = (\$68,000)(0.54) - \$175,000(0.2983) + \$5,000(0.1483)$$

$$+ \ 0.46(\$34,000)$$

$$= \$36,720 - \$52,203 + \$742 + \$15,640$$

$$= \$899$$

Conclusion: Alternative D is the acceptable investment, having the largest positive cash flow. The same conclusion may be arrived at using the present equivalent approach.

6.16.3 Evaluating Mutually Exclusive Alternatives Using Rate of Return As the Decision Criteria

The rate-of-return approach as the decision criteria for evaluating mutually exclusive alternatives is not normally recommended when MARR is known because of the added computation necessary in applying this method. However, this approach does have certain advantages. Mainly:

1. For sensitivity analysis with respect to the rate of return, knowing the actual rates of return on each overall investment and each increment of investment supplies more information to the analyst than is supplied by the annual or present equivalent approaches.
2. For capital budgeting purposes (where capital rationing is involved) ranking investments using the rate-of-return approach allows the cutoff rate of return in terms of funds available to be established.
3. Perhaps the most significant reason for the rate-of-return approach is that management most readily relates to investments in terms of return on investment. The main disadvantages of the rate-of-return approach are:
   a. Knowledge of revenues is essential to a rate-of-return calculation. Determining the portion of revenues attributable to a specific investment alternative (e.g., a single machine in an assembly line) may be difficult to accomplish.

b. Computational effort is usually significantly greater and the probability of an error in the analysis is increased if the analyst does not fully understand and appreciate the approach.

When using the rate-of-return method it is essential to do an incremental analysis. The relationship between the rate of return on total cash flows and the rate of return on incremental cash flows is not the same as the relationship between annual equivalent or present equivalent amounts. That is, for two alternatives A and B, $i_B - i_A$ does not necessarily equal i_{B-A}. An improper analysis leads to two possible common errors:

1. Accepting the investment with the highest rate of return on the overall investment
2. Accepting the alternative with the highest investment that meets MARR/BTRR

When the "do nothing" alternative is available and when three or more alternatives are under consideration $(X^2 + X)/2$, rates of return on the overall and incremental investments must be calculated. (X = the number of investments.)

When the do-nothing alternative is not available, this number reduces to $(X^2 - X)/2$. Therefore, when the number of alternatives being considered is large, the evaluation becomes tedious and subject to error if a systematic approach is not used. Smith* has developed an excellent system for the rate-of-return method that allows a thorough and effective analysis. This system is basically as follows:

1. Arrange the investments in increasing order of investment size.
2. Calculate the rate of return on each increment of investment.
3. Develop a choice table. The choice table allows the analyst to select the optimal alternative for any MARR or cutoff rate of return.

This choice table may be derived directly from the tabular format or one may use what Smith refers to as a network diagram. The network diagram is recommended until the analyst becomes thoroughly familiar with the procedure.

Using points to represent alternatives (such as Alternatives A, B, C), a network diagram showing rate of return relationships can be drawn as an x-sided figure to represent the x alternatives. Arrange the alternatives in a clockwise manner in ascending order of investment size, connecting

---

* Gerald W. Smith, *Engineering Economy: Analysis of Capital Expenditures*, 3rd ed., The Iowa State University Press, Ames, Iowa, 1979, p. 109.

all points in the diagram. Indicate the rate of return on each increment of investment on the diagram, and use arrows to show the direction of increasing investment size.

Figure 6.3 in example 6.22 demonstrates the use of the network diagram. Diagram 1 represents the case for the do-nothing alternative. Diagram 2 represents the case where investment in one of the alternatives is mandatory.

Example 6.22

Problem: We are given three mutually exclusive alternatives to consider. Sufficient funds are available for any investment that meets MARR. Assume that the do-nothing alternative is available.

	Alternative		
	A	*B*	*C*
First cost	$12,000	$40,000	$20,000
Life	5 years	5 years	5 years
Salvage value	$1,000	$4,000	$2,500
Operating costs	$28,000	$20,000	$25,000
Operating revenue	$33,000	$33,000	$33,000
Before-tax cash flow	$5,000	$13,000	$8,000

$t = 50\%$. Use the straight-line method (equation 4.2) to calculate the depreciation allowance for tax purposes.

Solution

Investment A

$$AE = (AEOR - AEOC)(1 - t) - P(A/P, i_a, 5) + NSV(A/F, i_a, 5)$$

$$+ t(AED)$$

Try 15%:

$$AE = \$5,000(0.5) - \$12,000(0.2983) + \$1,000(0.1483)$$

$$+ 0.5(\$2,200)$$

$$= \$2,500 - \$3,580 + \$148 + \$1,100$$

$$= \$168$$

Try 20%:

$$AE = \$5,000(0.5) - \$12,000(0.3344) + \$1,000(0.1344)$$

$$+ 0.5(\$2,200)$$

$$= \$-279$$

$$i_A = 15\% + 5(168/447) = 16.9\%$$

Invesment B

$$AE = \$13,000(0.5) - \$40,000(A/P, i, 5) + \$4,000(A/F, i, 5)$$
$$+ 0.5(\$7,200)$$

$$i_B = 10.7\%$$

Investment C

$$AE = \$8,000(0.5) - \$20,000(A/P, i, 5) + \$2,500(A/F, i, 5)$$
$$+ 0.5(\$3,500)$$

$$i_C = 16.0\%$$

Investment B − A

$$AE = \$8,000(0.5) - \$28,000(A/P, i, 5) + \$3,000(A/F, i, 5)$$
$$+ 0.5(\$5,000)$$

$$i_{B-A} = 8\%$$

Investment C − A

$$AE = \$3,000(0.5) - \$8,000(A/P, i, 5) + \$1,500(A/F, i, 5)$$
$$+ 0.5(\$1,300)$$

$$i_{C-A} = 15.0\%$$

Investment B − C

$$AE = \$5,000(0.5) - \$20,000(A/P, i, 5) + \$1,500(A/F, i, 5)$$
$$+ 0.5(\$3,700)$$

$$i_{B-C} = 5\%$$

Arranging data in a tabular format, we have

| Alternative | Investment | *ROR (%) on the Incremental Investment over:* | | |
		O	*A*	*C*
A	$12,000	16.9		
C	20,000	16.0	15.0	
B	40,000	10.7	8.0	5.0

Using network diagram 1 in Figure 6.3, we can now develop a choice table that allows us to select the optimal investment for any given MARR/ BTRR value.

(The dashed line shows the decision path.)

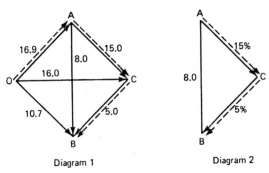

Diagram 1   Diagram 2

FIGURE 6.3 Network diagrams.

Procedure

1. Start with the minimum investment alternative and select the alternative that generates the highest rate of return on the incremental investment (note that the incremental investment over the do-nothing alternative is the total investment) over this minimum investment alternative.
2. Move to the alternative selected in the network diagram and select the alternative from the investments remaining that generates the highest rate of return on the incremental investment.
3. Continue this process until all increments of investment have been checked. *Proceed around the network diagram in a clockwise manner.*

   a. Using network diagram 1 (Figure 6.3):

      (1) Starting with the do-nothing alternative and checking the ROR on each incremental investment we have

$$i_A = 16.9\% \qquad i_C = 16.0\% \qquad i_B = 10.7\%$$

Select alternative A.

      (2) Proceed to investment A in the network diagram. Check the ROR on each increment of investment over alternative A:

$$i_{C-A} = 15.0\% \qquad i_{B-A} = 8.0\%$$

Select increment C − A.

      (3) Proceed to investment C in the network diagram. Check the ROR on each increment of investment:

$$i_{B-C} = 5.0\%$$

Select increment B − C.

This completes the network analyses. The choice table is

16.9% < MARR	choose 0 (do nothing)
15.0% < MARR ≤ 16.9%	choose A
5.0% < MARR ≤ 15.0%	choose C
MARR ≤ 5.0%	choose B

b. *Problem:* Assume that one investment must be selected. Using network diagram 2 (Figure 6.3):

(1) We check the rates of return on the incremental investments available and select the investment that pays the highest rate of return:

$$i_{C-A} = 15.0\% \qquad i_{B-A} = 8.0\%$$

Choose increment C − A.

(2) Proceed to investment C and check the rate of return on each incremental investment over C:

$$i_{B-C} = 5.0\%$$

This completes our analysis.
The choice table is

15.0% < MARR	choose A
5.0% < MARR ≤ 15.0%	choose C
MARR ≤ 5.0%	choose B

We cannot make a final choice regarding which alternative to select until we know the cutoff rate of return or a MARR value close to the cutoff rate of return.

6.17 MULTIPLE RATES OF RETURN

The general cash flow pattern dealt with in this text, and the normal one occurring in industry, is one where a large investment is followed by a series of receipts. However, there are situations where, in addition to the original investment, other large investments are required during the life of the project. Such situations include:

1. Enhanced recovery schemes such as water flooding, gas injection, or

thermal methods for an oil field currently producing under primary depletion

2. A cash flow cycle for nuclear fuel where the initial cost is followed by revenues from use or sale and finally by the high disposal costs of used nuclear fuel

3. A strip-mining operation which has a large initial investment followed by cash receipts and finally significant environmental penalties to restore the site

These situations can give rise to multiple rates of return. This is illustrated in the following example.

Example 6.23

Problem: Calculate the rate of return for the following cash flow.

Case (a)

Solution

$$PE = \$-5,000 + \$11,500(P/F, i, 1) - \$6,600(P/F, i, 2)$$

$$= \$-5,000 + \frac{\$11,500}{1 + i} - \frac{\$6,600}{(1 + i)^2}$$

i (%)	0%	5	10	15	20	25	30
PE ($)	-100	-34	0	+9.5	0	-24	-59

The data are graphed in Figure 6.4. Note that the rates $i = 10\%$ and $i = 20\%$ both make the present equivalent value equal to zero.

Problem: Calculate the rate of return for the following cash flow, which is case (a) flipped over.

Case (b)

Solution

$$PE = \$5,000 - \$11,500(P/F, i, 1) + \$6,600(P/F, i, 2)$$

i (%)	0%	5	10	15	20	25	30
PE ($)	+100	+34	0	-9.5	0	+24	+59

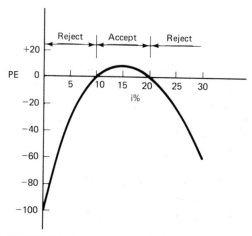

FIGURE 6.4 Present equivalent graph for Example 6.23, case (a).

Again, $i = 10\%$ and $i = 20\%$ both satisfy the criterion PE = 0 (see Figure 6.5).

The dual rates arise because the present equivalent formula is a polynomial of n degrees, and therefore is capable of several roots. The difficulty in an economic analysis is in determining the correct rate of return as a measure of whether or not to accept the project. Both the cases in Example 6.23 result in a PE = 0 at rates of return of 10% and 20%, but a present equivalent calculation using a MARR of 15% would result in acceptance of case (a) and rejection of case (b).

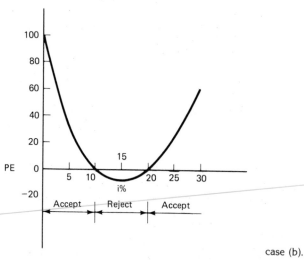

case (b).

Again, $i = 10\%$ and $i = 20\%$ both satisfy the criterion PE = 0 (see Figure 6.5).

There is no agreement in the literature as to the meaning, if any, that can be given to the multiple rates. However, the accept/reject decision can be made using the logic developed in Chapter 5 for the meaning of MARR.

According to Chapter 5, MARR is either the cost of capital (under conditions of no capital rationing) or the cut-off rate of return. This implies that in PE and AE calculations, receipts are treated as being reinvested at MARR in order to arrive at the final value. The following example applies this approach to a dual rate-of-return situation.

Example 6.24

Problem: Does the following cash flow diagram represent an acceptable investment at MARR = 15%? At MARR = 55%?

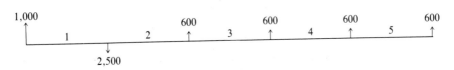

Solution: Calculate the rates of return:

$$PE = \$1,000 - \frac{\$2,500}{1 + i} + \frac{\$600}{(1 + i)^2} + \frac{\$600}{(1 + i)^3} + \frac{\$600}{(1 + i)^4} + \frac{\$600}{(1 + i)^5}$$

i (%)	0	10	15	20	30	40	41.5	50	55	60	70	80	86	90	100
PE ($)	900	456	316	211	77	7	0	−25	−31	−33	−27	−12	0	8	31

$$PE = 0 \quad at \quad i = 41.5\% \text{ and } 86\%$$

Assume that the initial receipt of $1,000 is invested at MARR for the first year and then calculate the rate of return on the modified project.

$MARR = 15\%$

First-year investment = $1,000(1.15) = \$1,150$

Modified project:

$$
\begin{array}{r}
2,500 \\
-1,150 \\
\hline
1,350
\end{array}
$$

$$PE = \$-1,350 + \$600(P/A, i, 4)$$

$i \approx 28\%$, which is greater than 15%; therefore, accept the investment. This is consistent with the 15% PE calculation, which is $316.

$MARR = 55\%$

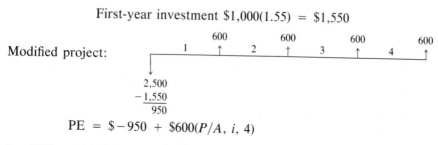

First-year investment $1,000(1.55) = \$1,550$

Modified project:

2,500
− 1,550
‾‾‾‾‾
950

$$PE = \$-950 + \$600(P/A, i, 4)$$

$i \simeq 51\%$, which is less than 55%; therefore, reject the investment. This is consistent with the 55% PE calculation, which is $\$-31$.

As is apparent in Example 6.24, the PE and AE criteria can be used in the multiple-rate-of-return situation as long as the MARR value is near the organization's reinvestment rate. If it is necessary to calculate a rate of return, then, as in the example, use the MARR to modify the cash flows into a conventional pattern, and calculate the rate of return using the modified diagram. This will give a result consistent with the PE and AE approaches.

Another problem created by the existence of multiple rates of return is that computer programs, designed to calculate rates of return, must contain a facility for checking the possibility of multiple rates of return. Otherwise, the programs, which usually have a numerical procedure for converging, could produce a unique and incorrect solution.

The following method is based on Descartes' rule of signs, which states the number of real positive roots of a nth-degree polynomial is less than or equal to the number of sign changes in the polynomial coefficients.

To determine if an investment project has multiple rates of return, or at least, more than one reversal in sign regarding the year-by-year cash flows, the following simple procedure may be used.

Step 1: Calculate the year-by-year net cash flows.

Step 2: Check the number of sign changes in these cash flows over the project life.

a. If there is no sign change, the project either has a negative rate of return (not economical as other profitability measures should also indicate), or it may have a large positive rate of return (usually for certain proposals with little or no investment).

b. If the sign changes only once, the project should have a unique rate of return.

c. If there is more than one sign change, it is possible that the project has more than one positive rate of return. Multiple sign changes are necessary but not sufficient for the existence of multiple rates of return. A sufficient condition rests with the relative magnitudes of these cash flows.

6.18 SUMMARY

As has been illustrated in this chapter, the decision-making power of the comparison methods of annual equivalent, present equivalent, and rate-of-return calculations has two aspects. First, the methodology requires an explicit presentation of the alternatives, with all the cash flows, salvage values, and tax effects specifically estimated. Second, strict adherence to the discounted cash flow formulas, together with the correct rate of return (MARR or BTRR), will identify the profit-maximizing alternative. The result is rational decision making based on economic criteria.

When a comparison of a group of mutually exclusive alternatives is under consideration and MARR/BTRR is known, the present equivalent or annual equivalent approach is recommended. These approaches minimize the calculations involved and will select the preferred alternative from the group. If MARR/BTRR is unknown and/or information is desired regarding the specific rate of return on each investment, the rate-of-return method is required.

PROBLEMS

6.1 The following two types of power plants are to be used for a mining operation. The expected period of service is 10 years. The before-tax rate of return is 15%.

	Steam Plant	Diesel Plant
Installed cost	$200,000	$300,000
Salvage value after 10 years	50,000	50,000
Annual operating costs	40,000	30,000

a. Compare these alternatives using the annual equivalent cost approach.
b. Convert the annual equivalent cost to a present equivalent cost.

6.2 Using an interest rate of 12%, find the annual equivalent cost for a proposed machine tool that has a first cost of $10,000, an estimated economic life of 8 years, and an estimated salvage value of $2,000. Annual maintenance will amount to $200 a year and periodic overhauls costing $600 each will occur at the end of the second, fourth, and sixth years.

6.3 Compare plan C and plan D on the basis of capitalized costs of perpetual service (PEC calculation with $n = \infty$) using an interest rate of 8%. Plan C calls for an initial investment of $500,000 and expenditures of $20,000 a year thereafter. It also calls for the expenditure of $100,000 at a date 10 years hence and every 10 years thereafter. Plan D calls for an initial investment of $800,000 followed by a single investment of $200,000, 25 years hence. It also involves annual expenditures of $20,000.

6.4 Compare investment A to investment B on the basis of capitalized cost of perpetual service (PEC calculation with $n = \infty$) using a BTRR = 15%. Investment A calls for:

1. An initial investment of $500,000
2. Expenditures of $20,000 a year for the first 10 years and $40,000 a year thereafter
3. Expenditures of $100,000, 10 years hence and every tenth year thereafter

   Investment B calls for:

1. An initial investment of $400,000
2. Annual expenditures of $40,000
3. An expenditure of $100,000 10 years hence

6.5 Compare the annual equivalent costs of two different machines designed to do the same job. The expected period of service is 10 years. BTRR = 15%.

	Machine A	Machine B
Installed cost	$20,000	$30,000
Salvage value after 10 years	1,000	5,000
Annual operating costs the first year	9,200	6,000
Annual operating costs will increase each year by:	500	500

6.6 The following estimates have been made for two alternatives; we must choose one of them. The before-tax rate of return required is 20%.

	A	B
Installed cost	$120,000	$150,000
Estimated useful life	10 years	10 years
Salvage at retirement	$20,000	$30,000
Annual operating costs	$20,000	$15,000

Try to minimize your computations as you determine which course of action to recommend.

6.7 The following estimates of cost apply to equipment alternatives A and B. The before-tax rate of return required is 20%.

	A	B
Installed cost	$100,000	$40,000
Operating costs	$5,000 at the end of year 1 and increasing by $1,000 per year for 20 years	$10,000 at the end of year 1 and increasing by $2,000 per year for 10 years
Overhaul costs every 5 years	$10,000	None req'd
Economic life	20 years	10 years
Salvage value at end of life (just overhauled)	$20,000	$10,000

a. Compare the present equivalent costs using a study period of 20 years.

b. Compare the annual equivalent costs.

6.8 The first cost of a new machine is $300,000. A major overhaul is (to be expensed not capitalized) expected to occur at the end of 10 years costing $25,000. Annual receipts are $60,000 and annual expenses are $10,000. The machine can be sold for $110,000 at the end of 15 years.

a. What is the prospective before-tax rate of return?

b. What is the prospective after-tax rate of return? Assume that $t = 46\%$, use the fast-write-off method, and assume 100% equity capital.

6.9 The first cost of a new milling machine is $200,000. Annual receipts from the machine are estimated to be $70,000. Annual expenses for the first year are estimated to be $25,000 and they are expected to increase by $500 per year every year thereafter. The salvage value is estimated to be $40,000 at the end of 10 years. $t = 46\%$ and $r = 0$.

a. What is the before-tax rate of return?

b. What is the after-tax rate of return?

Use the fast-write-off method.

6.10 A company wishes to compare the present equivalent, on an after-tax basis, of two investment opportunities.

	A	B
Initial cost	$75,000	$125,000
Salvage value in 10 years	25,000	75,000
BTCF	15,000	20,000

Use the declining-balance method for CCA and a 30% rate. $r = 0$, $t = 48\%$, and MARR $= 20\%$.

6.11 The following two machines are designed to perform the same job. Use an estimated life of 20 years. $r = 0$, MARR $= 15\%$, and $t = 46\%$. Use the fast-write-off method.

	Machine A	Machine B
First cost	$30,000	$20,000
Salvage value after 20 years	7,000	3,000
Annual operating cost	6,000	9,000

a. Compare these two machines using the present equivalent revenues (PER) approach.

b. Compare the two machines using the annual equivalent revenues (AER) approach.

6.12 Use an incremental analysis to compare the annual costs of two different machines designed to do the same job. The expected period of service is 10 years. Use the declining-balance method and a CCA rate of 20%. $r = 0$, $i_a = 12\%$, and $t = 40\%$.

	Machine A	Machine B
First cost	$20,000	$30,000
Salvage value after 10 years	1,000	5,000
Annual operating costs the first year	9,200	6,000
Annual operating costs will increase each year by	500	500
Additional operating cost in year 5	1,000	2,000

6.13 A company wishes to purchase either machine A or machine B to perform a specific operation. Compare these alternatives using the present-worth approach (present equivalent). Minimize your calculations.

	A	B
Initial investment	$75,000	$125,000
Estimated economic life	10 years	10 years
Estimated salvage value	$25,000	$75,000
Annual operating costs	$30,000	$20,000

Use the declining-balance method and a CCA rate of 30%. $r = 0$, $t = 40\%$, and MARR = 20%.

6.14 A machine has a first cost of $105,000, an estimated life of 10 years, and an estimated salvage value of $5,000. How much depreciation will be written off during the second year of life using:

a. The fast-write-off method?

b. The declining-balance method (rate = 10%)?

c. Compute the book value at the beginning of year 5 using the declining-balance method.

d. Assume a BTCF of $25,000 in year 3 and calculate the income tax payment to be made. Use a tax rate of 40%. Use the declining-balance method and a CCA rate of 10%. $r = 0$.

6.15 An investment with an estimated economic life of 20 years is under consideration.

	First Cost	Salvage Value
Building	$70,000	$ 10,000
Land	50,000	100,000
Equipment	40,000	10,000

Estimated annual operating costs	$20,000
Additional expense in year 3	$10,000

Use the declining-balance method and a CCA rate of 30%. $r = 50\%$, $i_d = 12\%$, $i_e = 18\%$, and $t = 50\%$. What is the annual revenue required to justify the project?

6.16 The first cost of a new machine is $300,000. A major overhaul is expected to occur at the end of 10 years costing $5,000. Estimated annual receipts are $60,000 and annual expenses are $16,100. Assume that the machine can be sold for $50,000 at the end of 15 years. Use the declining-balance method and a CCA rate of 20%. $r = 0$ and $t = 50\%$. What is MARR?

6.17 Find the after-tax rate of return on a project to which the following estimates apply: installed cost = $200,000; estimated resale value at the end of 10 years = $50,000; annual revenues = $100,000; annual operating costs = $40,000. $r = 0$ and $t = 50\%$.
a. Use the fast-write-off method.
b. Use the declining-balance method and a CCA rate of 20%.

6.18 As an engineer for the Big-T Corporation, you have recommended that the company spend $100,000 today for a small manufacturing facility. The company uses a 5-year time span for the evaluation of most investments. You estimate that the resale value of all assets in 5 years = $80,000. Operating costs = $300,000 per year. The effective tax rate = 46% and MARR = 15%. Use the fast-write-off method.
a. What is the annual revenue required to justify the investment?
b. What is the present equivalent of revenue requirements?
c. Should the investment be made if you can sell 300,000 units per year for $1 per unit?

6.19 You are part of the management team negotiating a labor contract with the union. A union representative remarks that any increase in wages costs the company only 50 cents on the dollar (assuming that $t = 50\%$), and therefore, the company should be much more receptive to wage increases than they actually are. How are you going to respond to this statement?
a. Is the statement true?

b. What happens to the rate of return on the investment if operating costs increase more than operating revenues? Complete the following table to verify your conclusions. (Assume that capital cost allowance = $0.10.)

Year	Operating Revenue	Operating Cost	D	TI	IT	ATCF
1	$1.00	$0.80				
2	1.00	0.85				

6.20 A $100,000 investment in machinery is proposed. It is anticipated that annual revenues from this investment will be $25,000 and annual expenses $10,050 a year for 10 years. A major overhaul (to be expensed not capitalized) is expected at the end of 5 years at a cost of $5,000. Use the fast-write-off method to calculate CCA. Assume a 10-year life and a resale value of $30,000. $r = 0$ and $t = 46\%$.
a. Calculate the annual revenue required to justify this investment using MARR $= 10\%$.
b. Calculate the actual after-tax rate of return on the investment.
c. On what basis would you accept the investment?

6.21 The first cost of a new machine is $300,000. A major overhaul is expected to occur at the end of 3 years costing $25,000. Annual receipts are $80,000 and annual expenses are $11,350. The machine can be sold for $200,000 at the end of 5 years. Use the declining-balance method and a CCA rate of 30%. $r = 0$ and $t = 40\%$.
a. What is BTRR?
b. What is MARR?

6.22 Compare the annual revenue requirements for two different machines designed to do the same job. The expected period of service is 10 years.

	Machine 1	Machine 2
Installed cost	$50,000	$80,000
Salvage value after 10 years	10,000	30,000
Annual operating costs the first year	30,000	24,000
Annual operating costs will increase each year by	2,000	2,000

$i_d = 10\%$, $r = 40\%$, $t = 46\%$, and $i_e = 13.1\%$.
a. Use the declining-balance method and a CCA rate of 20%
b. Use the fast-write-off method to calculate CCA

6.23 A company wishes to purchase either machine A or machine B to perform a specific operation. Compare these alternatives using the present-worth approach (present equivalent). Minimize your calculations.

	Machine A	Machine B
Initial investment	$100,000	$150,000
Estimated economic life	10 years	10 years
Estimated salvage value	$25,000	$75,000
Annual operating costs	$30,000	$20,000

Use the declining-balance method and a CCA rate of 30%. $r = 25\%$, $i_e = 12\%$, $i_d = 8\%$, $t = 50\%$.

6.24 The first cost of a new roll former is $300,000. A major overhaul is (to be expensed not capitalized) expected to occur at the end of 5 years costing $5,000. Annual receipts are $60,000 and annual expenses are $16,150. The estimated resale value for the machine at the end of 15 years is $50,000. Use the fast-write-off method to calculate CCA
a. What is the prospective before-tax rate of return?
b. What is the prospective after-tax rate of return?
c. What is the rate of return on the equity investment? $r = 40\%$, $i_d = 9\%$, and $t = 46\%$.

6.25 A company is considering the purchase of a new lathe for $100,000 which has an estimated useful life of 10 years and negligible salvage value. Estimated annual operating revenues $= \$60,000$. Estimated annual operating costs $= \$36,000$. Use the fast-write-off method to calculate CCA. $r = 0$, and $t = 46\%$. What is the after-tax rate of return on this investment?

6.26 The information is identical to Problem 6-25 except that the net salvage value in year 10 is $83,300. What is the after-tax rate of return on this investment?

6.27 A construction company is considering some new equipment with a first cost of $100,000 and an estimated salvage value in 5 years of $30,000. This equipment is expected to have operating costs of $25,000 per year. Use the declining-balance method and a 30% CCA rate. $r = 0$, MARR $= 12\%$, and $t = 40\%$. Determine the revenue requirements to justify this investment.

6.28 A taxi company is considering the purchase of five automobiles for a total cost of $60,000. The estimated useful life of these vehicles is 4 years and their total salvage value at that time is estimated to be $15,000. Estimated annual operating costs $= \$20,000$ per unit. Use the declining-balance method and a CCA rate of 30%. MARR $= 12\%$ and $t = 40\%$. What revenue is required to justify the investment?

6.29 The information is identical to Problem 6.28 except that the past performance indicates that each taxi will earn, on the average, $100 per day for 245 days each year.

a. What is the rate of return on this investment?

b. How sensitive is this investment to the revenues generated? What is the rate of return if revenues drop to $95 per day per unit?

6.30 A utility is considering the purchase of a new electric generating unit with an initial cost of $500,000. The estimated useful life of this unit is 20 years with an estimated salvage value of $40,000. Use the declining balance method and a CCA rate of 30%. Estimated annual operating costs = $50,000. $r = 60\%$, $i_d = 12\%$, $t = 50\%$, and $i_e = 16\%$.

a. What revenue is required to justify this investment?

b. What is the after-tax rate of return to the equity holder if annual revenues are $124,500?

6.31 A company is considering the purchase of some new equipment worth $200,000. The equipment has an estimated useful life of 8 years and a salvage value of $40,000. Use the fast-write-off method to calculate CCA. Estimated annual operating costs = $25,000. $r = 25\%$, $i_d = 8\%$, $t = 50\%$, and $i_e = 12\%$. What revenues are required to justify this investment?

6.32 You and a friend have an opportunity to go into a small business manufacturing circuit boards for a major electronics company. You estimate that you will need $30,000 capital. You each have $7,500 and you can borrow the balance from the bank at 12%.

1. Initial investment is $20,000 for equipment and $10,000 for working capital.
2. Estimated salvage value on the equipment in 5 years is negligible.
3. Estimated annual operating costs = $40,000 per year.
4. Estimated annual operating revenues = $51,750 per year.
5. Use the fast-write-off method for CCA purposes.
6. An equipment overhaul costing $2,000 is expected to occur in year 3.
7. $t = 40\%$.

a. Calculate the expected rate of return on equity capital.

b. How sensitive is the investment to a change in operating revenues or operating costs? What is the rate of return on equity capital if annual revenues drop to $49,800?

6.33 The first cost of a new punch press is $200,000. Annual receipts from the machine are estimated to be $84,450. Annual expenses for the first year are estimated to be $40,000 and they are expected to increase by $3,000 per year thereafter. The salvage value is estimated to be $120,000 at the end of 10 years.

a. What is the before-tax rate of return?

b. What is the after-tax rate of return? Use the fast-write-off method to calculate CCA. $r = 0$ and $t = 46\%$.

c. What is the rate of return on the equity investment in part (b) if $r = 30\%$ and $i_d = 10\%$?

6.34 Two enterprising engineering students have decided to start the Keepon Trucking Company. They put in $10,000 of their own money and borrowed another $10,000 from the bank at a 12% annual rate. $15,000 was used to purchase of a single-axle flat-deck truck. The remaining $5,000 was retained as working capital. Estimated salvage value for the truck in 3 years is $3,000. They estimated that operating costs (driver, gas, oil, maintenance) would come to $25,000 per year and the truck would earn (gross revenue) $32,450 per year. Use an income tax rate of 25%. Use the declining-balance method and a CCA rate of 30%.

a. Calculate the expected rate of return on equity capital.

b. What is the expected rate of return on equity capital if revenues increase to $34,750?

6.35 Global Industries is considering the following investment:

	First Cost	Salvage
Building	$100,000	$ 75,000
Working capital	100,000	100,000
Land	50,000	90,000
Equipment	40,000	5,000

Financing is 40% debt capital, $i_d = 10\%$, and $t = 50\%$. Annual operating costs are $50,000 and an additional expense of $25,000 occurs in year 5. The estimated life of this investment is 10 years. Use the declining-balance method. The CCA rate on the building is 10%; the CCA rate on equipment is 20%.

a. Calculate the annual revenue necessary to make this a feasible investment using MARR = 12%.

b. Calculate the rate of return on equity capital assuming that operating revenues are $115,850.

6.36 A contractor purchased $400,000 worth of equipment to build a dam. After 2 years and the completion of the dam, the contractor sold the equipment for $300,000. The contractor attributed the sale of the equipment at an amount higher than book value to the excellence of maintenance. He estimated that he spent $16,000 per year more for maintenance than necessary "just to keep the equipment running," and by so doing reduced his operating costs exclusive of maintenance on the contract by $12,000 per year. The contractor's income tax rate was 46%. The contractor's taxable income was $108,000 in each of the 2 years required to complete the contract. Use the fast-write-off method to calculate the CCA. $r = 0$ and MARR $= 15\%$.

a. Construct an after-tax cash flow diagram for the contractor.

b. Construct an after-tax cash flow diagram for the contractor if he had spent $16,000 per year less on maintenance, and as a result his operating costs increased by $12,000 per year and the equipment sold for its book value at the end of the 2-year period.

6.37 A new apartment block under consideration can be depreciated using the declining-balance method and a CCA rate of 10%.

	First Cost	Salvage Value (10 years)
Building	$250,000	$150,000
Land	50,000	

Estimated annual operating revenues $= \$50,000$. Estimated annual operating costs $= \$8,600$. An additional expense at the end of year $3 = \$10,000$. $r = 0$; $t = 40\%$, and $n = 10$ years.

a. Assume that the land can be sold for $175,000 at the end of 10 years. What is the after-tax rate of return?

b. If the investment can be sold for $300,000 at the end of 5 years (building, $200,000; land, $100,000), should the apartment block be sold at this time?

6.38 A project under consideration is estimated to have a 20-year useful life. Estimated annual operating costs are $100,000 with an additional expense of $100,000 in year 5. Use the declining-balance method. The CCA rate on the building is 5%; the CCA rate on equipment is 20%.

	Initial Investment	Salvage Values
Building	$200,000	$100,000
Land	50,000	100,000
Working capital	40,000	40,000
Equipment	100,000	5,000

$r = 25\%$, $i_d = 12\%$, $i_e = 18\%$, and $t = 50\%$. What is the annual revenue required to justify this investment?

6.39 A company is considering two alternatives for printing equipment. Use the fast-write-off method to calculate the CCA. MARR $= 15\%$ and $t = 40\%$.

	Alternative	
	A	B
Installed cost	$50,000	$40,000
Estimated salvage value in year 10	0	0
Estimated annual operating costs	22,000	25,000

Which alternative is preferable?

6.40 The Topdog Oil Company is considering an investment in an oilfield pumping station which has a useful life of 10 years. The estimated annual operating costs are $75,000 with an additional expense in year 3 of $30,000.

	Initial Cost	Market Value in Year 10
1. Buildings	$100,000	$80,000
2. Working capital	50,000	50,000
3. Land	50,000	90,000
4. Equipment	200,000	75,000
5. Additional equipment (a pump) will be added 5 years from today	50,000	0

6. Use the declining-balance method. The CCA rate on the building is 10%; the CCA rate on equipment is 30%.
7. $i_d = 10\%$, $t = 46\%$, $r = 60\%$, and $i_e = 16.9\%$.

Calculate the revenue requirements necessary to justify the investment.

6.41 Five years ago Global Developments invested in a rental property. The question today is whether to retain or sell this property.

	Original Cost	Market Value Now	Market Value in 10 Years
Land	$100,000	$150,000	$200,000
Building	300,000	250,000	150,000
Working capital	75,000	75,000	75,000

1. Use the declining-balance method and a CCA rate of 10%.
2. Operating revenue per year = $150,000.
3. Operating costs per year = $56,350. $r = 0$ and $t = 46\%$.

  a. Calculate the historic after-tax rate of return on the investment over the past 5 years.
  b. Find the prospective after-tax rate of return if the investment is retained for another 10 years.
  c. Which rate of return is relevant to the company's decision, and under what condition should management sell now?

6.42 A project under consideration is estimated to have a 10-year useful life. The estimated annual operating costs are $50,000 with an additional expense in year 5 of $25,000. Use the declining-balance method and a 10% CCA rate on the building. Use the fast-write-off method for equipment

	Initial Cost	Salvage Value
Building	$100,000	$25,000
Land	50,000	75,000
Working capital	90,000	90,000
Equipment	40,000	20,000

$r = 50\%$, $i_d = 10\%$, $i_e = 18\%$, and $t = 40\%$. What annual revenue must the project generate to justify this investment?

6.43 You are given a set of mutually exclusive alternatives from which you may choose one. Each alternative is expected to have an infinite life and produce annual after-tax cash flow savings as follows (the do-nothing alternative is also available):

Alternative	Initial Cost	After-Tax Cash Flow
A	$2,000	$ 450
B	4,000	800
C	1,000	200
D	3,000	750
E	5,000	1,060

a. Draw a network diagram and develop a choice table.
b. Which alternative is the best choice if:
   (1) MARR = 26%?
   (2) MARR = 18%?
   (3) MARR = 14%?
c. In what range of MARR is alternative C the proper choice?

6.44 You are given a set of five mutually exclusive alternatives from which you may choose one. They are listed in ascending order of investment size. The lives are identical and salvage values are negligible. The after-tax rates of return are:

Alternative	Overall ROR (%)	ROR on an Incremental Investment over: A	B	C	D
A	20				
B	36	50			
C	30	40	24		
D	26	34	20	14	
E	25	30	18	11	8

a. Draw a network diagram and develop a choice table.
b. which alternative is the best choice if:
   (1) MARR = 13%?
   (2) MARR = 35%?
   (3) MARR = 40%?
c. In what range of MARR is alternative C the proper choice?

6.45 You are given a set of five mutually exclusive alternatives from which you must choose one. They are listed in ascending order of investment size. The lives are identical and salvage values are negligible. After-tax rates of return are:

| Alternative | Overall ROR (%) | ROR (%) on Incremental Investment over: | | | |
		A	B	C	D
A	30				
B	32	37			
C	30	31	22		
D	27	26	18	16	
E	24	22	12	10	9

Which alternative is the best choice if:
a. MARR = 12%?
b. MARR = 25%?
c. MARR = 40%?

6.46 A syndicate of investors are considering buying a low-producing oil well from an oil company for $10,000. The well is expected to produce revenue of $12,500 in year 1 and $13,200 in year 2. In year 3 additional equipment will be required, so the cash flow will be an expenditure of $16,500. The well will produce $13,150 in revenue in year 4. In years 5 and 6 it will be necessary to use enhanced recovery techniques, resulting in a negative cash flow of $16,400 in year 5 and $17,300 in year 6. The well will produce revenue of $21,600 in year 7. All cash flows are after taxes. The investors must agree to operate the well for the full 7 years.
a. If the oil company's MARR is 10% and the investor's MARR is 20%, is this an acceptable transaction for both parties?
b. Is it still acceptable if the oil company's MARR is 30%?

6.47 A researcher with a promising invention is trying to raise money. Currently, the researcher's cash needs are small, but when the invention gets to the prototype and preproduction stages, in 5 years and 10 years, respectively, the need will be considerable. Therefore, the following investment package is conceived. If an investor purchases a

$7,800 share of the invention today, it can be paid for with $1,600 5 years hence and $6,190 10 years hence. Because the project requires R&D, by making the purchase the investor qualifies for an immediate $1,000 investment credit. It is estimated that the share can be sold at the end of year 15 for $10,000. Assume that all cash flows are after-tax cash flows. What would the MARR of an investor have to be for this to be an acceptable package?

APPENDIX 6A

Analysis of Investment Decisions Using a Microcomputer

The microcomputer revolution has made tremendous computing power readily available. Calculations that previously required much time and table searching can be readily solved with a microcomputer. Mortgage calculation programs are now part of the standard demonstration software that is supplied with the micros.

A very powerful software package now available for most microcomputers is the *electronic spreadsheet*. The first package of this type was Visicalc, developed in 1979. Since then there have been a host of similar packages, two of the more popular being Supercalc and Lotus 1-2-3. Although the packages differ in sophistication, the basic principal is the same. That is, they provide the user with the electronic equivalent of a large sheet of squared paper, with the columns labeled A, B, C, . . . across the sheet and the rows numbered down the sheet. Thus each square has a unique address or label, such as A25 (column A, row 25). The power of the packages is that one can make any square a function of other squares. For example, square A25 can be the sum of squares A1 to A24 inclusive. Then as you change the numbers in squares A1 to A24, the total in A25 also changes. Thus it is possible to develop a program to do complicated cash flow analysis with a minimum of effort. Further, such a program permits one to experiment quickly and simply with different assumptions and parameters, such as MARR, tax rates, and depreciation schedules.

The following microcomputer program is designed to do revenue requirements and rate-of-return calculations on either a before- or an after-tax basis. The program is written using Supercalc (1.12, November 1982) but can readily be adapted to any other spreadsheet program.

6A.1 PROGRAM DESCRIPTION

When the program is loaded, Table 6.A1 appears on the screen. Where the word "DATA" appears, the user inserts the problem data. Where the word "CODE" appears the user specifies the correct option code. Row 2 is where the basic parameters—years, MARR, debt ratio, and so on—are inserted; rows 6 to 9 are for the cash flows; rows 11 to 14 are for initial and terminal values; and the answers appear in rows 19 to 22.

Although there are options to simplify and speed up the calculations, the program is best introduced by a basic example.

Example 6A.1

Problem: Find the before-tax AER for the following investment in equipment: initial cost = $50,000, annual operating cost = $2,000, life = 11 years, MARR = 10%, and SV = 0.

Solution: To solve, load the program and enter the following data:

11 in A2	(life n)
11 in A4	(life for CCA calculation when using SL—equation 4.2)
0.1 in B2	(MARR)
0 in B7	(code for uniform costs)
2,000 in C7	(annual operating cost)
50,000 in E11	(initial cost)

The AER answer, $9,698, will appear in F22.

Problem: Use the same data but do an after-tax calculation with $t = 46\%$ and straight-line depreciation (equation 4.2).

Solution: Enter .46 in E2 and the AER answer, $12,383, will appear in F22.

TABLE 6A.1 Cash flow program for microcomputers using Supercalc 1.12.

	A	B	C	D	E	F	G	H	I
1	n1	i	r	id	t	ie	if	.1	A/P,i,n
2	5	.1	DATA	DATA	DATA	DATA	DATA		.263797
3	n2(cca)	DB(BLDG)	DB(EQUIP)			CTF-B	CTF-E	p/c,i,n	a/p,i,n
4	DATA	DATA	DATA	DATA	DATA			3.79079	ERROR
5	YEAR			CODE	CODE	3	4	5	6
6	OR	CODE	DATA	DATA	DATA	DATA	DATA	DATA	DATA
7	OC	CODE	DATA	DATA	DATA	DATA	DATA	DATA	DATA
8	CCA-B	CODE		DATA	DATA	DATA	DATA	DATA	DATA
9	CCA-E1	CODE		DATA	DATA	DATA	DATA	DATA	DATA
10	BLDG		LAND	WC	EQUIP	CTF-B	CTF-E	CCTF-B	CCTF-E
11	P(SL)	DATA	DATA	DATA	DATA	1	1	1	1
12	P(DB)	DATA	DATA	DATA	DATA				
13	SV	DATA	DATA	DATA	DATA				
14	BV	DATA	DATA	DATA	DATA				
15	NSV-SL	0	0	0	0				
16	NSV-DB	0	0	0	1				
17	CCA-SL	0	0	0	0				
18	PEOR		PEOC	P(SL)	P(DB)	P-SV-SL	P-SV-DB	PED	
19	0		0	0	0	0	.620921	0	
20									
21	PE-SL	AE-SL	PE-DB	AE-DB	PER-SL	AER-SL	PER-DB	AER-DB	
22	0	0	.620921	.163797	0	0	-.62092	-.16380	

6A.2 PROGRAM OPTIONS

The program contains several options that the user can select.

1. The discount rate to be used can be entered in cell B2 or values for r, i_d, t, i_e, and i_f can be entered into cells C2 to G2 and the discount rate will be calculated. (Place a 0 in cell B2.)

2. With respect to operating revenues (OR) and operating costs (OC), two options exist.

   a. If OR/OC are a uniform amount each year enter the value in C6/C7. Enter a 0 in B6/B7.

   b. If OR/OC are represented by a geometric progression (e.g., inflation) enter the OR/OC at the end of year 1 in C6/C7 and enter a 1 in B6/B7.

   c. In either part (a) or (b), if any additional OR/OC occurs in any specific year, enter the value into the cell for that specific year (e.g., if an additional OC occurs in year 3, place the value in cell F7). To handle a nonuniform series of cash flows or an arithmetic gradient series, the values must be entered in D6 to W6 and/or D7 to W7.

3. Capital investment values are entered under the column headings B10 to E10 (e.g., building values in column B).

   a. For straight-line depreciation (equation 4.2): $D = (P - SV)/n$
      (1) Place capital investments in B11 to E11.
      (2) Place salvage values (SV) in B13 to E13.
      (3) Place book values at disposal (BV) in B14 to E14.

   b. Depreciation allowance—straight line (equation 4.2). Enter life for depreciation purposes in cell A4.
      (1) Building: If B8 = 0, annual depreciation allowance = B17; if B8 = 1, annual depreciation values are entered into D8 to W8.
      (2) Equipment: If B9 = 0, annual depreciation allowance = E17; if B9 = 1, annual depreciation values are entered into cells D9 to W9.

   c. For declining-balance depreciation:
      (1) Place capital investments in cells B12 to E12.
      (2) Place salvage values (SV) in cells B13 to E13.
      (3) Place book values (BV) for land and working capital in cells C14 and D14. No values need to be entered in cells B14 and E14.

4. Capital tax factors:

   a. Enter declining-balance rate for buildings in B4.

   b. Enter declining-balance rate for equipment in C4.

    c. If the half-year rule applies to buildings, enter a 1 in D4.

    d. If the half-year rule applies to equipment, enter a 1 in E4.

The program will readily handle declining-balance or straight-line capital (equation 4.2) capital cost allowance. Any other depreciation method requires that the annual capital cost allowance values be entered for each year. For buildings place a 1 in cell B8 and the CCA values in cells D8 to W8. For equipment place a 1 in cell B9 and the CCA values in cells D9 to W8.

6A.3 DATA AND CODE ENTRY

Data may be entered wherever the word "DATA" appears plus in cells A2 and B2.

    Note: Cells A2 and B2 presently show values. Text (e.g., DATA) is treated as zeroes and if zeroes are stored on the disk in these two cells, division by zero occurs. If this should happen, load cells A1:I2. Place values in cells A2 and B2 and copy these values to the disk and reload the entire program.

Codes are entered where the word "CODE" appears.

Cell	Value to be Entered
A2	Enter life of the investment.
B2	Enter discount rate in B2 or leave B2 blank and enter values in C2, D2, F2, and G2.
C2	Enter debt ratio if no value placed in B2.
D2	Enter interest rate on debt if no value placed in B2.
E2	Enter income tax rate.
F2	Enter interest rate on equity if no value placed in B2.
G2	Enter inflation rate if no value placed in B2.
A4	Enter life for depreciation allowance calculation when using SL (equation 4.2). Enter 2 when using the fast-write-off method.
B4	Enter declining-balance rate for buildings.
C4	Enter declining-balance rate for equipment.
D4	Enter 1 if the half-year convention applies to buildings.
E4	Enter 1 if half-year convention applies to equipment.
B6	Enter 0 if OR values are uniform; enter 1 if (P/C) factor applies.
C6	Enter OR for year 1 if OR values are uniform.
D6–W6	Enter any additional revenues for specific years (e.g., gradient values start in year 2).
B7	Enter 0 if OC are uniform; enter 1 if (P/C) factor applies.
C7	Enter OC for year 1 if OC values are uniform.
D7–W7	Enter any additional operating costs for specific years.

B8	Enter 0 if depreciation allowance for buildings is straight-line (equation 4.2) or declining-balance; otherwise, enter 1 and enter depreciation allowance for each year in cells D8 to W8.
B9	Enter 0 if depreciation allowance for equipment is straight-line or declining-balance; otherwise, enter 1 and enter depreciation allowance for each year in cells D9 to W9.
B11	Enter initial investment in buildings if using SL.
C11	Enter initial investment in land if using SL.
D11	Enter initial investment in working capital if using SL.
E11	Enter initial investment in equipment if using SL.
B12	Enter initial investment in buildings if using DB.
C12	Enter initial investment in land if using DB.
D12	Enter initial investment in working capital if using DB.
E12	Enter initial investment in equipment if using DB.
B13	Enter salvage value on disposal for buildings.
C13	Enter salvage value on disposal for land.
D13	Enter salvage value on disposal of assets for working capital.
E13	Enter salvage value on disposal for equipment.
B14	Enter book value on disposal for buildings when using SL. [When using declining-balance (DB) no value needs to be entered.]
C14	Enter book value on disposal for land.
D14	Enter book value on disposal of assets for working capital.
E14	Enter book value on disposal for equipment. When using declining-balance, no value needs to be entered.

6A.4 CALCULATED VALUES

Calculated values appear in the cells listed below. If any individual data value is changed, all cells affected are automatically changed. Any comments below also refer to fast write-off, which is in effect SL with the ½ year rule.

H2	Discount rate
I2	$(A/P, i, n)$
H4	$(P/C, i, n)$
I4	$(A/P, i, n)$ for the value placed in A4 (the life for depreciation purposes)
F11–I11	Capital tax factors
B15–E15	Net salvage values (NSV) when using SL
B16–E16	Net salvage values (NSV) when using DB
B17–E17	Values for depreciation allowance (SL)
B19	PEOR

C19	PEOC
D19	P(SL): initial investment when using SL
E19	P(DB): initial investment when using DB
F19	$P - SV - SL$: present equivalent of the salvage values when using SL
G19	$P - SV - DB$: present equivalent of the salvage values when using DB
H19	PED: present equivalent of depreciation allowance
A22	PE — SL: present equivalent value when using SL
B22	AE — SL: annual equivalent value when using SL
C22	PE — DB: present equivalent value when using DB
D22	AE — DB: annual equivalent value when using DB
E22	PER — SL: present equivalent revenue when using SL
F22	AER — SL: annual equivalent revenue when using SL
G22	PER — DB: present equivalent revenue when using DB
H22	AER — DB: annual equivalent revenue when using DB

Rate-of-Return Calculations

The present equivalent and annual equivalent values in cells A22 to D22 are calculated for any interest rate given as input data. When the rate of return is the unknown value required, one has to adjust manually the interest rate entered in cell B2 until the values in cells A22 to D22 become zero.

Example 6A.2

Problem: Global Industries is considering the purchase of a new piece of equipment which has an estimated economic life of 5 years. MARR = 15% and t = 46%. Annual operating revenues = $100,000, annual operating costs = $40,000, initial cost = $170,000, and estimated salvage value in 5 years = $50,000. Determine the annual equivalent value and the revenue requirements. Assume that the equipment was purchased in 1979 (full depreciation allowance is taken in year 1). Use the declining-balance method and a CCA rate of 20%.

Solution: AE = $483 and AER = $99,105. The investment meets the MARR requirement. Table 6A.2 shows the data input and values calculated.

Problem: Assume that the equipment was purchased in 1984 (the half-year rule applies). Use the declining-balance method and a CCA rate of 20%.

Solution: AE = −$386 and AER = $100,715. The investment does not meet the MARR requirement. Table 6A.3 shows the data input and values calculated.

TABLE 6A.2

	A	B	C	D	E	F	G	H	I
1	n1	i	r	id	t	ie	if		A/P,i,n
2	5	.15	DATA	DATA	.46	DATA	DATA	.15	.298316
3	n2(cca)	DB(BLDG)	DB(EQUIP	CTF-B	CTF-E			p/c,i,n	a/p,i,n
4	DATA	DATA	.2	CODE	CODE	DATA	DATA	3.35216	ERROR
5	YEAR	0	1	1	2	3	4	5	6
6	OR	CODE	100000	DATA	DATA	DATA	DATA	DATA	DATA
7	OC	CODE	40000	DATA	DATA	DATA	DATA	DATA	DATA
8	CCA-B	CODE	DATA	DATA	DATA	DATA	DATA	DATA	DATA
9	CCA-E1	CODE	DATA	DATA	DATA	DATA	DATA	DATA	DATA
10	BLDG	DATA	LAND	WC	EQUIP	DATA	DATA	DATA	DATA
11	P(SL)	DATA	DATA	DATA	DATA	CTF-B	CTF-E	CCTF-B	CCTF-E
12	P(DB)	DATA	DATA	DATA	170000	1	.737143	1	.754286
13	SV	DATA	DATA	DATA	50000				
14	BV	DATA	DATA	DATA	DATA				
15	NSV-SL	0	0	0	1				
16	NSV-DB	0	0	0	36857.1				
17	CCA-SL	0	0	0	0				
18									
19		PEOR	PEOC	P(SL)	P(DB)	P-SV-SL	P-SV-DB	PED	
20		335216.	134086.	0	125314.	.497177	18324.5	0	
21	AE-SL		PE-DB	AE-DB	PER-SL	AER-SL	PER-DB	AER-DB	
22	32400.1		1620.05	483.287	134085.	39999.7	332215.	99105.0	

PE-SL 108610.

TABLE 6A.3

	A	B	C	D	E	F	G	H	I
1	n1	i	r	id	t	ie	if	0	A/P,i,n
2	5	.15	DATA	DATA	.46	DATA	DATA	.15	.29316
3	n2(cca)	DB(BLDG)	DB(EQUIP	CTF-B	CTF-E			p/c,i,n	p/c,i,n a/p,i,n
4	DATA	DATA	.2	1	2	3	4	3.35216	ERROR
5	YEAR	0	1	DATA	DATA	DATA	DATA	5	6
6	OR	CODE	100000	DATA	DATA	DATA	DATA	DATA	DATA
7	OC	CODE	40000	DATA	DATA	DATA	DATA	DATA	DATA
8	CCA-B	CODE		DATA	DATA	DATA	DATA	DATA	DATA
9	CCA-E1	CODE		DATA	DATA	DATA	DATA	DATA	DATA
10	BLDG	DATA	LAND	WC	EQUIP	CTF-B	CTF-E	CCTF-B	CCTF-E
11	P(SL)	DATA	DATA	DATA	170000	1	.737143	1	.754286
12	P(DB)	DATA	DATA	DATA	50000				
13	SV	DATA	DATA	DATA	DATA				
14	BV	DATA	DATA	DATA	0				
15	NSV-SL	0	0	0	36857.1				
16	NSV-DB	1	0	0	0				
17	CCA-SL	0	0	0					
18	PEOR		PEOC	P(SL)	P(DB)	P-SV-SL	P-SV-DB	PED	
19	335216.		134086.	0	128229.	0	18325.0	0	
20									
21	PE-SL	AE-SL	PE-DB	AE-DB	PER-SL	AER-SL	PER-DB	AER-DB	
22	108610.	32400.0	-1293.7	-385.94	134086.	40000.0	337611.	100715.	

Problem: Assume that the equipment was purchased in 1984. Use the fast-write-off method.

Solution: AE = \$4,055 and AER = \$92,490.60.

Note: In the first problem, a 0 is placed in cell E4. In the second, a 1 is placed in cell E4. All other data are identical. In the third, a 1 is placed in cell B9 and the CCA values are placed in D9, D10, and D11. Zeroes are placed in cells C4 and E4. A 2 is placed in cell A4 and \$70,000 is moved from cell E12 to E11.

6A.5 PROGRAM LISTING

The cell numbers containing equations and the program listing are given in Table 6A.5.

TABLE 6A.4

	A	B	C	D	E	F	G	H	I
1	n1 5	i .15	r DATA	id DATA	t .46	ie DATA	if DATA	.15	A/P,i,n .29316
2	n2(cca) 2	DB(BLDG) DATA	DB(EQUIP) DATA	CTF-B CODE	CTF-E CODE				
3								p/c,i,n 3.35216	p/c,i,n .437977
4	YEAR 0			1	2	3	4	5	6
5	OR CODE								
6	OC CODE		100000						
7	CCA-B CODE		40000						
8	CCA-E1 1								
9				42500	85000	42500			
10		BLDG	LAND	WC	EQUIP 170000	CTF-B 1	CTF-E 1	CCTF-B 1	CCTF-E 1
11	P(SL)	DATA	DATA	DATA	170000				
12	P(DB)	DATA	DATA	DATA	0				
13	SV	DATA	DATA	DATA	50000				
14	BV	DATA	DATA	DATA	0				
15	NSV-SL	0	0	0	31304.3				
16	NSV-DB	1	0	0	0				
17	CCA-SL	0	0	0	34000				
18	PEOR 335216.		PEOC 134086.	P(SL) 170000	P(DB) 0	P-SV-SL 16155.9	P-SV-DB .497177	PED 129173.	
19									
20									
21	PE-SL 14185.4	AE-SL 4055.09	PE-DB 108610.	AE-DB 32400.1	PER-SL 310043.	AER-SL 92490.6	PER-DB 134085.	AER-DB 39999.7	

TABLE 6A.5 Program listing.

H1		P = (C1*D1)*(1 − E1)
H2		P = (C2*D2)*(1 − E2) + (1 − C2)*((1 + F2)*(1 + G2) − 1) + B2
I2		P = (H2*(1 + H2)^A2)/((1 + H2)^A2 − 1)
H4		P = (1 − (((1 + G2)/(1 + H2))^A2))/(H2 − G2)
I4		P = (H2*(1 + H2)^A4)/((1 + H2)^A4 − 1)
F11		P = 1 − (E2*B4)/(H2 + B4)
G11		P = 1 − (E2*C4)/(H2 + C4)
H11		P = 1 − (E2*B4)/(H2 + B4)*(1 + H2/2)/(1 + H2)
I11		P = 1 − (E2*C4)/(H2 + C4)*(1 + H2/2)/(1 + H2)
B15	TR	P = IF(B11 = 0,0,IF(B13< = B11,B13 − (B13 − B14)/A2*E2*(1/I2),B13 − ((B13 − B11)*(E2/2) + (B11 − B14)/A2*E2*(1/I2))))
C15	TR	P = IF(C11 = 0,0,C13 − (C13 − C11)*(E2/2))
D15	TR	P = D11
E15	TR	P = IF(E11 = 0,E15 = 0,IF(E13< = E11,E13 − (E13 − E14)/A4*E2*(1/I4),E13 − ((E13 − E11)*(E2/2) + (E11 − E14)/A4*E2*(1/I4))))
A16		P = ■NSV-DB
B16		P = IF(B12 = 0,B16 = 0,IF(B13< = B12,B13*F11,((B13 − B12) − (B13 − B12)*(E2/2)) + B12*F11))
C16		P = IF(C12 = 0,0,C13 − (C13 − C12)*(E2/2))
D16		P = IF(D12 = 0,0,D13)
E16		P = IF(E12 = 0,E16 = 0,IF(E13< = E12,E13*G11,(E13 − E12) − ((E13 − E12)*(E2/2)) + E12*G11))
A17		P = ■CCA − SL
B17		P = IF(B11 = 0,0,(B11 − B14)/A2)
C17		P = IF(C11 = 0,0,C13 − (C13 − C11)*(E2/2))
D17		P = D11
E17		P = IF(E11 = 0,0,(E11 − E14)/A2)
B18		P = ■ PEOR
C18		P = ■ PEOC
D18		P = ■ P(SL)

224

E18	P= ▪ P(DB)
F18	P= ▪P-SV-SL
G18	P= ▪P-SV-DB
H18	P= ▪ PED
B19	P= IF(B6=0,C6*(1/I2)+NPV(H2,D6:W6),C6*H4+NPV(H2,D6:W6))
C19	P= IF(B7=0,C7*(1/I2)+NPV(H2,D7:W7),(C7*H4)+NPV(H2,D7:W7))
D19	P= B11+C11+D11+E11
E19	P= IF(D4=0,B12*F11,B12*H11)+C12+D12+IF(E4=0,E12*G11,E12*I11)
F19	P= (B15+C15+D15+E15)*(1/(1+H2))^A2
G19	P= (B16+C16+D16+E16)*(1/(1+H2))^A2
H19	P= IF(B8=0,B17*(1/I2),NPV(H2,D8:W8))+IF(B9=0,E17*(1/I2),NPV(H2,D9:W9))
A21	P= ▪ PE-SL
B21	P= ▪ AE-SL
C21	P= ▪ PE-DB
D21	P= ▪ AE-DB
E21	P= ▪▪ PER-SL
F21	P= ▪▪ AER-SL
G21	P= ▪ PER-DB
H21	P= ▪ AER-DB
A22	P= (B19−C19)*(1−E2)−D19+F19+E2*H19
B22	P= A22*(I2)
C22	P= (B19−C19)*(1−E2)−E19+G19
D22	P= C22*(I2)
E22	P= C19+(D19−F19−(E2*H19))/(1−E2)
F22	P= E22*(I2)
G22	P= C19+(E19−G19)/(1−E2)
H22	P= G22*(I2)

225

CHAPTER SEVEN

Inflation Recession

North America and most areas of the world have experienced many years of inflation. Since inflation can have a major effect on the feasibility of a project, it is necessary to incorporate inflationary effects into an economic analysis. The most visible sign of inflation is a general increase in prices. People are apt to complain, quite correctly, that "the dollar doesn't go as far as it used to."

In an inflationary economy the purchasing power of the dollar drops. The many explanations as to the cause or causes of inflation are not discussed in detail in this chapter. However, in simple terms it is reasonable to say that inflation results when aggregate demand exceeds aggregate supply within the economy.

Two basic theories of what causes this to happen are referred to as:

1. *Demand pull*
   a. Increase in the stock of money at existing prices and interest rates creates excessive demand for nonmonetary assets, such as land, equipment, and consumer goods.

227

   b. Increasing expenditures open an inflationary gap after full employment is attained.

2. *Cost-push (or income) inflation*

   a. Cost-push inflation arises when income groups attempt to raise their real incomes by raising their monetary incomes. This increases the cost of the goods they produce, creating a spiraling situation of rising incomes and rising prices.

   b. If output cannot expand to meet expectations, prices rise and all groups become somewhat frustrated.

7.1 MONEY AS A SCALE OF MEASUREMENT

In Canada the dollar is used as a scale of measurement to measure accounting and other economic activities. When the dollar changes in value, say during an inflationary or recessionary period, the scale of measurement changes.

The measurement of an object requires the selection of one of its attributes and expressing this attribute in terms of some scale. For example, one may wish to measure the length or weight of various objects. The chosen attribute of length may be measured in feet or metres. We can readily transform from one scale to another, and we must convert all measurements to one scale or the other before performing any calculations. For example, adding meters to feet gives us a meaningless expression.

One of the functions of money is to serve as a common denominator for exchange. Measurements using different scales (metres and feet, pounds and newtons, etc.) cannot be directly compared, multiplied, or added. The same is true if the exchange ratio of money has changed between two points in time. Although the measurement unit (e.g., dollar) is denoted by the same name as before, its size has changed. Two monetary values (e.g., dollar values) used as a unit of measurement at two different points in time are as different from each other as metres and feet. Therefore, it is essential to transform or adjust these dollar values to values that represent the same purchasing power at any point in time before one can use this unit of measurement in an economy study.

7.2 PRICE INDICES

A change in the money scale is reflected by a reciprocal change in the general price level. A change in the money scale cannot be measured directly; however, the reciprocal change in the general price level can easily be measured by the use of index numbers. This represents an indirect measure.

Indirect measures are widely used in practice; for example, the height of a column of mercury in a thermometer can be used indirectly to measure the temperature of a specific object or the surrounding environment. Similarly, the general price level index provides an indirect measure of the purchasing power or exchange value of money.

The *Consumer Price Index* (CPI) is calculated by the government by adding together the costs of a large list of items that are regularly purchased by wage earners. The total is expressed as a percentage of a base-year total. Thus this index is one measure of the change in the purchasing power of money. It is analogous to taking the same shopping list to the grocery store each week and recording how the total bill varies.

For example, suppose that the CPI increased during a 5-year period from 100 points to 150 points. This means that the prices of goods and services measured by the index increased by 50%. The index thereby measures the relationship of goods to money. The reciprocal of this index measures the relationship of money to goods (e.g., $100/150 = 0.67$). The money unit (e.g., dollar) that served as a measuring scale shrank by 33% during the given period (e.g., from 1.00 to 0.67). The average annual inflation rate, i, during the period can be calculated from the equations of Chapter 3.

$$100(1 + i)^5 = 150$$

$$i = 8.4\%$$

An *index* is a ratio between two measurements. A *price index* is the ratio between the prices of a commodity (or a group of commodities) during the period in question.

7.3 SELECTING THE RIGHT PRICE INDEX

Companies are faced with the problem of trying to determine how much their costs are really increasing. If a company is using the wrong measure, it could have a serious impact on the financial stability of the firm. The widely used CPI, for instance, may not be appropriate under all conditions. For example, the lumber industry (including sawmills, veneer, plywood, and sash door products) has had price fluctuations over the years significantly different from the CPI quoted in Canada. According to the Industry Selling Price Index (ISPI), in 1972 lumber prices rose 22%, whereas the CPI rose 5%; in 1974 lumber prices dropped 2% and the CPI rose by 11%; and in 1978 lumber prices increased by 16% and the CPI increased by 9%.

These variations clearly demonstrate the need to select the right index from published statistics. The availability and content of various indices can be obtained from government sources in the United States and Canada.

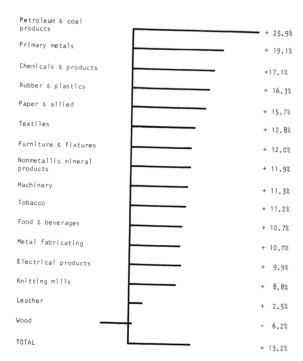

FIGURE 7.1 Change in Canadian Industry Selling Price Index from 1979–1980.

There is probably a published index available that suits your individual company's needs.

The CPI is the best guide to what employees are going to be demanding in wage increases. Therefore, if wages constitute the majority of costs, the CPI will represent a reasonable measure of inflation. In Canada the Industry Selling Price Index (ISPI) is the best guide to increases with respect to input costs for products that are produced domestically and have gone through some manufacturing processing. Figure 7.1 shows the change in the Canadian Industry Selling Price Index for several products from 1979 to 1980.

7.4 THE IMPACT OF INFLATION ON INVESTMENT DECISIONS

To appreciate more fully the impact of inflation on the after-tax dollar, we shall consider a series of examples that address two basic questions:

1. What impact does inflation have on investment decisions?
2. How may the inflationary effects be incorporated into an economy study? (How is the dollar transformed or adjusted?)

To examine these questions, it is first necessary to have several explicit definitions. They are:

Current dollars or actual dollars: These are the actual amounts that are exchanged. They are the numbers that appear on a cash flow diagram. For example, when a mortgage is negotiated, a repayment scheme is agreed upon. The monthly amount is in actual or current dollars. A salary is paid in actual dollars, as are utility bills and grocery bills.

Constant dollars or real dollars: These are dollars that have the same, or constant, purchasing power relative to some point in time. Usually, they are prefixed by some date, such as 1984 dollars. Estimates are often made in constant-dollar amounts to ensure a valid basis for comparison with other projects.

Responsive receipts and responsive expenditures: The word "responsive" is used to indicate that the amounts change with the change in price levels. For example, if the operating costs were inflating at 10% per year, the fully responsive situation would have the operating revenues increasing at 10% per year. If you were to negotiate a loan at the bank, your loan rate would probably be tied to the bank prime rate, say 2% over the bank prime rate. Thus the interest rate you pay is fully responsive to changes in the bank prime rate.

The examples that follow illustrate the effect of inflation on investments.

Example 7.1

Problem: Assume that you make two equal investments of $2,991 for a period of 5 years. Investment A pays a fixed before-tax rate of return of 20%. Investment B guarantees you a fixed before-tax rate of return of 20% and compensates you for the decrease in purchasing power of the dollar due to a 10% inflation rate (receipts are fully responsive).

Solution

$$\$2,991(A/P,\ 20\%,\ 5) = \$1,000$$

The following table shows the impact of inflation on investments A and B.

Year	Investment A: Fixed Receipts			Investment B: Fully Responsive Receipts		
	BTCF [current (actual) dollars]	10% (P/F)	BTCF [constant (real) dollars]	BTCF [current (actual) dollars]	10% (P/F)	BTCF [constant (real) dollars]
0	$-2,991		$-2,991	$-2,991		$-2,991
1	1,000	0.9091	909	1,100	0.9091	1,000
2	1,000	0.8265	826	1,210	0.8265	1,000
3	1,000	0.7513	751	1,331	0.7513	1,000
4	1,000	0.6830	683	1,464	0.6830	1,000
5	1,000	0.6209	621	1,611	0.6209	1,000
	BTRR = 9%			BTRR = 20%		

Conclusion

1. Inflation reduces the real return on an investment with unresponsive receipts.
2. The before-tax rate of return remains unchanged if receipts are fully responsive to inflation.

Since most corporations and individual investors are faced with the realities of income tax payments, we need to assess the impact of inflation on the after-tax cash flows in an inflationary economy.

Example 7.2

Problem: Evaluate the responsive investment of Example 7.1 on an after-tax basis, first in a noninflationary economy, and then with a 10% inflation rate and fully responsive receipts. Assume that $t = 50\%$ and 100% equity capital. Use the straight-line method (equation 4.2). SV = 0.

Solution: $D = \$2,991/5 = \$598$ per year. The following tables illustrate the impact of inflation on the after-tax rate of return on 100% equity capital.

For cash flows in a noninflationary economy (with no inflation there is no difference between responsive and fixed receipts):

Year	BTCF	AED	TI	IT	ATCF
0	$-2,991	0	0	0	$-2,991
1-5	1,000	598	402	201	799
	After-tax rate of return on equity capital: $i_e = 10.5\%$				

For cash flows in a 10% inflationary economy, before-tax cash flows are fully responsive to inflation:

Year	BTCF	AED	TI	IT	ATCF (current dollars)	10%(P/F)	ATCF (constant dollars)
0	$-2,991	$ 0	$ 0	$ 0	$-2,991		$-2,991
1	1,100	598	502	251	849	0.9091	772
2	1,210	598	612	306	904	0.8265	747
3	1,331	598	733	367	964	0.7513	724
4	1,464	598	866	433	1,031	0.6830	704
5	1,611	598	1,013	507	1,104	0.6209	686

Real after-tax rate of return on equity capital: $i_e = 7.0\%$

Conclusion: The real after-tax rate of return on 100% equity capital is reduced in an inflationary economy even when receipts are fully responsive because the depreciation allowance does not respond to inflation. As a result, the income tax payments are more than fully responsive.

Example 7.3

Problem: Example 7.3 is intended to show the effect of introducing debt capital into the financial structure. It is assumed that the debt capital is in the form of bonds with a fixed interest rate over the life of the bonds. In periods of rapid inflation and fluctuating interest rates, bond issues will probably have fixed interest rates only for relatively short periods. Adjustment clauses will probably be included with the issue to allow periodic adjustment of the interest rate. $P = \$2,991$, $SV = 0$, $r = 30\%$, $i_d = 12\%$, and $n = 5$ years.

$$\text{Equity capital} = \$2,991 \times 70\% = \$2,094$$
$$\text{Debt capital} = \$2,991 \times 30\% = \underline{\quad 897 \quad}$$
$$\text{Total} = \$2,991$$

Solution: Assuming that the debt is reduced uniformily over the 5-year period, the interest on debt will be calculated as follows:

Year	Debt Principal	Interest on Debt (12%)
1	$897.00	$107.64
2	717.60	86.11
3	538.20	64.58
4	358.80	43.06
5	179.40	21.53

The impact of a 10% inflation rate on the after-tax rate of return with 30% debt capital in the structure is as follows.

For cash flows in a noninflationary economy:

Year	BTCF	AED	Interest on Debt	TI	IT	ATCF	Debt. Princ. + Int.	Bal. for Equity
0	$-2,991	$ 0	$ 0	$ 0	$ 0	$-2,991		
1	1,000	598	108	294	147	853	$287	$566
2	1,000	598	86	316	158	842	265	577
3	1,000	598	65	337	169	831	244	587
4	1,000	598	43	359	180	820	222	598
5	1,000	598	22	380	190	810	201	609

After-tax rate of return on equity capital: $i_e = 12.4\%$

For cash flows in a 10% inflationary economy assuming that the before-tax cash flows are fully responsive to inflation:

Year	BTCF	AED	Int. on Debt	TI	IT	ATCF (current dollars)	Debt. Princ. + Int.	Balance for Equity Curr. Dollars	Const. Dollars
0	$-2,991	$ 0	$ 0	$ 0	$ 0	$-2,991			
1	1,100	598	108	394	197	903	$287	$616	$560
2	1,210	598	86	526	263	947	265	682	563
3	1,331	598	65	668	334	997	244	753	566
4	1,464	598	43	823	412	1,052	222	830	567
5	1,611	598	22	991	496	1,115	201	914	568

Real after-tax rate of return on equity capital based on constant dollars: $i_e = 11.7\%$

Conclusion: The after-tax rate of return receives some protection when debt capital enters into the capital structure. Whether or not the equity capital will receive full protection from the debt capital in an inflationary economy depends on the debt ratio. This ratio, for most industrial firms, does not exceed 25 to 30%; therefore, full protection is unlikely.

Figure 7.2 illustrates the debt ratio required, at various inflation rates, to give the equity holder full protection in an inflationary economy. For example, with an inflation rate of 6% the debt ratio required is 24%. The debt ratio increases to 36% for an inflation rate of 12%.

This figure was developed using the following data: P = $2,991; SV (in year 5) = 0; BTCF = $1,000 (zero date) and is fully responsive to inflation; the straight-line method (equation 4.2) was used; t = 50%, i_e = 10%, i_d = 5%; and the inflation-adjusted rates i_{ef} and i_{df} were calculated based on the specific inflation rate used for each curve.

The equity holder's return on investment drops because the depreciation allowance is unresponsive to inflation and income tax is more than fully responsive.

Figure 7.3 illustrates the significant increase in taxes that occurs in an inflationary economy. For example, with a 6% inflation rate over the 5-

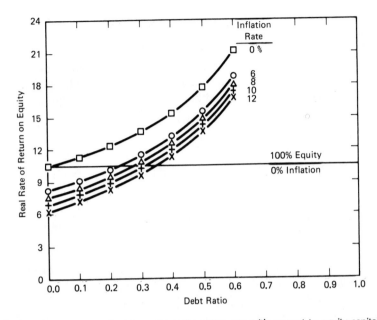

FIGURE 7.2 Debt ratio required to maintain the status quo with respect to equity capital in an inflationary economy.

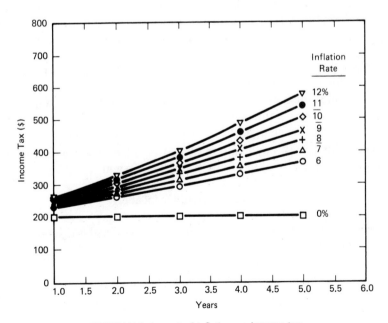

FIGURE 7.3 Impact of inflation on income tax.

year period, income taxes increase by 240%, whereas the before-tax cash flow only increases by 34%.

This figure was developed using the following data: P = $2,991; SV (in year 5) = 0; BTCF = $1,000 (zero date) and is fully responsive to inflation; the straight-line method (equation 4.2) was used; t = 50%; r = 0; i_e = 10%; and i_{ef} was calculated based on the specific inflation rate used for each curve.

7.5 ACCOUNTING FOR INFLATION IN AN ECONOMIC ANALYSIS

There are two generally accepted approaches that may be used with respect to discounting cash flows involving inflation.

1. Discount the actual cash flows (in current dollars) with a discount rate adjusted for inflation (i_{af}).
2. Convert the actual after-tax cash flows to constant dollars (zero date dollars) and discount these cash flows using (i_a) the MARR value for a noninflationary economy.

Great care must be taken with the constant-dollar conversion method. The conversion must be applied only to after-tax cash flows, or to public projects where there are no income taxes. Converting before-tax cash flows to constant-dollar amounts could introduce two possible sources of error:

1. When the income tax is calculated using actual (current) before-tax cash flows, the income tax payment is more than fully responsive to inflation. If the before-tax cash flows are converted to constant dollars, the calculated income tax payment is not responsive and thus will be significantly lower than the actual income tax payment.
2. These lower tax payments may also result in a lower marginal tax rate than actually applies due to the tax system.

7.5.1 Adjusting the Discount Rate for Inflation

When 100% equity capital is involved, the adjusted discount rate in an inflationary economy may be calculated using

$$i_{af} = i_{ef} = (1 + i_e)(1 + i_f) - 1 \qquad (7.1)$$

where i_{af} = adjusted discount rate considering inflation

i_{ef} = adjusted discount rate using 100% equity capital

i_e = minimum after-tax rate of return on equity (real or true rate) in an inflation-free economy

i_f = expected inflation rate

To apply this equation the analyst must establish an inflation-free value for (i_e) at some base year and estimate the inflation rate to be expected over the life of the investment analysis. The minimum after-tax rate of return (real or true rate) on equity capital in an inflation-free economy will vary somewhat from industry to industry and to some degree within segments of the same industry. However, this value will probably fall in the range 6 to 10%.

In times of double-digit inflation, 10% or more, one would expect discount rates on equity capital to be in the range 16 to 23%.

This was the situation in the late 1970s and early 1980s when inflation was running close to 10% per year. Assume that i_e = 6% and i_f = 10%. Then

$$i_{ef} = (1 + i_e)(1 + i_f) - 1$$

$$= (1.06)(1.10) - 1 = 16.6\%$$

Assume that i_f = 10% and i_e = 12%. Then

$$i_{ef} = (1.10)(1.12) - 1 = 23.2\%$$

The discount rate (i_{df}) to be used with respect to debt capital to arrive at an adjusted MARR (i_{af}) value will be the fixed rate on current bond issues where bond financing is used. If funds are borrowed from a lending institution (e.g., bank), the rate will normally be tied to the prime rate and expected fluctuations in the prime rate should be considered in calculating i_{df}. If interest rates remain high and continue to experience wide fluctuations over relatively short time spans, bond issues may be forced to reevaluate the interest rate paid periodically.

The adjusted MARR (i_{af}) value, including both the effect of debt and taxes, to be used as a discount rate may be calculated as follows:

$$i_{af} = (ri_{df})(1 - t) + (1 - r)[(1 + i_e)(1 + i_f) - 1] \qquad (7.2)$$

Assume that i_{df} = 12%, i_e = 8%, i_f = 10%, r = 30%, and t = 40%. Then

$$i_{af} = (0.3)(0.12)(0.6) + (0.7)[(1.08)(1.10) - 1]$$

$$= 2.16\% + 13.16\% = 15.3\%$$

7.5.2 Converting After-Tax Cash Flows to Constant Dollars

The second method of accounting for inflation is to reduce the after-tax cash flows to constant or real dollars. That is, to dollars of constant purchasing power. This can be accomplished by using the P/F factor and the average annual inflation rate. Thus, if an automobile costs $10,000, actual dollars, in 1985, and the inflation rate, i_f, has been 9% from 1980 to 1985, the automobile costs

$$\$10,000(P/F, 9\%, 5) = \$6,500 \text{ in 1980 dollars}$$

It is interesting to apply these formulas to the stories of parents and others regarding prices in the 1950s when cigarettes were 35 cents, whiskey $2.50, a car $3,000, and the minimum wage was 60 cents/hour. If we assume a 4% annual inflation rate from the 1950s to the 1980s, the $(F/P, 4\%, 30)$ factor equals 3.2434 and the 1980 figures are

$$\$0.35 \times 3.2434 = \$1.13$$
$$\$2.50 \times 3.2434 = \$8.10$$
$$\$3000 \times 3.2434 = \$9,730$$
$$\$0.60/\text{hour} \times 3.2434 = \$1.95 \text{ per hour}$$

Such a calculation provides some perspective on whether things are actually getting better or worse.

Example 7.4 will demonstrate that the two methods of accounting for inflation in an economy study are equivalent.

Example 7.4

Problem: The initial investment = $10,000, estimated salvage value = 0, estimated life = 5 years, 100% equity capital, income tax rate t = 50%, the CCA is calculated using the straight-line method, i_e = 8%, and i_f = 10%. Operating revenues and operating costs are fully responsive to inflation. Annual operating revenues and operating cost in constant (today's) dollars are $6,000 and $2,000, respectively.

Solution

$$i_{af} = (1 + i_e)(1 + i_f) - 1 = 18.8\%$$
$$\text{AED} = (\$10,000)/5 = \$2,000 \text{ per year}$$

Year	Oper. Revenues	Oper. Costs	BTCF	AED	TI	IT	ATCF Curr. Dollars	ATCF Const. Dollars
0			$-10,000				$-10,000	$-10,000
1	$6,600	$2,200	4,400	$2,000	$2,400	$1,200	3,200	2,909
2	7,260	2,420	4,840	2,000	2,840	1,420	3,420	2,826
3	7,986	2,662	5,324	2,000	3,324	1,662	3,662	2,751
4	8,785	2,928	5,857	2,000	3,857	1,929	3,928	2,683
5	9,663	3,221	6,442	2,000	4,442	2,221	4,221	2,621

Discounting the after-tax cash flows in current (actual) dollars using the discount rate adjusted for inflation (18.8%), the net present value of the investment is

$$PE = \$-12,000 + \$3,200(0.8418) + \$3,420(0.7085)$$
$$+ \$3,662(0.5964) + \$3,928(0.5020) + \$4,221(0.4226)$$
$$= \$1056$$

Discounting the after-tax cash flows in constant (zero date) dollars using the discount rate (8%) for a noninflationary economy the net present value of the investment is

$$PE = \$-12,000 + \$2,909(0.9259) + \$2,826(0.8573)$$
$$+ \$2,751(0.7938) + \$2,683(0.7350) + \$2,621(0.6806)$$
$$= \$1056$$

Thus we see that either method of discounting the after-tax cash flows gives us an identical answer.

7.6 COMPOSITE INFLATION RATES

The Consumer Price Index (CPI) is a measure of the average rate of inflation that affects wage earners. Specific items or commodities can be subject to rates that are either higher or lower than the rates reflected by the CPI. For example, energy costs increased little during the 1950s and 1960s and then increased significantly during the 1970s. When evaluating projects it is important to use an inflation rate that reflects price changes associated with resources to be used in the project. A highly labor intensive project should probably use a different inflation rate than that used for an energy-intensive project. Furthermore, two or three different inflation rates are often used within a single project analysis: possibly one inflation rate for labor costs, another for materials, and a third for the energy component. It is also possible to calculate a weighted average for the different resources used and derive a composite inflation rate that is specific to a project.

Example 7.5

Problem: Assume that operating costs are composed of $50,000 in direct labor costs (DL), $100,000 in direct material costs (DM), and $50,000 in overhead costs (OH). These costs are assumed to increase by 8%, 3%, and 6%, respectively. What is the composite inflation rate to apply to operating costs?

Solution

$$\text{Total operating costs} = \$200,000$$

$$\text{Composite inflation rate} = \frac{\text{DL}(8\%) + \text{DM}(3\%) + \text{OH}(6\%)}{\text{total operating costs}}$$

$$= \frac{\$4,000 + \$3,000 + \$3,000}{\$200,000}$$

$$= 5\%$$

7.7 REVENUE REQUIREMENTS AND INFLATION

Considering inflation with respect to operating costs and/or operating revenues represents a situation where the growth is exponential and the progression is geometric.

Example 7.6

Problem: Find the annual equivalent operating costs (AEOC) given that $i_f = 6\%$, $i_{af} = 20\%$, and $n = 10$ years. AEOC is $100 today and are expected to be fully responsive to inflation.

Solution: Using equation 3.19, we have

$$C = \$100(1.06) = \$106$$

$$P = C\{1 - [(1 + i_f)^n/(1 + i_{af})^n]/(i_{af} - i_f)\}$$

$$= \$106\{[1 - (1.06)^{10}/(1.20)^{10}]/(0.20 - 0.06)\}$$

$$= \$106(5.0769) = \$538$$

$$\text{AEOC} = P(A/P, 20\%, 10) = \$538(0.2385) = \$128$$

An alternative method of calculating AEOC considering inflation in operating costs is given by the equation below. This equation is used in computer models to calculate AEOC when operating costs are fully responsive to inflation. The (F/P) term calculates the current dollar amount for year z, the P/F term moves it to time zero, and the A/P term converts the present value to a uniform series.

$$\text{AEOC} = \left[\sum_{z=1}^{n} OC_z(F/P, i_f, z)(P/F, i_{af}, z) \right] (A/P, i_{af}, n)$$

where OC_z represents the cash operating costs for the zth year with no inflation (constant-dollar amount).

Example 7.7

Problem: Determine the annual before-tax cash flows necessary to maintain a constant (real) rate of return of 10% on a 100% equity capital investment of $3,791. Assume an inflation rate of 9.1%.

Solution

$$i_{af} = (1 + i_e)(1 + i_f) - 1$$

$$= (1.10)(1.091) - 1$$

$$= 20\%$$

The cash flow diagram assuming no inflation is

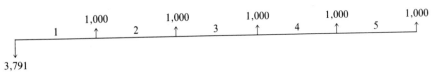

3,791

The cash flow diagram assuming 9.1% inflation is

3,791

With no inflation:

$$P = \$1,000(P/A, 10\%, 5) = \$3,791$$

With 9.1% inflation and fully responsive cash flows:

$$P = (\$1,091)(P/F, 20\%, 1) + (\$1,190.28)(P/F, 20\%, 2)$$
$$+ (\$1,298.60)(P/F, 20\%, 3) + (\$1,416.77)(P/F, 20\%, 4)$$
$$+ (\$1,545.70)(P/F, 20\%, 5)$$
$$= \$3,791$$

Using equation 3.19, we have $i_f = 9.1\%$, $i_e = 10\%$, $i_{af} = (1.091)(1.10) - 1 = 20\%$, and $C = \$1,000(1.091) = \$1,091$.

$$(P/C, i, k, n) = [1 - (1 + i_f)^5/(1 + i_{af})^5]/(i_{af} - i_f)$$
$$= [1 - (1.091/1.20)^5]/(0.20 - 0.091)$$
$$= 3.47541$$
$$P = C(P/C, 20\%, 9.1\%, 5) = \$1,091(3.47541) = \$3,791$$

Example 7.8

Problem: An investment under consideration has an initial cost of $20,000, an estimated salvage value of $5,000, and a useful estimated life of 5 years. The declining-balance method and a CCA rate = 30% will be used to calculate depreciation allowance. Annual operating costs are estimated to be $2,000 today (constant dollars) and are fully responsive to inflation. $r = 0$, $t = 50\%$, $i_e = 8\%$, and $i_f = 11.1\%$. What is the annual revenue required to justify this investment?

Solution

$$i_{af} = i_{ef} = (1 + i_e)(1 + i_f) - 1 = 20\%$$

Using equation 3.19, we have

$$(P/C, i, k, n) = [1 - (1.111/1.20)^5]/(0.20 - 0.111) = 3.59278$$
$$AEOC = C(P/C, i, k, 5)(A/P, 20\%, 5)$$
$$= \$2,222(3.59278)(0.3344) = \$2,669$$

$$AER = AEOC + [P(A/P, i_{af}, n)CTF - SV(A/F, i_{af}, n)CTF]/(1 - 0.5)$$

$$CTF(SV) = 0.7000 \qquad CTF(P) = 0.7250$$

$$AER = \$2,669 + [\$20,000(0.3344)0.7250 - \$5,000(0.1344)0.700]/0.5$$
$$= \$2,669 + (\$4,849 - \$470)/0.5$$

$$= \$2,669 + \$8,757$$

$$= \$11,426$$

This value can be compared to the revenue requirements in the no-inflation situation using a discount rate of 8%.

$$\text{CTF(SV)} = 0.6053 \qquad \text{CTF}(P) = 0.6199$$

$$\text{AER} = \$2,000 + [(\$20,000)(0.2505)0.6199$$

$$- \$5,000(0.1705)0.6053]/0.5$$

$$= \$7,179$$

Thus we see how inflation can affect an investment.

Example 7.9

Problem: Assume that the conditions in Example 7.8 have not changed except that financing is by 36% debt capital at a cost of 12%. What is the annual equivalent revenue requirement?

Solution: $i_e = 8\%$, $i_f = 11.1\%$, $i_{df} = 12\%$, $r = 36\%$, and $t = 50\%$. When there is debt in the structure, use equation 7-2 to calculate i_{af}.

$$i_{af} = ri_{df}(1 - t) + (1 - r)[(1 + i_e)(1 + i_f) - 1]$$

$$= (0.36)(0.12)(0.5) + (0.5)[(1.08)(1.111) - 1] = 15\%$$

$$(P/C, i_{af}, i_f, n) = [1 - (1.111/1.15)^5]/(0.15 - 0.111) = 4.0615$$

$\text{AEOC} = \$2,222(4.0615)(A/P, 15\%, 5) = \$2,692$

$\text{CTF(SV)} = 0.6667 \qquad \text{CTF}(P) = 0.6884$

$\text{AER} = \text{AEOC} + [P(A/P, 15\%, 5)\text{CTF} - \text{SV}(A/F, 15\%, 5)\text{CTF}]/(1 - t)$

$$= \$2,692 + [\$20,000(0.2983)0.6884$$

$$- \$5,000(0.1483)0.6667]/0.5$$

$$= \$9,917$$

Summary

$$\text{AER with no inflation} = \$7,179$$

$$\text{AER with 36\% debt} = \$9,917$$

$$\text{AER with 100\% equity} = \$11,416$$

Assuming that revenues are \$7,179 per year in a noninflationary economy and that these revenues are fully responsive to inflation:

$$C = \$7{,}179(1.111) = \$7{,}976$$

$$\text{AEOR} = C(4.0615)(A/P, 15\%, 5) = \$9{,}663$$

Thus we see that a debt ratio of close to 36% and fully responsive revenues are required to maintain an 8% return to the equity holder.

Example 7.10

Problem: This example converts the current (actual) cash flows to constant (real) dollars before calculating the real rate of return. First cost = \$10,000 and estimated salvage = 0. Use the straight-line method to calculate depreciation allowance. Operating revenue = \$6,000 per year, operating cost = \$2,000 per year, life = 5 years, $r = 0$, $t = 50\%$, and $i_f = 10\%$. Determine the real after-tax rate of return on the investment.

Solution

$$D = (P - SV)/n = (\$12{,}000)/5 = \$2{,}000 \text{ per year}$$

Year	BTCF	AED	TI	IT	ATCF (current dollars)	ATCF (constant dollars)
0	\$ – 10,000	—	—	—	\$ – 10,000	\$ – 10,000
1	4,000	\$2,000	\$2,000	\$1,000	3,000 × 0.9091 =	\$2,723
2	4,000	2,000	2,000	1,000	3,000 × 0.8265 =	2,479
3	4,000	2,000	2,000	1,000	3,000 × 0.7513 =	2,254
4	4,000	2,000	2,000	1,000	3,000 × 0.6830 =	2,049
5	4,000	2,000	2,000	1,000	3,000 × 0.6209 =	1,863

$$\text{PE} = \$ - 10{,}000 + \$2{,}723(P/F, i, 1) + \$2{,}479(P/F, i, 2)$$
$$+ \$2{,}254(P/F, i, 3) + \$2{,}049(P/F, i, 4) + \$1{,}863(P/F, i, 5)$$

Try 4%:

$$\text{PE} = \$ - 10{,}000 + \$10{,}197 = + \$197$$

Try 5%:

$$\text{PE} = \$ - 10{,}000 + \$9{,}934 = \$ - 66$$

$$i_e = 4.7\%$$

The rate of return with no inflation = 15.2%.

7.8 SUMMARY

Inflation does have a significant impact on investment decisions, and recognition of this impact is critical to the ongoing economic health of the corporation. The credibility of the analysis is subject to using an inflation rate that is representative of the investment under study. As suggested, this may require the use of more than one inflation rate or a composite rate that is representative. If a constant-dollar study is considered, the after-tax cash flows should be converted to constant dollars to ensure that the true effects of taxation are measured. The inclusion of inflation in the economic analysis and the determination of a real rate of return reflected by the financial statements is essential to the economic health of the firm.

PROBLEMS

7.1 When 100% equity capital is involved the discount rate in an inflationary economy is

$$(1 + i_e)(1 + i_f) - 1$$

$$i_{ef} = i_{af}$$

Determine the discount rate (i_{af}) to be used for cash flow calculations in an inflationary economy when using current dollars and when debt capital is involved. (This discount rate should also work for 100% equity capital.)
a. Show how to calculate i_{af}.
b. Show the final form of the equation for AE and PE.

7.2 Assume that the following conditions exist: $r = 0$, $i_e = 12\%$, $i_f = 7.1\%$, and $n = 5$ years. Operating costs with no inflation equal $1,000 per year and are fully responsive to inflation (100% equity capital). Determine the present equivalent value:
a. Ignoring inflation.
b. Considering inflation.

7.3 An investment is under consideration which has a first cost of $40,000, an estimated salvage value of $6,000, and an estimated useful life of 5 years. Estimated annual operating revenue is $22,400 with no inflation (zero date). Estimated annual operating costs are $8,000 with no inflation (zero date). Use the declining-balance method and a CCA rate of 30% $r = 0$, $t = 50\%$, and $i_f = 9.1\%$. Determine the after-tax rate of return:
a. Ignoring inflation.

b. Considering inflation and assuming that operating revenues and operating costs are fully responsive to inflation.

c. Calculate i_{ef} using equation 7.1 and the i_e value from part (a).

d. Why is the answer to part (c) different from that of part (b)?

7.4 An investment is under consideration which has a first cost of $50,000, an estimated salvage value of zero, and an estimated useful life of 5 years. Estimated annual operating revenues in year 1 including inflation are $26,550. Estimated annual operating costs in year 1 are $10,000. Use the fast-write-off method for capital cost allowance. $r = 0$, $t = 50\%$, and $i_f = 6.5\%$. Determine the after-tax rate of return:

a. Ignoring inflation.

b. Considering inflation and assuming that operating revenues and operating costs are fully responsive to inflation.

c. Calculate i_{ef} using equation 7.1 and the i_e value from part (a).

d. Why is the answer to part (c) different from that of part (b)?

7.5 An investment is under consideration which has a first cost of $50,000, an estimated salvage value of $20,000, and an estimated useful life of 5 years. Estimated annual operating costs are $25,000 with no inflation and are fully responsive to inflation. Use the declining-balance method and a CCA rate of 20% $r = 0$, $t = 50\%$, $i_e = 12\%$, and $i_f = 7.1\%$.

a. Calculate the annual revenue requirements to justify the investment assuming no inflation.

b. Calculate the annual revenue requirements to justify the investment assuming inflation.

7.6 Assume that the operating revenues in Problem 7.5 are $40,200 per year, at zero date, and are fully responsive to inflation. All other data are as stated in the problem. Determine the after-tax rate of return (i_{af}).

7.7 An investment is under consideration which has a first cost of $60,000, an estimated salvage value of $20,000, and an estimated useful life of 5 years. Use the fast-write-off method for capital cost allowance. Operating costs = $30,000 per year (no inflation) and are fully responsive to inflation. $r = 0$, $t = 50\%$, $i_e = 9\%$, and $i_f = 10.1\%$. Determine the necessary revenue to justify the investment:

a. Ignoring inflation.

b. Considering inflation and fully responsive operating costs.

7.8 Assume that the conditions in Problem 7.7 have not changed except that financing is by 40% debt capital (i_{df}) at a cost of 15%. Determine the necessary revenue to justify this investment.

7.9 Operating costs for a specific project under consideration are composed of $40,000 in direct labor costs, $20,000 in overhead costs, and $80,000 in direct material costs. These costs are based on no inflation and at zero date. The inflation rates associated with these operating costs are 10%, 4%, and 6% respectively.

a. What composite inflation rate should be applied to operating costs for this study?

b. Assume that the Consumer Price Index is used as a basis to predict inflation on this job and that this index suggests using a rate of 10%. What error is introduced in annual equivalent operating costs? Base the calculation on a 5-year life. $r = 0$, $i_e = 9.1\%$.

7.10 The following investment is under consideration.

Initial investment	$400,000
Economic life	10 years
Salvage value	$110,000
Annual operating cost with no inflation (zero date)	$50,000
Annual operating revenues with no inflation (zero date)	$140,000

$r = 50\%$, $i_{df} = 8\%$, $t = 50\%$, $i_e = 11.1\%$, and $i_f = 8\%$.

Use the declining-balance method and a CCA rate of 20%. Operating costs and operating revenue are fully responsive to inflation.

a. What annual revenue is necessary to justify this investment?

b. Should the investment be made?

7.11 The following data pertain to a computerized honing machine. Operating costs are fully responsive to inflation.

Initial investment	$600,000
Economic life	5 years
Salvage value	$200,000
Operating costs (zero date)	$15,000

$r = 20\%$, $i_{df} = 12\%$, $t = 48\%$, $i_e = 9.2\%$, and $i_f = 7.3\%$. Use the fast-write-off method for capital cost allowance. There is an additional maintenance expense of $5,000 at the end of the second year. What is the required revenue to justify this investment?

7.12 A company is considering an investment in a new fabrication plant. The following data pertain to this investment.

	Initial Cost	Salvage Value in 10 Years
Building	$300,000	$175,000
Land	100,000	150,000
Working capital	200,000	200,000
Equipment	250,000	100,000

$r = 50\%$, $i_{df} = 12\%$, $t = 40\%$, $i_e = 9\%$, and $i_f = 7.2\%$. The economic life of the project is 10 years. Annual operating costs are $70,000 (zero date) and are fully responsive to inflation. Use the fast-write-off method for capital cost allowance on equipment and a 10% declining-balance rate for the building. What are the annual revenue requirements to justify this investment?

7.13 ABC Corporation is considering an investment in a project with the following data:

Initial cost	$300,000
Economic life	5 years
Salvage value	$75,000
Operating costs are fully responsive	
(zero date)	$12,000

$r = 20\%$, $i_{df} = 12\%$, $t = 48\%$, $i_e = 9.2\%$, and $i_f = 7.3\%$. There is an additional capital disbursement of $15,000 at the end of the second year. Use the declining-balance method and a CCA rate of 30%. What is the required revenue to justify this investment?

7.14 Evaluate the following investment proposal:

Initial investment	$125,000
Salvage value in 5 years	$35,000
Economic life	5 years
Annual operating costs with no	
inflation (zero date)	$20,000
Annual operating revenue with no	
inflation (zero date)	$55,000

$r = 0$, $t = 48\%$, $i_e = 9\%$, and $i_f = 10.1\%$. Use the declining-balance method and a CCA rate of 25%. Operating costs and operating revenue are both fully responsive to inflation. Should this investment be made?

7.15 ABC Corporation is considering two mutually exclusive machines to perform an operation. The duration of the project is 8 years, at the end of which time the machine will be sold.

	Machine A	Machine B
Initial cost	$100,000	$150,000
Salvage value	20,000	30,000
Annual operating costs with no inflation (zero date)	25,000	15,000

$r = 20\%$, $i_{df} = 12\%$, $t = 48\%$, $i_e = 9.2\%$, and $i_f = 7.3\%$. Operating costs are fully responsive to inflation. Use the fast-write-off method for the CCA. Which machine should be selected?

7.16 The duration of a project is 4 years, for which a machine with the following data is being considered:

Initial cost	$30,000
Salvage value	5,000
Annual operating cost (zero date)	15,000
Annual operating revenues (zero date)	26,400

$r = 0$, $t = 46\%$, and $i_f = 7.3\%$. Use the fast write-off method for the CCA. What is the expected rate of return on this investment?

7.17 The first cost of a new milling machine under consideration is $100,000. It has an estimated salvage value of zero and an estimated useful life of 10 years. Use the fast-write-off method for the CCA. Annual operating costs with no inflation = $40,000 (zero date) and are fully responsive to inflation. $r = 0$, $t = 46\%$, $i_e = 8\%$, and $i_f = 6.5\%$. What annual revenue is required to justify this investment?

7.18 An investment is under consideration which has a first cost of $60,000 and an estimated salvage value of zero in 5 years. Use the declining-balance method and a 20% CCA rate. Annual operating costs with no inflation = $15,000 (zero date) and are fully responsive to inflation. $r = 50\%$, $l_{df} = 9\%$, $t = 40\%$, $i_e = 12\%$, and $i_a = 8.7\%$. What annual revenue is required to justify this investment:
a. In a noninflationary economy? $(A/P, 8.7\%, 5) = 0.2551$.
b. In an inflationary economy? $i_f = 5.9\%$ and $P/C = 4.00377$.

7.19 An investment in a new office building is under consideration which has an estimated useful life of 20 years. Use the declining-balance method and a CCA rate of 10%.

	P	SV in 20 Years
Building	$309,000	$250,000
Land	90,000	150,000

Estimated annual operating revenues = $100,000 (no inflation). Estimated annual operating costs = $45,000 (no inflation). Both revenues and costs are fully responsive to inflation. $r = 0.4$, $i_{df} = 10\%$, $t = 50\%$, and $i_f = 8.7\%$.
a. What is the after-tax rate of return (i_{af}) on this investment?
b. What is the rate of return on equity capital (i_e)?

7.20 The following investment, which has an estimated useful life of 15 years is under consideration. Use the declining-balance method and a 10% rate for the building and use the fast-write-off method for equipment.

	P	SV in 15 Years
Building	$300,000	$100,000
Land	100,000	200,000
Working capital	75,000	75,000
Equipment	200,000	0

Annual estimated operating costs with no inflation are $150,000. Operating costs are fully responsive to inflation. $r = 0.25$, $i_{df} = 10\%$, $t = 50\%$, $i_e = 12\%$, and $i_f = 5.7\%$. What is the present equivalent of revenue requirements to justify this investment?

7.21 The CanGrow Corporation is considering an investment in a new fabrication plant. The following conditions exist:

	Initial Cost	Salvage Value in 10 Years
Buildings	$200,000	$150,000
Land	50,000	100,000
Working capital	100,000	100,000
Equipment (installed cost)	250,000	100,000

Annual operating costs are $50,000 per year with no inflation (zero date) and are fully responsive to inflation. Use the declining-balance method for CCA. The rate on the equipment $= 20\%$; the rate on the building $= 10\%$. $r = 50\%$, $i_{df} = 12\%$, $t = 40\%$, $i_e = 9\%$, and $i_f = 7.2\%$. Calculate the annual revenue requirements necessary for this investment.

7.22 An investment is under consideration which has a first cost of $100,000, an estimated salvage value of zero, and an estimated useful life of 5 years. Use the fast-write-off method to calculate the CCA. Annual operating costs with no inflation (zero date) are assumed to be $10,000 per year and fully responsive to inflation. $r = 0.5$, $l_{df} = 8\%$, $t = 50\%$, $i_e = 11.1\%$, and $i_f = 8\%$. Determine the annual equivalent revenue requirement.

7.23 Does inflation usually lower or raise the return on investments to the equity holder? Why? (Answer in $\frac{1}{2}$ page or less.)

7.24 A manufacturing plant is being considered to manufacture television circuit boards. The following estimates have been developed:

	P	SV in 10 years
Building	$300,000	$100,000
Land	100,000	150,000
Working capital	127,000	127,000
Equipment	200,000	0

Annual operating revenues with no inflation (zero date)	$240,000
Annual operating costs with no inflation (zero date)	100,000
A major overhaul in year 5 is expected to cost	25,000

$r = 0.40$, $i_{df} = 10\%$, $t = 46\%$, and $i_f = 8\%$. Assume that operating revenue and operating costs are fully responsive to inflation. Calculate the rate of return on the equity capital (i_e) to the nearest percent. Use the fast write-off-method for equipment and the declining-balance method for the building using a 5% rate.

7.25 The following costs have been estimated for a facility to manufacture a new product which is expected to have a good market for 10 years.

	P	SV in 10 Years
Buildings	$500,000	$200,00
Land	$150,000	250,000
Working capital	175,000	175,000
Equipment	400,000	0

Operating revenues (zero date—no inflation)	$700,000
Operating costs (zero date—no inflation)	350,000
A major expense is expected in year 3	50,000

$r = 0.2$, $i_{df} = 12\%$, $t = 42\%$, $i_f = 7.3\%$, and $i_e = 9\%$. Use the fast-write-off method for equipment. Use the declining-balance method and a 10% rate for the building. Assume that the operating revenues and the operating costs are fully responsive to inflation.

a. Calculate the present equivalent revenue required to justify this investment.

b. Should the investment be made?

7.26 Sparkall Corporation is considering a facility for the production of automobile spark plugs. The following economic data pertain to this proposal, which has an estimated useful life of 10 years.

	Initial Cost	Salvage Value in 10 Years
Land	$200,000	$350,000
Building	300,000	150,000
Working capital	250,00	250,000
Equipment	500,000	0

$r = 50\%$, $i_{df} = 12\%$, $t = 40\%$, $i_e = 9\%$, and $i_f = 7.1\%$. Annual operating costs are $75,000 with no inflation (zero date) and are fully responsive to inflation. There is an additional operating expense of $20,000 at the end of the fourth year. Use the declining-balance method and a 10% rate for the building. Use the fast-write-off method for equipment. What is the present equivalent revenue requirement to justify this investment?

CHAPTER EIGHT
Break-Even Analysis

Break-even analysis is usually associated with the process of determining the level of sales necessary for a company to recover all of its costs incurred in providing a product to the market. In this analysis, profit is initially often assumed to be zero. Although this assumption may not be favorable to many managers, it establishes the base for the evaluation. A company can then determine if sales of this level are achievable. If this sales level is not achievable, the company will incur a loss, and if sales exceed break-even, the company will realize a profit. However, break-even analysis has much broader applications than the concept above and the analysis may be applied to calculate the break-even point at any specified rate of return on investment. For example, break-even analysis may apply to:

1. Revenue requirements calculations, the objective being to determine the revenue required in a break-even sense to meet all costs, including profit associated with each alternative

2. The increased sales necessary to offset the investment in additional production equipment

3. The level of sales and production that produces equivalence between two alternative investment proposals

Thus many of the problems previously discussed throughout this text may be thought of in a break-even sense. In this chapter the format more specifically associated with break-even analysis is introduced and applied.

8.1 THE CONCEPT OF FIXED AND VARIABLE COSTS

A company uses market reports to determine the level of sales forecast. From this information the company can estimate a set selling price and the multiple of the two represents sales revenue. The cost side of the organization is separated into two components: fixed costs and variable costs. Fixed costs include those costs which are fixed or do not vary with the level of production in the short run. Fixed costs include such costs as rent, insurance, and administrative costs. Variable costs cover such items as material, supplies, and labor. The variable costs will vary with the level of production of the plant. Note that variable costs are fixed per unit of output and fixed costs vary per unit of output. As output increases, fixed costs are spread over more items.

There are different methods of solving break-even analysis problems and the base for these methods is the algebraic approach. Revenue is the multiple of volume of sales and sales price. Costs are the sum of fixed costs, and the product of variable costs and sales volume. Break-even in its simplest form occurs at the point where total revenue equals total costs.

$$\text{Total revenue (TR)} = \text{total cost (TC)} \tag{8.1}$$

$$\text{TR} = \text{SP} \times S \qquad \text{TC} = \text{FC} + \text{VC} \times S$$

where SP = selling price per unit
  S = number of units sold
  FC = fixed costs
  VC = variable cost per unit

Equation 8.1 assumes that inventory is zero and that all production goes to sales immediately. This leads us to

$$S \times \text{SP} = \text{FC} + \text{VC} \times S \tag{8.2}$$

The concept of fixed costs and variable costs will now be more fully explained with reference to the corporate cash flow diagram (Figure 8.1).

The after-tax cash flow, income tax, administrative costs, utilities, insurance, and property taxes tend to be unaffected by the number of units produced. That is, they are fixed costs in the short run. Material and labor costs tend to vary according to the number of units produced and are generally referred to as variable costs. However, many costs, such as maintenance, utilities, and selling costs, have a fixed and a variable component,

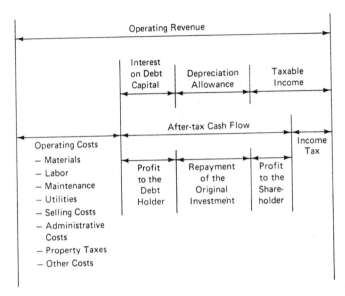

FIGURE 8.1 Fixed and variable costs within the organization.

depending on the number of units produced. These costs are often treated as being fixed but may be split into fixed and variable costs, depending on the circumstances and accuracy desired. All costs are variable over time; thus the term in the "short run" must be fully understood.

Note that this definition of fixed costs includes profit. This is the normal approach, and permits direct calculation of break-even volume as illustrated in Example 8.1. Sometimes it is desirable to separate the profit component out of the fixed costs. This approach is explained in Section 8.2.

The short run, by definition, represents that time span over which management is not able to alter the resources of the organization (e.g., equipment) but is able to change the firm's output by varying such variable factors of production as labor and/or materials (usually considered to be one year).

Example 8.1

Problem: Returning to the small-company example, Global Industries, discussed in Chapter 2, assume that Global would like to know the sales volume required over the next year to meet the following annual fixed and variable costs.

1. Fixed costs: ATCF = $20,000, income tax = $5,000, administrative costs = $4,000, rent = $6,000, utilities and insurance = $4,000. Therefore FC = $39,000.

2. Variable cost per unit: material = \$10, labor = \$5, selling expense = \$2. Therefore VC = \$17/unit.

3. Selling price per unit = \$25.

Solution: Using equation 8.2, we have

$$S(SP) = FC + (VC)S$$

$$S(\$25) = \$39,000 + (\$17)S$$

$$S = \$39,000/\$8 = 4,875 \text{ units}$$

This example may be represented graphically as shown in Figure 8.2.

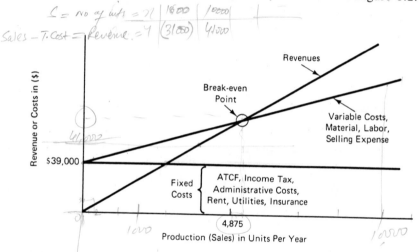

FIGURE 8.2 Break-even with linear revenue and cost curves.

Problem: Using the fixed and variable costs computed above, find the average cost at the annual output levels of 1,000, 2,000, 3,000, 4,000, 5,000, and 6,000 units.

Solution

Annual Output	Total Costs	Average Cost	Incremental Cost per Unit
1,000	\$ 56,000	\$56.00	\$17.00
2,000	73,000	36.50	17.00
3,000	90,000	30.00	17.00
4,000	107,000	26.75	17.00
5,000	124,000	24.80	17.00
6,000	141,000	23.50	17.00

8.2 CONTRIBUTION MARGIN APPROACH

The most common question asked in break-even analysis is at what level of sales will break-even be reached? Equation 8.2 may be restated in terms of sales volume (S).

$$S(SP) = FC + (VC)S$$

$$S(SP - VC) = FC$$

$$S = FC/(SP - VC) \tag{8.3}$$

CM = SP - VC
= 25 - 17
= 8

The divisor of equation 8.3 $(SP - VC)$ is of interest. This factor is known as the *contribution margin* (CM) and the contribution margin approach is another method for solving break-even problems. This is simply the difference between the selling price per unit and the variable costs per unit. If the contribution margin is negative, the company should consider shutting down production since not even variable costs are being met. The difference should be positive and it is this difference per unit of sales that is available to cover fixed costs including profit. If the contribution margin is positive, there is a level of sales at which fixed costs can be recovered and contributions can be directed toward profit. A company may wish to set a level of profit each year to know at what sales level this profit is achievable. Under these conditions profit becomes part of the fixed costs. To differentiate between break-even at zero profit and break-even with a profit contribution, the before-tax profit (P) in equation 8.3 is shown as a separate term:

$$S = (FC + P)/(SP - VC) \tag{8.4}$$

Example 8.2

Problem: A company produces an item with a fixed cost of $10,000 (zero profit) and a variable cost of $0.60 per unit. The item currently sells for $1.00. What is the contribution margin (CM), and what is the break-even sales volume at zero profit?

Solution

$$CM = SP - VC$$

$$= \$1.00 - \$0.60 = \$0.40$$

$$S = FC/CM$$

$$= \$10,000/\$0.40 = 25,000 \text{ units}$$

S(SP) = FC + (VC)S
25,000(1.00) = 10,000 + (0.60)25,000
25,000 = 10,000 + 15,000
25 = 25

These results are illustrated in Figure 8.3.

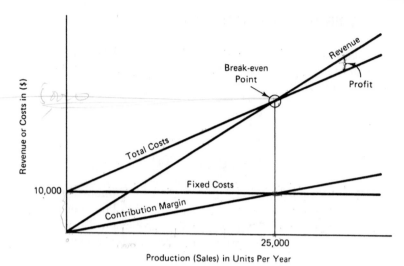

FIGURE 8.3 Contribution margin and fixed costs.

Problem: Assume that the variable costs rise by 10% from $0.60 to $0.66.

Solution

$$CM = \$1.00 - \$0.66 = \$0.34$$

Break even in units $\longleftarrow$ $S = FC/CM = \$10,000/\$0.34 = 29,412$ units ✓ SP

This represents an increase in sales volume of 17.65%.

Problem: Assume that fixed costs increase by 10% and variable costs remain at $0.60.

Solution

$$S = FC/CM = \$11,000/\$0.40 = 27,500 \text{ units}$$

This represents an increase in sales volume of 10%.

Problem: Assume that the sales price increases by 10%.

Solution

$$CM = SP - VC = \$1.10 - \$0.60 = \$0.50$$
$$S = FC/CM = \$10,000/\$0.50 = 20,000 \text{ units}$$

This result shows a decrease in sales volume of 20%.

Problem: Assume that SP, VC, and FC all increase by 10%.

VC ↑ —— ↑ in units but not at same rate

FC ↑ —— ↑ at same rate

SP ↑ —— ↓ but not at same rate

Solution

$$CM = SP - VC = \$1.10 - \$0.66 = \$0.44$$

$$S = FC/CM = \$11,000/\$0.44 = 25,000 \text{ units}$$

The net increase in sales volume is zero.

Example 8.3

Problem: In Example 8.2, the sales volume of 25,000 units at $1.00 per unit generates a revenue of $25,000. Fixed costs at zero profit equal $10,000. Variable costs are equal to $0.60 per unit. Assume that sales are 35,000 units and that fixed costs remain at $10,000. Determine the profit generated.

Solution

$$CM = SP - VC = \$0.40 \text{ per unit}$$

Contribution to FC + profit

$$CM = 35,000(\$0.40) = \$14,000$$

$$\text{Fixed cost} = \$10,000$$

$$\text{Profit} = \$14,000 - \$10,000 = \$4,000$$

Problem: From the equations or graphical solutions the company can evaluate its position. As in Example 8.2, if costs increase by 10%, the break-even sales volume increases by 17.65% to 29,412 units. With this information management has several alternative strategies open for consideration. For example, (1) raise the selling price; (2) initiate an advertising campaign costing $2,000 and raise the price by 5%; or (3) do nothing.

Solution

1. If the selling price is raised by 10%:

$$CM = \$1.10 - \$0.66 = \$0.44$$

$$S = FC/CM = \$10,000/\$0.44 = 22,727 \text{ units}$$

The sales volume could drop 2,273 units and the company could break even with a 10% increase in selling price and variable cost.

2. If the advertising campaign and 5% price rise are employed:

$$CM = \$1.05 - \$0.66 = \$0.39$$

$$S = (\$10,000 + \$2,000)/\$0.39$$

$$= 30,769 \text{ units}$$

If the advertising campaign is employed, the sales volume would have to increase by 5,769 units.

3. The do-nothing alternative with variable cost up 10% requires a sales volume of 29,412 units. Sales would have to increase by 4,412 units.

The company is now in a position to evaluate the alternatives. Although the first alternative may appear to be a good choice, a 10% increase in price may drive away more buyers than the company can afford to lose. A market survey may be necessary to indicate which alternative is the best.

8.3 OPERATING LEVERAGE

With the advance of modern technology, companies have become more capital intensive, with a resulting reduction in variable costs. The ratio of capital or fixed costs to variable costs is called *operating leverage* (see Figure 8.4). Usually, the higher the fixed cost or operating leverage, the higher the break-even point. However, if sales are above the break-even point, the profits will be greater; and if sales are below the break-even point, the losses will increase. Profits and losses are very sensitive to the level of operating leverage.

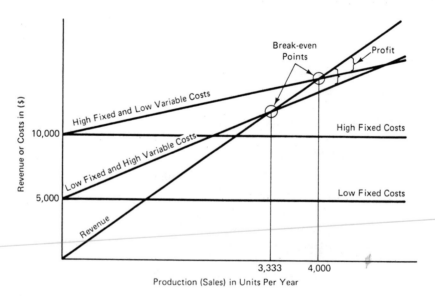

FIGURE 8.4 Operating leverage.

Example 8.4

Problem: A product has a revenue of $6.00 per unit. There are two alternatives open to the company to produce this product. Alternative A has a fixed cost of $5,000 and a variable cost of $4.50 per unit. Alternative B has a fixed cost of $10,000 and a variable cost of $3.50 per unit.

Solution

Alternative A

$$CM = \$6.00 - \$4.50 = \$1.50$$

$$S = FC/CM = \$5,000/\$1.50 = 3,333 \text{ units}$$

Alternative B

$$CM = \$6.00 - \$3.50 = \$2.50$$

$$S = FC/CM = \$10,000/\$2.50$$

$$= 4,000 \text{ units}$$

The following table shows the sensitivity of profit to sales volume as fixed costs increase.

Sales Volume	Profit Alt. A	Alt. B
2,000	$(2,000)	$(5,000)
3,333	0	(1,667)
4,000	1,000	0
6,000	4,000	5,000
7,000	5,500	7,500

Capital-intensive organizations take a higher risk with a greater possibility of higher losses should sales not materialize. But if sales do materialize, the high-technology, low-variable-cost organizations will realize a greater profit.

Example 8.5

Problem: Companies Y and Z both produce computer component parts for a very popular personal computer system. One of the components can be produced by either organization. Company Y has fixed costs of $1,250,000 and variable costs of $6.50 per unit of output, whereas company Z has fixed costs of $2,000,000 and variable costs of $4.00 per unit of output. Both companies have a production capacity limit of 500,000 units. If the selling price is $10.00, determine the break-even sales levels for both firms and express these values as percentages of the stated full capacity.

Solution

Company Y

$$S = FC/CM$$

$$= \$1,250,000/(\$10.00 - \$6.50)$$

$$= 357,143 \text{ units}$$

$$\text{Percent of full capacity} = 357,143/500,000$$

$$= 71.4\%$$

Company Z

$$S = \$2,000,000/(\$10.00 - \$4.00)$$

$$= 333,333$$

$$\text{Percent of full capacity} = 333,333/500,000$$

$$= 66.7\%$$

Problem: At what percentage of the production capacity are total costs for both companies equal?

Solution

$$FC + VC(S) = FC + VC(S)$$

$$\$1,250,000 + \$6.50(S) = \$2,000,000 + \$4.00(S)$$

$$\$2.50(S) = \$750,000$$

$$S = 300,000 \text{ units}$$

$$\text{Percent of full capacity} = 300,000/500,000$$

$$= 60\%$$

Problem: Assume that inventory cannot be stocked. At what sales price would both firms shut down in the short run?

Solution: When CM is less than or equal to zero,

$$CM = (SP - VC)$$

For firm Y, shutdown is at a selling price of $6.50. For firm Z, shutdown is at a selling price of $4.00.

Problem: Which of the two firms is more sensitive to changes in sales?

Solution: From Figure 8.5, company Z is more sensitive to changes in sales, because as sales go up or down, the profit of company Z correspondingly

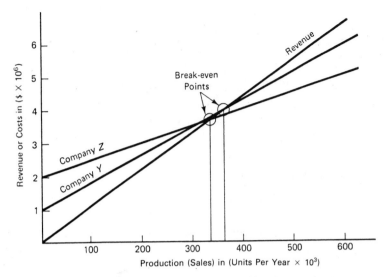

FIGURE 8.5 Operating leverage and break-even.

responds much more than that of company Y. This is due to the flat curve, characterized by high fixed costs and low variable costs.

8.4 DISCRIMINATION PRICING

Discrimination pricing can easily be evaluated using break-even analysis. Discrimination pricing is selling the same product at two different prices. Products may be sold at a lower price in discount stores with labels removed, or sold to a foreign market with reasonable assurance that the products will not reenter the country and displace sales of the same model at a higher price.

A company may have a product that is near the break-even point which yields only a small amount of profit. The company then receives a proposal for a large-volume purchase of the same product at a reduced price, with reasonable assurance that the sale will not affect the current market volume of their existing product. The company should proceed with this sale provided that the contribution margin of the added volume is still positive.

Example 8.6

Problem: The current product of a company has a selling price of $5.00, variable costs of $3.00, and fixed costs of $5,000. The current market volume is 3,000 units. Determine the amount of profit that this product will return to the company.

Solution

$$CM = SP - VC = \$5.00 - \$3.00$$
$$= \$2.00 \text{ per unit}$$
$$\text{Contribution} = (3,000 \text{ units})(\$2.00/\text{unit})$$
$$= \$6,000$$
$$\text{Fixed cost} = \$5,000$$
$$\text{Profit} = \$1,000$$

Conclusion: The product is profitable.

Problem: The company now receives an offer for an additional 1,500 items at $4.00 per unit with assurance that current volumes will not be reduced. If fixed costs remain unchanged, should the company accept the offer?

Solution

$$CM = \$4.00 - \$3.00 = \$1.00 \text{ per unit}$$
$$\text{Contribution} = (1,500 \text{ units})(\$1.00/\text{unit})$$
$$= \$1,500$$
$$\text{Profit} = \$1,500 + \$1,000 = \$2,500$$

Conclusion: Accept the offer.

8.5 BREAK-EVEN ANALYSIS FOR MULTIPRODUCT ORGANIZATIONS

Break-even analysis can make a major contribution in multiproduct organizations. Many organizations use unsubstantiated profit analysis to justify discontinuing production of a product that appears to be unprofitable. The result often transfers fixed costs to the remaining products and significant losses to the organization. Many products which may appear unprofitable are carrying a large amount of overhead and fixed costs which have to be carried by other products if discontinued. Therefore, as in a single-product organization, break-even analysis can be performed for multiproduct organizations. However, variable and fixed costs require a much greater in-depth investigation to determine specific costs by product. Many costs are common to more than one product.

The allocation of fixed costs that are common to more than one product in a multiproduct organization may be difficult. The question arises: Should the costs be allocated on the basis of sales value, number of units produced, number of labor hours used, or number of machine hours used?

This problem is discussed more fully in Chapter 13. Accountants generally allocate on the basis of labor and machine hours, but other allocation methods can be used. However, regardless of the allocation method used, the problem is that to discontinue a product, these allocated fixed costs have to be reallocated, as the product discontinued will no longer be supporting its share. In general, if a product is providing a positive contribution margin, the product should be retained since the product is supporting a share of overhead cost which would otherwise have to be carried by other products.

The method of allocation may not be important, but the fact that the fixed costs will remain even if the product does not is a very important point. This is why knowing the contribution margin of each product is very important.

The variable costs within a multiproduct organization should be allocated to each product. In this manner all the variable costs associated with a product are known and the contribution margin can be easily calculated. If the contribution margin is negative, the product should be carefully reviewed and possibly dropped.

Example 8.7

Problem: A company produces four products, A, B, C, and D.

Product	A	B	C	D	Total
Sales	$200,000	$ 300,000	$350,000	$150,000	$1,000,000
Costs					
Fixed	50,000	200,000	100,000	50,000	400,000
Variable	125,000	110,000	200,000	95,000	530,000
Total cost	$175,000	$ 310,000	$300,000	$145,000	$ 930,000
Profit	$ 25,000	$ − 10,000	$ 50,000	$ 5,000	$ 70,000

CM   75,000   190K   150 K   55 K

Profit analysis alone indicates that product B is losing money. Should product B be dropped?

Solution: If the fixed costs of $200,000 associated with product B do not vanish with the product, they must be reallocated to the remaining products. Assuming the following reallocation the result would then be:

Product	A	C	D	Total
Sales	$ 200,000	$ 350,000	$ 150,000	$ 700,000
Costs				
Fixed	100,000	200,000	100,000	400,000
Variable	125,000	200,000	95,000	420,000
Total cost	$ 225,000	$ 400,000	$ 195,000	$ 820,000
Profit	$ − 25,000	$ − 50,000	$ − 45,000	$ − 120,000

If product B were dropped, this company would probably collapse.

Product B is carrying 50% of the fixed costs of the organization. When product B is evaluated on the basis of contribution margin, the answer is very clear. With sales of $300,000 and a variable cost of only $110,000, the product is contributing $190,000 to fixed costs. This value represents a greater contribution than any of the other products. From the contribution margin approach all products are contributing and should be retained.

8.6 THE EFFECT OF INCOME TAX ON BREAK-EVEN ANALYSIS

Referring to Figure 8.1, we can see that:

$$\text{Taxable income} = \text{income tax} + \text{profit to the shareholder}$$

and

$$IT = TI \times t$$

so

$$IT = t(IT + \text{profit to the shareholder})$$

$$IT(1 - t) = t(\text{profit to the shareholder})$$

Therefore, income tax (IT) is equal to

$$IT = [t/(1 - t)](\text{profit to the shareholder}) \tag{8.5}$$

Example 8.8

Problem: The following information is available with respect to a specific product line. Fixed costs, excluding profit and income tax, are equal to $20,000. Profit required by the shareholder = $10,000, $t = 48\%$, VC = $2.50 per unit, and SP = $5.00 per unit. What sales volume is required at break-even?

Solution

$$IT = t/(1 - t)(\text{net income})$$

$$= (0.48)/(0.52)(\$10,000) = \$9,231$$

$$CM = SP - VC = \$2.50 \text{ per unit}$$

$$FC = \$20,000 + \$10,000 + \$9,231 = \$39,231$$

$$S = FC/CM = (\$39,231)/\$2.50$$

$$= 15,692 \text{ units}$$

The sales volume required at break-even is 15,692 units.

Example 8.9

Problem: Western Industries produces one model of garden tractor. The fixed costs of assets are $500,000, with an estimated salvage value of $100,000 in 5 years. Assume the straight-line depreciation method is used (equation 4.2). The variable costs of production are $1,500 per unit, with other fixed costs of $50,000 per year. An after-tax rate of return on investment of 15% is required with a tax rate of 41%. Determine the selling price for a production of 3,000 units per year.

Solution

$$D = (\$500,000 - \$100,000)/5 = \$80,000 \text{ per year}$$

Using equation 6.3 to calculate the revenue requirement:

$$AER = \$1,500/\text{unit} + \$50,000 + [\$500,000(A/P, 15\%, 5)$$
$$- \$100,000(A/F, 15\%, 5) - 0.41(\$80,000)]/0.59$$
$$= \$1,500/\text{unit} + \$50,000 + [\$500,000(0.2983)$$
$$- \$100,000(0.1483) - 0.41(\$80,000)]/0.59$$
$$= \$1,500/\text{unit} + \$222,068$$

Then the sales revenue must equal the revenue requirement:

$$SP(S) = AER$$
$$SP = \$1,500/\text{unit} + \$222,068/3,000 \text{ units}$$
$$= \$1,574.02 \text{ per unit}$$

8.7 THE MAKE VERSUS BUY DECISION

Firms that produce more than one product often continue to produce products without ever questioning whether any of the products or components should be produced for the organization by an external manufacturer. The most important factor in evaluating such a proposal is what in-house costs will be deferred if the component is produced elsewhere. The variable costs, such as labor and material, are of course deferred, and these are easily identified. The costs that are not so easily identified are the ones associated with fixed costs or overhead. Questions that will have to be asked include: If the component is produced elsewhere, what overhead costs can be reduced? Is space at a premium, and could this be used for other product lines?

Example 8.10

Problem: Company Y produces three products, A, B, and C. The costs are as follows:

	Product		
	A	B	C
Variable labor	$ 6.00	$ 3.00	$ 4.00
Cost of material	5.00	5.00	8.00
Variable costs	$11.00	$ 8.00	$12.00
Overhead costs			
Milling machine	$ 2.00	$ 1.25	$ 1.75
Space rent	3.00	0.50	1.50
Others	2.00	1.00	3.00
Total	$18.00	$10.75	$18.25

Company Z makes an offer to produce the items for $12.00, $10.00, and $11.50, respectively. Which of the items should be produced by company Z for company Y? Assume that only the variable cost will be deferred: $11.00, $8.00, and $12.00 for products A, B, and C, respectively.

Solution

Saving = deferred costs − costs of acquiring from company Z

Saving product A = $11.00 − $12.00 = $ − 1.00

Saving product B = $8.00 − $10.00 = $ − 2.00

Saving product C = $12.00 − $11.50 = $0.50

Therefore, have product C produced by company Z.

Problem: Company M offers the company the going in-house rate for use of space and the milling machine. What should the company do if the offer from company Z still stands?

Solution: The deferred costs now are not only the variable costs but also the overhead costs associated with rent and the milling machine.

Saving A = $11.00 + $5.00 − $12.00 = $4.00

Saving B = $8.00 + $1.75 − $10.00 = $ − 0.25

Saving C = $12.00 + $3.25 − $11.50 = $3.75

Therefore, have products A and C produced by company Z and rent space and milling machine capacity to company M.

The examples as given are simplifications of real-life situations. There are many other factors that must be considered in the make versus buy decision. Two of the most important factors in the make versus buy decision are the security of supply and the quality of the product supplied. If the company is not sure of the integrity of the supplier, this can greatly influence the decision-making process as to whether to make or to buy the product.

The company may have moral obligations of contracts with suppliers and unions which may not allow or could make it very difficult for them to discontinue in-house production. These are just a few of the constraints that will have to be considered in the make versus buy decision.

8.8 THE PATTERN OF CORPORATE DEVELOPMENT

Growth within an organization may or may not be desirable, depending on the specific objectives to be accomplished. Nevertheless, the development of appropriate policy decisions for the organization requires some understanding of how and why organizations grow and decline. Such change may be viewed in terms of goods and/or services and their markets as they relate to the specific organization in question.

There is considerable evidence to support the theory that goods, services, and technologies have a finite life. That is, they are introduced, they grow, reach a maturity level, and then decline. Figure 8.6 shows a typical life cycle for a product or service.

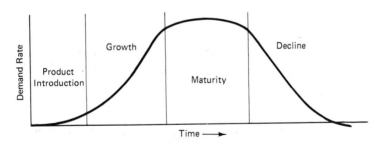

FIGURE 8.6 A life cycle for a product or service.

Organizations also tend toward a type of life cycle in their development. This development is characterized by four distinctive stages of development. These are:

1. The owner-entrepreneur who markets a single product or service and operates within an informal framework.
2. A single unit (e.g., plant) managed by a team of departmental managers (e.g., production, marketing, engineering, and service). Sales represent a single line of related products through a single distribution channel.

3. Several regional units reporting to a corporate headquarters, each unit being structured similar to stage 2 above. Each unit produces and markets the same line of products through multiple channels.

4. Several semiautonomous units reporting to corporate headquarters, each with structure 2 or 3. Each unit produces different lines of products for separate markets with multiple channels.

Growth as a means for avoiding goods or service obsolescence is implicit in the organizational development. Consideration of growth needs and opportunities represents the decision-making process within the firm. This process requires that marketing strategies, product development policies, and questions of diversification be reviewed periodically. Break-even analysis becomes a very useful tool within this review process.

Example 8.11

Problem: Global Industries has experienced rapid growth since initially organized. The company is still basically a one-person, single-product organization. Sales of hand carts have started to level off and the cost of closing a sale has increased.

1. Fixed costs:
   a. Equipment has an estimated value equal to book value today $ 40,000
      Salvage value in 5 years = $10,000
   b. Working-capital requirements   60,000
          Total fixed capital costs $100,000
   c. Administration expenses $ 20,000
   d. Utilities and insurance   5,000
   e. Selling costs   10,000
   f. Maintenance   5,000
   g. Rent   10,000
          Total fixed operating costs per year $ 50,000
2. Variable costs:
   a. Labor and material = $15.00 per unit
3. Maximum annual sales and production = 6,000 units
4. Selling price = $30.00 per unit
5. Present sales level = 4,500 units

In reviewing operations the company feels that sales of 5,000 units per year can be reached and maintained over the next 5 years but that after 5 years sales in the present market area will probably start to decrease.

The company is considering two alternative strategies. One involves an expansion of the existing marketing area for the same product (alternative B). The other alternative is to start production of a second product, a grass-seed spreader (alternative C). Either alternative requires:

1. Additional equipment costs $20,000
2. Salvage value = $10,000 in 5 years
3. Working capital $40,000
   Total additional fixed capital costs $60,000
4. Other fixed costs increase by $25,000 per year
   Total additional fixed operating costs $25,000
5. Variable costs per yard cart decrease by $3.00/unit
6. Variable costs per grass-seed spreader equals $14.00/unit
7. Selling price per grass-seed spreader equals $25.00/unit
8. $t = 30\%$ and $i_a = 20\%$
9. Depreciation allowance based on the straight-line method (equation 4.2)

Sales for alternative B are estimated to be 7,000 units per year. Sales for alternative C are estimated to be 5,000 yard carts and 2,000 grass-seed spreaders per year.

Solution: Find AER for alternative A (existing operation).

$$D = (\$40,000 - \$10,000)/5 = \$6,000 \text{ per year}$$

$$AER = \$15/\text{unit} + \$50,000 + [\$100,000(A/P, 20\%, 5)$$

$$- \$70,000(A/F, 20\%, 5) - 0.30(\$6,000)]/0.7$$

$$= \$15/\text{unit} + \$50,000 + [\$100,000(0.3344)$$

$$- \$70,000(0.1344) - \$1,800]/0.7$$

$$= \$15/\text{unit} + \$50,000 + (\$33,440 - \$9,408 - \$1,800)/0.7$$

$$= \$15/\text{unit} + \$81,760$$

For a sales and production volume of 5,000 units:

$$AER = \$15.00/\text{unit}(5,000 \text{ units}) + \$81,760 = \$156,760$$

$$AE = \text{total revenue} - \text{total costs}$$

$$= \$150,000 - \$156,760 = \$-6,760$$

The present operation is not meeting the 20% return desired on the investment. Find AER for alternative B.

$$D = (\$60,000 - \$20,000)/5 = \$8,000/\text{year}$$

$$AER = \$12/\text{unit} + \$75,000 + [\$160,000(0.3344)$$

$$- \$120,000(0.1344) - 0.30(\$8,000)]/0.7$$

$$= \$12/\text{unit} + \$75,000 + \$49,966$$

$$= \$12/\text{unit} + \$124,966$$

For a sales and production volume of 7,000 yard carts per year:

$$AER = \$84,000 + \$124,966 = \$208,966$$
$$AE = TR - TC = \$210,000 - \$208,966$$
$$= \$1,034$$

Alternative B meets the MARR requirement.

Find AER for alternative C. Let the yard cart be product X and the grass-seed spreader be product Y.

$$D = (\$60,000 - \$20,000)/5 = \$8,000/year$$
$$VC = \$12/X \text{ unit} + \$14/Y \text{ unit}$$
$$AER = \$12/X \text{ unit} + \$14/Y \text{ unit} + \$75,000$$
$$+ [\$160,000(0.3344) - \$120,000(0.1344)$$
$$- 0.30(\$8,000)]/0.7$$
$$= \$12/X \text{ unit} + \$14/Y \text{ unit} + \$124,966$$

For a sales and production volume of 5,000 units of (X) and 2,000 units of (Y):

$$AER = \$60,000 + \$28,000 + \$124,966 = \$212,966$$
$$AE = TR - TC$$
$$= \$150,000 + \$50,000 - \$212,966 = \$-12,966$$

Alternative C does not meet the MARR requirement.

Problem: For alternative C, assume that you are satisfied with the market projection of 5,000 yard carts but you would like to know how many grass-seed spreaders you would have to sell for alternative B and alternative C to be equivalent.

Solution

Selling price for yard carts	$30/unit
Selling price for grass-seed spreaders	$25/unit
Variable cost for grass-seed spreaders	$14/unit

Alternative B

$$TR - TC = \$1,034$$

Alternative C

Sales equal: 5,000 yard carts + Y grass-seed spreaders

$$TC = AER = \$60,000 + (\$14/unit)Y + \$124,966$$

$$= \$184,966 + (\$14/unit)Y$$

$$TR = (\$5,000)(\$30/unit) + Y(\$25/unit)$$

$$= \$150,000 + Y(\$25/unit)$$

For alternative B to be equivalent to alternative C:

$$TR - TC \text{ (alternative B)} = \$1,034$$

$$\$1,034 = \$150,000 + Y(\$25/unit)$$

$$-\$184,966 + Y(\$14/unit)$$

$$= -\$34,966 + Y(\$11/unit)$$

$$Y = \$36,000/\$11$$

$$= 3,273 \text{ units}$$

8.9 BREAK-EVEN ANALYSIS TO DETERMINE THE OPTIMUM NUMBER OF PRODUCTION UNITS TO INSTALL

All manufacturing equipment has a production capability limit; that is, the equipment is capable of producing only so many items before its physical limits are reached. A company may wish to evaluate how many machines of a new type of production equipment should be purchased. This analysis will assume that all additional units of production equipment have fixed and variable costs identical to the first piece of equipment.

The first unit of production equipment can be evaluated independent of additional pieces of equipment. The first piece of equipment has a production capability limit, fixed costs, and variable costs. Given the selling price the break-even point can be evaluated. For a constant selling price, the point of maximum profit will occur at the production capability limit. Therefore, the range where profit will be earned is between the break-even point and the production capability limit and the profit increases as sales approach the production capability limit. If another unit of production equipment is added with identical costs, the same factors will hold true for this unit.

The important point is that an additional piece of production equipment should not normally be added unless sales are greater than the maximum production of the existing units of production plus sufficient sales to meet breakeven on the additional component. If this requirement cannot be satisfied, the company will make a greater profit by dropping sales back to the point where sales can just be met by the last installed unit of production equipment.

$$PCL = \text{production capability limit/unit}$$

$$n = \text{number of production equipment units}$$

If

$$n(PCL) + BE < S < (n + 1)(PCL)$$

where BE is break-even sales volume for unit $n + 1$, install $n + 1$ units. But if

$$n(PCL) < S < n(PCL) + BE$$

install n units.

From Figure 8.7 the dotted areas are ranges of profit. Note that the maximum profit in each sector is at the production capability limit. Once an additional unit of equipment is installed and sales do not reach the next break-even point, the company will operate in one of the hatched areas. Although a profit is being made on all equipment installed previously, the

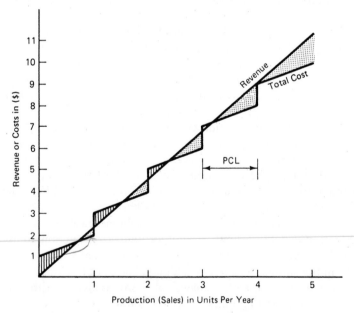

FIGURE 8.7 Optimum number of units of productions.

company is losing money on the installation of the last unit, and the profit of this organization will be reduced by the amount of the loss on this unit. The profit or loss is the vertical distance between the sales line (revenue) and the costs line. This approach is very helpful in selecting the number of production units to install.

Volume discounts on the purchase of equipment and materials may also be a major consideration in determining the number of production units to install. Table 8.1 indicates an example of volume discounts. Labor is not included in the table, but it should be noted that if each machine required an operator, an experienced operator could be hired first and as more equipment is added, this operator could supervise less experienced and lower-cost operators on the additional equipment. If this were to hold true, labor could also be included as a volume discount item in the table.

TABLE 8.1 Volume discounts.

Units of Production Equipment	Incremental Fixed Cost/Unit	Material
1	F.C.	$XXX/unit
2	98% F.C.	98% $XXX/unit
3	96% F.C.	96% $XXX/unit
4	94% F.C.	94% $XXX/unit
5	92% F.C.	92% $XXX/unit
6	90% F.C.	90% $XXX/unit
7	88% F.C.	88% $XXX/unit
8	86% F.C.	86% $XXX/unit

8.10 BREAK-EVEN UNDER NONLINEAR CONDITIONS

To this point we have considered only linear relationships. Within certain ranges and limitations the linear approximations are suitable for solving many problems. However, the nonlinear case should also be considered.

In many cases, as the manufacturing process expands, so does the efficiency of the firm up to a point; after that point, as the firm expands, inefficiency enters into the firm's operations. These stages of growth and efficiency are known as *increasing returns to scale* and *decreasing returns to scale*. This phenomenon is best represented by quadratic equations. For examples of increasing returns to scale, assume that one operator could run three machines but that only one machine is purchased. Therefore, as more machines (up to three) are added, labor cost per unit of output will decrease. But now assume that a fourth machine is added and that overtime and idle machine time waiting for the operator are uneconomical to the firm. Or assume a maintenance staff that as more machines are added becomes more efficient to a point. After this point the maintenance section becomes too large and inefficiencies become the norm.

The nonlinear aspect also holds true for revenue. If we assume perfect competition, one supplier is small in comparison to the market size and will not have any effect on the market price. But for the firm that supplies a large percentage of the total market, an increase in the amount supplied and sold will necessitate a reduced price in order to sell the extra supply. Therefore, this revenue curve would also be nonlinear (Figure 8.8).

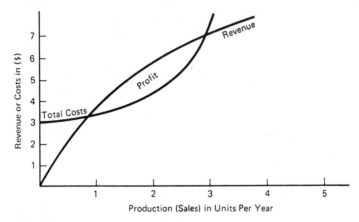

FIGURE 8.8 Break-even under nonlinear conditions.

From calculus it is known that there is a maximum or a minimum at the point where the derivative is equal to zero. The cost curve will take the general quadratic form of $ax^2 + bx + c$. This is also true for the revenue curve in a nonlinear case.

Since profit is revenue minus costs as long as either costs or revenue is in a quadratic form, so also will be the profit function. Profit will also take the general form of $ax^2 + bx + c$.

$$dP/dx = 2ax + b$$

A maximum or minimum value is then found by setting this term equal to zero.

$$0 = 2ax + b$$

$$x = -b/2a$$

Substituting this value of x back into the general quadratic equation to find the maximum profit yields

$$\text{Maximum profit} = ax^2 + bx + c$$

$$= a(-b/2a)^2 + b(-b/2a) + c$$

$$= b^2/4a - b^2/2a + c$$

$$= -b^2/4a + c$$

Profit is revenue minus costs, the first derivative of which is marginal revenue minus marginal cost. Then set marginal revenue minus marginal cost equal to zero to find the point of maximum profit. Therefore, at the point of maximum profit, marginal revenue equals marginal cost. This is a well-known equation used by economists to find the point of maximum profit.

Example 8.12

Problem: Given that costs are a function of $0.1x^2 + 0.3x + 3$ and revenues are a function of $-0.25x^2 + 3x$, find the point of maximum profit and the amount of profit at that point.

Solution: Using method 1:

$$MC = dC/dx = 0.2x + 0.3$$

$$MR = dR/dx = -0.5x + 3$$

Maximum profit occurs when

$$MC = MR$$

$$0.2x + 0.3 = -0.5x + 3$$

$$0.7x = 2.7$$

$$x = 3.857$$

$$Costs = 0.1x^2 + 0.3x + 3$$

$$= 0.1(3.857)^2 + 0.3(3.857) + 3$$

$$= 1.488 + 1.157 + 3 = \$5.645$$

$$Revenue = -0.25x^2 + 3x$$

$$= -0.25(3.857)^2 + 3(3.857)$$

$$= -3.7191 + 11.571 = \$7.852$$

$$Profit = revenue - costs = \$2.207$$

Using method 2,

$$Profit = revenue - costs$$

$$= -0.35x^2 + 2.7x - 3$$

At maximum

$$x = -b/2a$$
$$= -2.7/2(-0.35) = 3.857$$

Maximum profit will occur at sales of 3.857. Maximum profit will be

$$P = -b^2/4a + c$$
$$= -(2.7)^2/4(-0.35) - 3$$
$$= \$2.207$$

8.11 MINIMUM-COST ANALYSIS

The cost components of an item or project are often related to a common variable. Some situations where this occurs are:

1. A bridge, where a change in the span length changes the cost of the span and the number of piers required and their total cost
2. A dike, where an increase in the height decreases the expected losses from flooding but increases the cost of the dike
3. A power line, where the increased area of the conductor decreases the power loss but increases the installation cost
4. Insulation, where increased thickness decreases heat loss but increases the installation cost
5. Production, where longer production runs decrease the fixed costs per unit but increase the storage costs

The general mathematical expression of these situations is

$$\text{Total cost} = Ax + \frac{B}{x} + C \qquad (8.6)$$

where x is the common variable and A, B, and C are constants.

The value of the variable that gives the minimum total cost can be found by calculus. Setting the first derivative equal to zero, we have

$$\frac{d(\text{cost})}{dx} = A - \frac{B}{x^2} = 0$$

$$x = \sqrt{\frac{B}{A}}$$

The second derivative,

$$\frac{d^2(\text{cost})}{dx^2} = 2\frac{B}{x^3}$$

is positive, so the value of x is a minimum. Substituting this minimum value back into the total cost curve yields

$$\text{Minimum total costs} = A\sqrt{\frac{B}{A}} + \frac{B}{\sqrt{B/A}} + C$$

$$= \sqrt{BA} + \sqrt{BA} + C$$

shows that the minimum is achieved when the directly proportional cost Ax is equal to the inversely proportional cost, B/x. This is also apparent when the components are plotted (Figure 8.9). The minimum total cost occurs at the value where the cost-component curves intersect.

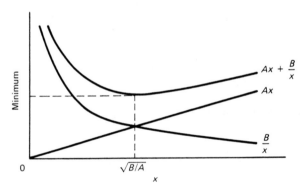

FIGURE 8.9 Minimum cost: general case.

8.11.1 Production Lot Size

A classical problem of minimum cost analysis is the determination of the number of a specific product to make in a single production run. High machinery setup costs encourage long production runs; high storage costs encourage short production runs. The balance between these costs can be determined as follows. Let

C_i = unit production cost (labor, materials, machine time) of product i

D_i = annual demand for product i (units/year)

Q_i = lot size (order quantity) of product i

S_i = setup costs for product i

H = holding costs as a fraction of unit cost for storing products for one year

P_i = production rate for product i

N = number of setups (production runs) per year

Assuming that the on-hand inventory goes down to zero just as the new production run is started, the inventory can be diagrammed (Figure 8.10).

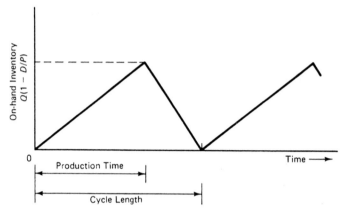

FIGURE 8.10 Plot of on-hand inventory versus time.

From the diagram the following relationships are derived:

$$\text{Cycle length in years} = \frac{Q}{D} \text{ years}$$

$$\text{Number of cycles per year} = N = \frac{D}{Q} \text{ cycles/year}$$

$$\text{Production time} = \frac{Q}{P} \text{ years}$$

$$\text{Maximum inventory} = \text{production time} \times (P - D)$$

$$= \frac{Q}{P}(P - D) = Q\left(1 - \frac{D}{P}\right)$$

$$\text{Average inventory} = \frac{1}{2} \text{ of maximum inventory}$$

$$= \frac{1}{2}Q\left(1 - \frac{D}{P}\right)$$

The total costs are the sum of the production costs, the setup costs, and the holding costs.

The *production costs* are the annual demand times the unit production costs:

$$D \times C$$

The *setup costs* are the cost of a single setup times the number of production runs per year:

$$N \times S = \frac{D}{Q}S$$

The *holding costs* are the holding fraction times the value of the average inventory:

$$H \times C \times \frac{1}{2}Q\left(1 - \frac{D}{P}\right)$$

So

$$\text{Total cost} = D \times C + \frac{D \times S}{Q} + \frac{H \times C(1 - D/P)}{2}Q$$

This is identical to the general mathematical expression (equation 8.6) with the lot size, Q, as the common variable. Solving for the minimum cost Q, we obtain

$$\frac{d(\text{total cost})}{dQ} = -\frac{(D)(S)}{Q^2} + \frac{(H)(C)(1 - D/P)}{2} = 0$$

gives

$$\text{Optimum } Q = \sqrt{\frac{(2D)(S)}{(H)(C)(1 - D/P)}}$$

$$\text{Optimum } N = \sqrt{\frac{(D)(H)(1 - D/P)}{2S}}$$

and substituting the optimum Q into the total cost equation gives

$$\text{Minimum total cost} = D \times C + \frac{(D)(S)}{\sqrt{\dfrac{(2D)(S)}{(H)(C)(1 - D/P)}}} + \frac{(H)(C)(1 - D/P)}{2}$$

$$\times \sqrt{\frac{(2D)(S)}{(H)(C)(1 - D/P)}}$$

$$= D \times C + \sqrt{\frac{(D)(S)(H)(C)(1 - D/P)}{2}}$$

$$+ \sqrt{\frac{(D)(S)(H)(C)(1 - D/P)}{2}}$$

$$= D \times C + \sqrt{(2D)(S)(H)(C)\left(1 - \frac{D}{P}\right)}$$

As noted before, at the minimum the variable cost components are equal.

Example 8.13

Problem: Global Industries manufactures a block valve which it sells at a constant rate of 500 per year. The direct production cost of the valve is $1,100 per unit, and it costs $460 to set up the machinery to manufacture the valve. The equipment can produce 12 valves per day and the company works 240 days per year. The company estimates that the costs of holding an item in inventory, cost of space, insurance, and of money tied up amounts to 25% of the cost of the item.

How many valves should Global produce in a single run?

Solution

$D = 500$ units/year

$C = \$1,100$

$S = \$460$

$P = 12$ units/day $\times$ 240 days/year $= 2,880$ units/year

$H = 0.25$

$$Q = \sqrt{\frac{(2D)(S)}{(H)(C)(1 - D/P)}} = \sqrt{\frac{2 \times 500 \times 460}{0.25 \times \$1,100 \times (1 - 500/2880)}} = 45 \text{ units}$$

45 units is an awkward number, requiring less than a normal shift. Therefore, choose as a lot size 48 units, an even 4 days of production.

The minimum cost at optimum Q is

$$500 \times \$1,100 + \sqrt{2(500)(460)(0.25)(\$1,100)\left(1 - \frac{500}{2,880}\right)}$$

$$= 550,000 + 10,224$$

$$= \$560,224$$

The cost to make in lots of 48 units is

$$500 \times \$1,100 + \frac{500}{48}(460) + \frac{1}{2}(48)(0.25)(\$1,100)\left(1 - \frac{500}{2,880}\right)$$

$$= 550,000 + 10,245$$

$$= \$560,246$$

The total cost curve is fairly flat in the region of the optimum Q and small changes, such as from 45 to 48, result in only minor cost changes.

A situation often encountered in a production environment is where a single facility or piece of equipment can be set up to produce several different products. In this situation it is usually best to have a fixed product sequence. That is, the machine produces a run of product A, then B, then C, and so on until the end product, where it then repeats the sequence. The fixed sequence makes scheduling easier and also permits the selection of low cost setups. For example, it may be more economical to go from product A to C, than from A to B. In this situation the lot sizes for all the products must be calculated together. To calculate them separately using the individual lot size formula will result in no fixed sequence, plus stockouts, since the machine will be busy with one product when it is needed for another.

The lot size formula for m products to be done in sequence on a single facility is developed as follows. H (holding cost) is common to all products N (number of setups per year) is common to all products; therefore, $Q_i = D_i/N$.

$$\text{Total cost} = \sum_{i=1}^{m} D_i C_i + N \sum_{i=1}^{m} S_i + \frac{H}{2N} \sum_{i=1}^{m} C_i D_i \left(1 - \frac{D_i}{P_i}\right)$$

$$\frac{d\,(\text{total cost})}{dN} = \sum_{i=1}^{m} S_i - \frac{H}{2N^2} \sum_{i=1}^{m} C_i D_i \left(1 - \frac{D_i}{P_i}\right)$$

$$\text{Optimum } N = \sqrt{\frac{H \sum_{i=1}^{m} C_i D_i (1 - D_i/P_i)}{2 \sum_{i=1}^{m} S_i}}$$

$$\text{Minimum cost} = \sum_{i=1}^{m} D_i C_i + \sqrt{2H \sum_{i=1}^{m} S_i \sum_{i=1}^{m} C_i D_i \left(1 - \frac{D_i}{P_i}\right)}$$

Example 8.14

Problem: On the same machine as Example 8.13, Global Industries also makes pipe fittings and globe valves. Calculate the lot sizes for each product.

Data: H = 0.25.

	D (units/year)	C	S ($/setup)	P (units/year)
Block valves	500	$1,100	460	2,880
Pipe fittings	1,000	375	200	2,400
Globe valves	250	1,500	700	720

Solution

$$N = \sqrt{\frac{0.25[500 \times 1,100 \times (1 - 500/2,880) + 1,000 \times 375(1 - 1,000/2,400) + 250 \times 1,500(1 - 250/720)]}{2(460 + 200 + 700)}}$$

$$= \sqrt{\frac{0.25[454,514 + 218,750 + 244,792]}{2,720}}$$

$$= \sqrt{\frac{0.25(918,056)}{2,720}} = \sqrt{\frac{229,514}{2,720}} = \sqrt{84.38}$$

$$= 9.186$$

Block Valves

$$Q = 500/9.186 = 54 \text{ units}$$

Pipe Fittings

$$Q = 1,000/9.186 = 109 \text{ units}$$

Globe Valves

$$Q = 250/9.186 = 27 \text{ units}$$

Note: Because of the interaction, the optimum Q value for the block valves changed from 45 to 54 units.

$$\text{Minimum cost} = (500 \times \$1,100 + 1,000 \times \$375 + 250 \times \$1,500)$$
$$+ \sqrt{2,720 \times 229,513}$$
$$= \$1,300,000 + \$24,985$$
$$= \$1,324,985$$

8.12 SUMMARY

Break-even analysis is a tool to assist the company in the evaluation of the level of sales required to break even either with or without a profit. In using the concepts of fixed costs, variable costs, and contribution margin, the company can evaluate whether to start production of a new product, enter a market, or evaluate a shutdown price.

Break-even analysis is very useful in evaluating alternatives. Alternatives such as high fixed costs with low variable costs versus low fixed cost and high variable costs can be evaluated within the given sales range and the company can then make the best decision for their needs.

In the latter part of the chapter the tax implications are evaluated with the effect they will have on a required level of sales. Section 8.10 is the evaluation of the nonlinear case and examination of those studies that require more accuracy and detail than the constraints imposed by the linear case.

Break even is a useful tool to evaluate the profit provided by individual products within the marketplace and the profit that may be generated by the acquisition of other products.

PROBLEMS

8.1 Company A is going into the business of digging ditches. The company will have a fixed cost of $17.98 for a shovel and pays an experienced operator $9.20 per hour. The company currently charges $11.00 per hour for the service.

a. Calculate the hours of operation required to break even.

b. Now assume that a $5.00 profit is required. Calculate the hours required to break even.

8.2 Global Industries annual sales volume is estimated at 100,000 units. The annual fixed costs are $100,000 and variable operating and maintenance costs are $5.00 per unit. Other annual fixed charges are insurance and utilities $3,000, income tax $4,000, and profit $20,000. Determine the selling price.

8.3 You are the manager of the local water treatment plant with an annual budget requiring revenues of $8,500,000. In the budget is included $500,000 (profit) to be returned to the city's general revenue fund. The average sales are 60,000,000 gallons per day. If a program is initiated to soften the water, a $300,000 fixed cost per year will be required, which will result in a 10% increase in sales and revenue. How much can be spent on additional chemicals and still maintain exactly the $500,000 going back into the city's general revenues?

8.4 You are a marketing consultant for a local brewery. An upgrade is required in one of your popular lines of beer. The contribution margin is now $1.00 per case. Current sales are 216,666 cases per month. The cost of the upgrade is $2,275,000. If the current contribution margin is to be maintained, a price increase is required. The price increase is expected to reduce sales by 15% immediately but will recover 1% per month over the next 20 months. After 20 months the sales are expected to remain at 105% of the pre-upgrade sales volume. How many months will it take to recover the cost of the upgrade and the lost sales during the lower demand?

8.5 Apex Plastics manufactures egg cartons. They currently have two injection machines, each with a fixed cost of $40,000 per year, and variable costs to produce an egg carton is $0.25. Each machine has a production capability limit of 500,000 cartons per year. If more machines are added, the maximum sales potential is estimated at 2,300,000 units per year. Make all calculations on a before-tax basis. The current sales price of the cartons is $35.00 per 100 cartons.
a. How many machines should be added, and what would the profit be?
b. What would the profit be if an additional unit were added?

8.6 Old East Fabricators has realized the potential for metal siding sales in the local area. A siding roll-forming machine costs $100,000, and material costs and labor = $1.00 per square foot. The siding sells for $1.25 per square foot. Assume that the life of the machine is 1 year, sales are unlimited, and the production capability limit of the roll former is 370,000 square feet per year.
a. Should the firm enter the business, and what would the profit be on one machine? On ten machines?
b. Now assume that the second machine can be purchased for 90% of the first and a further saving of $2,000 on each additional unit

up to a limit of 10 machines. How many units are required to break even?

8.7 A company is evaluating two different machines for producing the same product. Current annual sales are 100,000 units. Machine A has a first cost of $2,000,000 and a salvage value of $100,000 in 5 years. Machine B has a first cost of $600,000 and a salvage value of $100,000 in 6 years. The variable labor and materials per item for machines A and B are $4.50 and $8.00, respectively. Annual overhead costs for the company are $75,000. Use the straight-line method (equation 4.2) with an income tax rate of 31% and $i = 12\%$. Determine the sales price required for both machines.

8.8 Tell All Corporation sells portable air conditioners. Owing to economic conditions and high energy costs, sales have been lagging. The annual fixed costs with profit and tax are $200,000, with variable costs per unit of $200. The current selling price is $395 per unit.
a. Calculate the level of sales at the break-even point.
b. With sales at the break-even point the company has two choices available. Initiate an advertising campaign costing $25,000 and increase sales by 20%, or sell to a foreign distributor 400 units at $100 off full retail price but have added expenses to export of $10,000. Which of the two alternatives will return the greater profit to the firm?

8.9 ABC Corporation produces four products, A1, A2, A3, and A4. The costs are as follows:

	A1	A2	A3	A4
Labor	$ 4.00	$ 5.00	$ 8.00	$ 3.00
Material	6.00	4.00	2.00	7.00
	$10.00	$ 9.00	$10.00	$10.00
Space	$ 3.00	$ 0.50	$ 4.00	$ 0.75
Other overhead	5.00	6.00	4.00	4.50
Total	$18.00	$15.50	$18.00	$15.25

ABC is short of space and can use all the available space that becomes free. XYZ Company has made a bid of $12.00, $10.00, $13.50, and $12.00 to produce products A1, A2, A3, and A4, respectively. Which products, if any, should be produced by XYZ Company for ABC Company? (Other overhead remains under either alternative.).

8.10 A company produces products A, B, and C. The labor costs for the firm are $10.00 per hour and materials are $25.00 per square meter.

	Cost per unit		
	A	B	C
Materials (square meter)	0.25	0.65	0.59
Labor hours	0.10	0.50	0.50
Warehouse space costs	$1.00	$2.00	$3.00
Production equipment	$0.25	$0.50	$0.45
Overhead costs	$1.00	$2.00	$0.50

a. The selling price of the three products is $10.00, $27.00, and $25.00, respectively. Determine the contribution margin for each product assuming materials and labor are the only variable costs.

b. A local jobber has made the following offer to produce the products for $9.00, $20.00, and $23.00. Assuming all fixed costs remain, which of the items should the jobber produce?

c. Warehouse space and production equipment time can be used by other products at the going in-house rates. However, the overhead costs remain. Which of the items should now be produced by the jobber?

8.11 Acme Industries produces small construction equipment. The current value of their production equipment is $560,000 with an expected salvage value of $111,000 in 7 years. Annual administration charges are $80,000; labor and material, $43,000 per year; insurance and utilities, $5,000 per year; maintenance contract, $3,000 per year; depreciation is by the straight-line method (equation 4.2); $t = 33\%$; and $i_a = 20\%$. Estimated sales are 300 units. Determine the sales price at break-even.

8.12 The Big Mouth Advertising Company has approached Acme Industries of Problem 8.11 with a proposal that they can increase sales by 15% for the low annual charge of $29,995. Should Acme accept the offer? (Only labor and material costs will increase for Acme.)

8.13 Mal-Trust Trust Company is building a new 20-floor office tower. The capital cost of the tower is $20,000,000 and after 20 years the expected salvage value is $15,000,000. Operating costs per year are maintenance $35,000, insurance $20,000, and utilities $45,000. Assume that the depreciation for income tax is straight line (equation 4.2) and the tax rate is 41%. The square footage of the building is 5,000 ft^2 per floor. The current going rate for office space rent is $41.12 per square foot per year. Calculate the rate of return for the investment assuming 100% occupancy.

8.14 Your firm has a revenue curve of $-0.35x^2 + 4x$ and a total cost curve of $0.15x^2 + 0.33x + 4$, where x is the level of sales. Determine the point of maximum profit, the amount of profit at that point, and by how much profit will fall if sales decrease by 5%.

8.15 Air Rocky Ride is considering purchasing a new Boeing 767 for $28,000,000. This plane would be used only on a new route. The demand for the flight is estimated at 1,000 flights per year to be $-1.75x + 400$ (where $x =$ dollar cost of each ticket). The variable costs are mainly fuel expenses but include other costs, such as booking and ticketing expenses, and are estimated at $21,000 per flight. The life of the asset is estimated at 20 years with zero salvage value. Do not consider interest and all calculations are before taxes.
  a. Find the average cost of the fare per passenger which will result in the maximum profit for the firm.
  b. Calculate the average profit per year.

8.16 Engineers designing a bridge are trying to determine the most economical span length. The costs of the abutments and deck do not change with span length, but the cost of the superstructure girders and the number of piers do.

   Given:

   Weight of steel girders in metric tons per span $= 0.8 \times$ (span in meters)^2

   Length of bridge $= 350$ meters

   Cost of erected steel $= \$150$ per metric tonne

   Cost of one concrete pier $= \$400,000$

   Determine the economical span length.

8.17 a. The Classyhome Manufacturing Company manufactures a line of designer wastebaskets. Annual sales are constant at 10,000 baskets per year. The cost of the baskets is $5.30 per unit and the company can manufacture baskets at a rate of 400 per day. It costs $300 to set up and adjust the machinery to make baskets. If the company works 250 days per year and the costs of holding inventory are estimated to be 24%, what production lot size should they use?
   b. The Classyhome Company has decided to have its wastebaskets made by a supplier who has agreed to provide them at $5.30 per unit. This means no setup costs, but there is a delivery van cost of $50.00 per order. The supplier is in a different city and so, to save on delivery costs, the entire order will be shipped as a single

unit. This is equivalent to having a production rate of infinity. What lot size should the company order from the supplier?

8.18 a. The Flipfloppy Disk Company is planning its yearly production schedule. One of its products is the 5¼-inch DSDD floppy disk, which has a production cost of $0.85 per unit. Monthly sales are estimated to be 2,000,000 units. The equipment costs $7,000 to set up and is capable of producing disks at a rate 20,000 disks per hour. If the inventory holding cost is $0.01 per disk per month, and the company works 8 hours per day, 21 days per month, what size production lots should be scheduled?

b. On the same machinery the company can also make 5¼-inch SSDD, 3-inch SSDD, and 3-inch DSDD disks. The costs and rates are given in the following table.

Product	Setup Cost	Unit Cost	Demand (units/month)	Production (units/hour)
51/4-in. DSDD	$ 7,000	$0.85	2,000,000	20,000
51/4-in. SSDD	5,000	0.75	1,000,000	26,000
3-in. DSDD	14,000	0.50	1,500,000	18,000
3-in. SSDD	9,000	0.45	400,000	24,000

The company has decided to run the machine 12 hours per day, 21 days per month. Storage costs are $0.01 per disk per month. Calculate a production schedule for the company.

8.19 In a 1-kilometer power transmission line the energy loss in kilowatt-hours due to resistance is equal to $I^2R \times$ number of hours $\div$ 1,000. Given the following data, find the minimum-cost cross-sectional area for a copper conductor that operates 24 hours per day, 365 days per year.

Resistance (R) of a copper conductor of area 1 mm² and length 1 km equals 20 ohms and is inversely proportional to the area

Installation cost = $30,000.00 + $1.80 × (weight of copper in kilograms)

Life = 20 years

Salvage value = $0.95 × (weight of copper in kilograms)

Weight of copper conductors = 8.9×10^{-4} kg/mm³

Cost of lost energy = $0.05/kilowatt-hour

Current = I = 500 amperes

MARR = 15%, $t = 0$

CHAPTER NINE

The Impact of Commitment on Investment Decisions

There is a measure of risk associated with commitment to any investment. One must realize that most investment decisions are irreversible to some degree. That is, an investment involves commitment and some risk that there will be a cost associated with owning the asset even for a very short period of time. You may purchase a new automobile today for $12,000 and find that 1 week from today on having to sell this vehicle that its market value has dropped by $1,000. If we momentarily ignore any tax effects due to the sale, the value for the automobile to be used in any decision to replace the automobile is $11,000.

The more specialized the function of the asset, the more risk involved with a commitment. General-use type of equipment, such as vehicles and standard machine shop equipment, is normally readily disposed off without a major loss in book value to the company. However, very specialized equipment may have virtually zero market value even if disposed off shortly after the acquisition date.

Technological change can also have a significant impact on the market value of an existing asset. Many of us have been somewhat reluctant to admit that our treasured calculator, purchased less than a year ago for $400, may now have a market value of only $75. Similarly, the analyst who recommended the purchase of an in-house computer system a few months ago for $100,000 now realizes that the market value of this system today may be only $25,000. However, the company that is going to survive and grow in a competitive marketplace is the company that is continually on the search for a product or machine that is capable of improving productivity and thereby profits.

9.1 SUNK COSTS AND FIRST COSTS

The basis of decision analysis in this text is that money should be invested so that it earns at least MARR. Furthermore, the focus is always on the future, and the analysis is directed toward how to obtain the maximum from that which is currently owned. Past mistakes, or successes, are relevant only insofar as they affect the future.

9.1.1 Sunk Costs

Suppose, for example, that several years ago you purchased a high-quality, Swiss-made, clockwork wristwatch for $1,500. Friction has taken its toll and the watch now loses 1 or 2 minutes per day, so you take it to the repair shop to be adjusted. You receive an estimate that the cleaning and adjustment will cost $150 and the repairs will be guaranteed for 1 year. In the same shop it is possible to purchase a new electronic wristwatch, accurate to within 2 seconds per month, guaranteed for 2 years, for $129.95. If the objective is to have an accurate timepiece at the lowest cost, the correct decision is to dispose of the old watch for whatever salvage value that can be obtained and purchase the electronic watch. The $1,500 originally spent on the old watch is not relevant to the decision. It is a sunk cost. That is, the money is gone and no present action can recover it.

Sunk costs represent past actions. They are the result of decisions made in the past. Because of the very human tendency to explain, justify, rationalize, or blame, they often figure prominently in discussions relating to present decisions. In the example above, it would be natural for the repair shop to suggest that for only $150 you receive a "good as new" $1,500 watch. The reality, however, is that for $150 you receive a watch that may, or may not, perform as well as a $129.95 watch.

Companies, organizations, and governments often become involved in sunk-cost arguments. Much time, effort, and money is spent trying to rectify an uneconomic situation because of money already invested. A good example is the supersonic transport airplane, the SST. The American government was able to cancel additional investment in the SST when it became obvious that for environmental reasons the project could not be a success.

On cancellation the American government had $800 million sunk into development expenses. The British and French governments were not able to ignore the sunk cost arguments and so developed the Concorde, which is an engineering marvel but an economic failure.

Because it is not possible to predict the future, costly mistakes can be made. During the 1960s when petroleum was inexpensive, many large plants were built using gas and oil as fuel and feedstock. The petroleum price increases of the 1970s made these decisions appear unwise. However, the rational approach, taken by most companies, is not to worry about previous investment decisions that cannot be changed. They prefer to devote their energies to organizing and utilizing their resources and capabilities as best they can to ensure that current and anticipated projects are successful.

9.1.2 First Cost

Closely allied to sunk cost, and equally misleading, is the concept of an asset that has "paid for itself." Consider, for example, a machine shop which purchased a lathe 25 years ago. The machine has been well maintained and has given dependable service over the quarter century. On the books of accounts it has been depreciated to a zero book value. The company is considering the manufacture of a certain component on the lathe and is doing a revenue requirements calculation in accordance with equation 6.3.

$$AER = AEOC + [P(A/P, i_a, n) - NSV(A/F, i_a, n) - t(AED)]/(1 - t)$$

The question is: What value should be used as a first cost P for the lathe?

Again, it is not the past costs or services that are relevant. The book value of the machine is zero, but this does not mean that the machine has no value. Decision making deals with present alternatives, so these are what must be evaluated.

One obvious alternative is to sell the machine. This would generate a certain amount of cash that could then be invested by the organization at MARR or greater. Retention of the machine implies forgoing this alternative. Thus, to be consistent, if the machine is retained, it must generate as much revenue as can be generated by its equivalent cash value. This concept is included in the analysis by using the cash equivalent of the owned machine as the value of P in equation 6.3. This cash equivalent value is referred to as the *first cost* of an owned asset, and is the value to use in revenue requirement and capital recovery calculations. Therefore, it is necessary to have a method of establishing the true market value to be used as a first cost for all assets presently owned or being considered for purchase in an economy study. One approach is referred to as the *outsider viewpoint* by Thuesen and Fabrycky.*

9.1.3 The Outsider Viewpoint

Viewing an investment decision from the standpoint of not owning an existing asset ignores any tax effects that may arise due to the sale of the asset and thus only applies to a before-tax analysis. However, using this framework in which to evaluate the alternative eliminates any gains or losses associated with the original purchase of the "asset" and thus any need to consider a sunk cost. Therefore, in order to establish a valid first cost to be used in an economy study on a before-tax basis, an example using the outsider viewpoint is presented.

Example 9.1

Problem: Machine A was purchased for $10,000 a year ago and at that time was assumed to have an economic life of 5 years and an estimated salvage value of $2,000. Its market value today is $6,000 and operating costs are $8,000 per year. Machine B can be purchased for $12,000, has an estimated economic life of 4 years, and an estimated salvage value of $4,000. Operating costs are estimated to be $5,000 per year. The before-tax interest rate is 20%. Should the new machine be purchased?

Solution: The outsider viewpoint assumes that the company *does not own* machine A. Therefore, the first cost for machine A is its current market value, $6,000.

Machine A

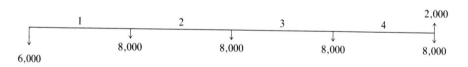

Machine B

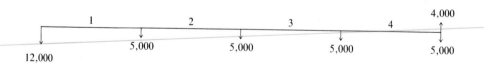

* G. J. Thuesen and W. J. Fabrycky, *Engineering Economy,* 6th ed., Prentice-Hall, Inc., Englewood Cliffs, N.J., 1984, p. 239.

$$AER = \$6,000(A/P,\ 20\%,\ 4) + \$8,000 - \$2,000(A/F,\ 20\%,\ 4)$$

$$= \$6,000(0.3863) + \$8,000 - \$2,000(0.1863)$$

$$= \$2,318 + \$8,000 - \$373 = \$9,945$$

$$AER = \$12,000(0.3863) + \$5,000 - \$4,000(0.1863)$$

$$= \$4,636 + \$5,000 - \$745 = \$8,891$$

Conclusion: Purchase machine B, because it has the lower revenue requirements.

The outsider viewpoint clearly establishes, on a before-tax basis, that the present market value of the existing asset is the value to be used as a first cost in an economy study.

In an economy study involving replacement, the trade-in alternative almost always exists and represents a feasible alternative to be carefully considered. However, the trade-in value offered for the existing asset and the purchase price of the new asset under consideration are not usually true market values for either contender. A classical example occurs when you start shopping for a new automobile. The salesperson, using good psychology, is usually prepared to give you top value for your old automobile as a trade-in on a new car.

Assume that you have advertised your present car and have established that the best price available is $3,000 (true market value). However, the salesperson has offered you $4,000 as a trade-in value on a new car priced at $10,000. This offer may be difficult to resist, but before accepting you should establish the price of the new car without any trade-in. The true market value of the new car is probably somewhat less than $10,000 and may be less than $9,000.

Example 9.2

Problem: Last week you purchased a beautiful new automobile for $12,000. Today you have started a new sales job that requires you to have a station wagon. You neither need nor can afford both units. Your supervisor has told you that the company is prepared to split any loss that you incur in selling your present automobile. However, you have to establish the true market value of the automobile. The following alternatives are open to you:

Alternative A:

Trade the car in on a station wagon. Trade-in value equals $11,000 on a station wagon with a purchase price of $14,000.

Alternative B:

Sell the car outright. The best price you have found is $10,000. The purchase price of the station wagon without a trade-in is $13,000.

With respect to the two alternatives determine:

1. What should the company pay you on disposing of your automobile?
2. What is your additional cash outlay, ignoring any tax effects due to a loss on disposal?

   Solution: For alternative A, the company reimburses you for

$$($12,000 - $11,000)0.50 = $500$$

Your additional cash outlay is

$$$14,000 - $11,000 - $500 = $2,500$$

For alternative B, the company reimburses you for

$$($12,000 - $10,000)0.50 = $1,000$$

Your additional cash outlay is

$$$13,000 - $10,000 - $1,000 = $2,000$$

Outsider Viewpoint: The outsider viewpoint, which approaches the problem using alternative B, clearly eliminates any confusion as to which is the proper method of analysis. The existing automobile can be purchased for $10,000, which represents the true market value. Therefore, the company should reimburse you for

$$($12,000 - $10,000)0.50 = $1,000$$

Your additional cash outlay is $2,000.

9.2 THE IMPACT OF TAXATION ON A DECISION TO RETAIN OR REPLACE AN ASSET

When taxes are considered in an analysis to replace, or just sell or retain an asset, it is essential to consider any gain or loss on disposal and the income tax effect of this gain or loss. Two questions arise:

1. What P value (referred to as P_v) should one use as a first cost for capital recovery purposes in the analysis?
2. What P value (referred to as P_t) should one use as a basis to calculate depreciation allowance for tax purposes?

Under certain conditions the P_v value may differ from the P_t value used in an economy study. For example, on large mega projects expenditures may be made over a period of years before the facility becomes operational. For economy study purposes all cash flows are usually adjusted to the operational date as a zero date. Under these circumstances *interest lost during construction* (ILDC) would become part of the P_v value used in the economy study; however, the P_t value for depreciation allowance purposes would not include the ILDC. In addition, when we are considering the sale of an asset, the book value usually differs from the market value. The book value represents the P_t value, whereas the market value represents the basis to determine the P_v value.

Smith* suggests that the P_v value to be used in the analysis must consider any capitalized costs to make the asset operational, the present market value of the asset, and the tax effects resulting from a market value that differs from book value for the asset.

$$P_v = \text{cost to make operational} + \text{market value forgone} \qquad (9.1)$$

$$+ \text{ income tax considerations}$$

Example 9.3

Problem: Global Industries own a lathe that was damaged in a fire. The economic analysis to determine repair or replacement requires that a P_v value be determined. The lathe can be sold in an "as is" condition for $25,000. The capitalized cost to repair the lathe is estimated to be $5,000. Assume that the sale would lead to a tax saving of $10,000.

Solution

$$P_v = \text{cost to make operational} + \text{market value forgone}$$

$$+ \text{ income tax considerations}$$

$$= \$5,000 + \$25,000 + \$10,000$$

$$= \$40,000$$

Example 9.4

Problem: Assume that you have had the misfortune to ruin the motor in your company automobile and a $2,000 expenditure (capitalized) is required to place the car in operating condition. The car has a market value of $3,000 in an "as is" condition. The book value of the car is $5,000 and assume

---

* Gerald W. Smith, *Engineering Economy: Analysis of Capital Expenditures*, 3rd ed., The Iowa State University Press, Ames, Iowa, 1979, p. 347.

that the loss on disposal amounts to a tax saving of $800. Calculate the P_v value to be used for your car in a replacement study.

Solution

$$P_v = \text{cost to make operational} + \text{market value forgone}$$
$$+ \text{ income tax considerations}$$
$$= \$2,000 + \$3,000 + \$800$$
$$= \$5,800$$

Example 9.5

Problem: Information pertaining to a presently owned asset to be used in a replacement cost study is as follows: original cost 4 years ago = $100,000, remaining book value today = $0, and market value today = $50,000. The capital cost allowance was based on the fast-write-off method. $i_a = 15\%$ and $t = 46\%$. Find the value P_v to be used in the replacement cost analysis.

Solution

$$\text{Cost to make operational} = \$0$$
$$\text{Market value} = \$50,000$$

Tax effects are as follows. The additional taxes in future years may be determined by spreading the recapture of $50,000 in capital cost allowance over the next 2 years, using the fast-write-off rates, and calculating the present equivalent of the additional taxes paid.

$$\text{Recaptured CCA each year} = \$50,000/2 = \$25,000$$
$$\text{Additional Taxes each year} = 25,000(.46) = \$11,500$$

```
                    11,500                    11,500
 ._____1___↑_____2_____↑
```

Present equivalent of additional taxes:

$$\text{PE} = +\$11,500(P/A, 15\%, 2)$$
$$= \$18,696$$
$$P_v = \text{cost to make operational} + \text{market value forgone}$$
$$+ \text{ income tax considerations}$$
$$= 0 + \$50,000 - \$18,696 = \$31,304$$

$31,304 represents the P_v value of the existing asset to be used in the economy study.

It is appropriate at this time to define two terms introduced by Terborgh:* defender and challenger. The *defender* is the asset that is presently owned and for which a replacement study is being considered. The *challenger* is the asset, model, or option that is being considered as the defender's replacement.

Example 9.6

Problem: The following information pertains to a replacement study to be undertaken by Global Industries. The defender (existing asset) was purchased 1 year ago. Assume that the economic life for both alternatives, as of today, is 9 years.

	Defender	*Challenger*
Original installed cost	$100,000	$150,000
Estimated salvage value in 9 years	0	0
Market value today	70,000	
Estimated annual operating costs	50,000	30,000

$t = 40\%$, MARR $= 15\%$, and $r = 0$. Use the declining-balance method and a CCA rate of 20%.

Solution: AER for alternative A will be solved by two methods.
Method 1:

$$\text{Loss on disposal} = BV - \text{selling price} = \$90,000 - \$70,000$$
$$= \$20,000$$

Determine the tax consideration due to this loss on disposal.
Present equivalent of tax savings in future years due to the loss on disposal:

$$\$20,000(td)/(i + d) = \$20,000)(0.22857) = \$4,571$$

P_v = cost to make operational + market value forgone

+ income tax considerations

$= 0 + \$70,000 + \$4,571$

$= \$74,571$

---

* George Terborgh, *Dynamic Equipment Policy,* McGraw-Hill Book Company, New York, 1949; George Terborgh, *Business Investment Management,* Machinery and Allied Products Institute, Washington, D.C., 1967.

Determine the present equivalent of all future capital cost allowance on the BV of $90,000.

PE of future tax savings due to capital cost allowance:

$$\text{BV}(td)/(i + d) = \$90,000 \ (0.22857) = \$20,572$$

PE of capital cost allowance in future years

$$= (\$20,572)/t = \$20,572/0.4 = \$51,429$$

$$\text{AER} = \text{AEOC} + [P(A/P, 15\%, 9) - \text{SV}(A/F, 15\%, 9)$$

$$\quad - t(\text{AED})]/0.6$$

$$= \$50,000 + [\$74,571(0.2096) - 0 - 0.4(\$51,429)0.2096]/0.6$$

$$= \$50,000 + (\$15,630 - \$4,312)/0.6$$

$$= \$68,864$$

Method 2:

$$\text{CTF(SV)} = 1 - (td)/(i + d) = 0.77143$$

$$\text{AER} = \$50,000 + [(\$70,000)(0.2096)(0.77143)]/0.6$$

$$= \$68,864$$

Using the CTF simplifies the calculation.

For Alternative B:

$$\text{CTF(P)} = 1 - \{td/(i + d) \ [(1 + i/2)/(1 + i)]\} = 0.7863$$

$$\text{AER} = \$30,000 + [\$150,000(0.2096)0.7863 - 0]/0.6$$

$$= \$71,202$$

Conclusion: Retain the defender.

Example 9.7

Problem: Last week the Deejay Corporation purchased a new milling machine for $100,000. Its salvage value today is $50,000. The salvage value in 10 years is zero. A new government contract received today means operating an additional shift on the milling machine, raising operating costs to $50,000 per year, or replacing the milling machine. An automatic milling machine can be purchased for $150,000 that will handle the job and has annual operating costs of $25,800 per year. The salvage value in 10 years is zero. Use the fast-write-off method. The economic life of both units is 10 years. $r = 0$ and $t = 50\%$. At what rate of return are these two investments equivalent? Should the automatic milling machine be purchased?

Solution: Try $i_a = 15\%$:

$$\text{Loss on disposal} = \$100,000 - \$50,000 = \$50,000$$

Spread the loss over 2 years using the fast-write-off method:

$$\text{CCA} = \$25,000 \text{ per year}$$

Tax Saving   12,500   12,500

$$\text{PE} = +\$12,500(P/A, 15\%, 2)$$

$$= \$20,321$$

$$P = 0 + \$50,000 + \$20,321 = \$70,321$$

	Defender	Challenger	C − D
$P_v(C - D)$	$-70,321	$-150,000	$-79,679
Operating costs	-50,000	-25,800	24,200
SV	0	0	0
CCA base	100,000	150,000	50,000

$$\text{PED} = \$12,500(P/F, 15\%, 1) + \$25,000(P/F, 15\%, 2)$$

$$+ \$12,500(P/F, 15\%, 3) = \$37,992$$

$$\text{PE}(C - D) = \$24,200(0.5)(P/A, 15\%, 10) - \$79,679 + 0.5(\$37,992)$$

$$= \$9$$

$$i_a = 15\%$$

Fifteen percent represents the rate of return on the incremental investment. Therefore, if MARR $\leq 15\%$, purchase the new machine; otherwise, retain the defender.

9.3 COMMITMENT TO AN INVESTMENT IN STAGES

Today there are many large projects throughout the world at some phase of completion. Many of these projects have been under study or construction for several years (e.g., tarsands, petrochemical plants, power dams, nuclear energy systems). The commitment to investment in these large projects occurs at several stages before the facility finally becomes operational. In these circumstances to simplify calculations all cash flows for an economy study should be moved to the operational date (which becomes the zero date for the analysis). The P_v value for economy study purposes under these circumstances will normally be different than the P_t value used to calculate capital cost allowance for tax purposes. Two terms need clarification before proceeding:

1. *Operational date:* represents the date the facility goes into operation. This date becomes the zero date for the analysis.
2. *Observation date:* represents the point in time that a decision has to be made as to whether the project should proceed or be abandoned.

Consider the following sequential decisions that result in an increasing degree of commitment.

Example 9.8

Problem: Global Petroleums is interested in building a plant to surface-mine tarsands. The company is evaluating the major material-handling equipment to use to bring the raw tarsand to the processing plant. The stages of evaluation and commitment considered are as follows:

Stage of Design	Time Span (years)
Conceptual design	2.0
Placing the order	0.5
Inspect and accept	3.0
Transport to the site	0.5
Assemble/install	0.5
Debug and test	0.5
	7.0

MARR $= 20\%$, $t = 50\%$, and $r = 0$. Assume that *the tax effect of any loss on disposal is realized in the year the loss occurs.* (In practice, the company may not have any profits until the plant becomes operational. However, assume that profits are generated from other operations of the company that can be used to offset any tax effects due to a loss on disposal.)

In actual practice, commitment will occur on a continuing basis over the total time span from the beginning of conceptual design through to the point when the equipment becomes operational. For simplicity we have assumed major commitments at specific points in time which represent the beginning of each stage of evaluation (all values are in millions of dollars):

-7.0	-5.0	-4.5	-3.0	-2.0	-1.5	-1.0	-0.5	0
50.0	100.0	100.0	70.0	50.0	5.0	15	10	
Conceptual design	Place order	Inspect	Inspect	Inspect	Transport	Assemble	Debug	Operational date

303

1. *Conceptual design:* The evaluation of alternatives such as draglines, bucket conveyors, tractor trailer units, conveyor systems, and so on. This stage represents many thousands of hours involved in travel, group discussions, and cash flow analyses before selecting an alternative and placing an order with a supplier. Assume that a 2-year time span is required to complete this stage. The estimated cost of the conceptual design is $50,000,000.

2. *Placing the order:* Assume that a decision is made to use draglines to remove the overburden and tarsand and that a conveyor system will be used to transport the tarsand to the processing plant.

   a. Estimated installed and debugged cost for three
      draglines $300,000,000

   b. Estimated installed and debugged cost for the
      conveyor system $50,000,000

   On placing the order for such specialized equipment, Global is committed to an initial investment of $100,000,000. This $100,000,000 represents a bond that is forgone if the order is canceled.

3. *Inspect and accept the product at the manufacturer's plant:* Inspection and commitment would actually occur at several stages of completion throughout the 3-year period to manufacture the equipment. Three inspection points and their associated commitment are shown on the cash flow diagram.

4. *Transport to the site:* Transportation to the site would occur at several stages (e.g., on completion of the first dragline it would be disassembled and transported to the site).

5. *Assemble/install:* Assembly of the draglines and conveyors would actually commence shortly after components begin to arrive on site.

6. *Debug and test:* A final sign-off and final payment is usually made after the completion of debugging and testing and the plant is placed on production. In our example the final commitment is indicated at -0.5 year.

Following are the stages of investment commitment:

Date[a] (years)	Stage of Commitment	Actual Cash Flow (millions)	P_v at Zero Date (millions)	Estimated Cumulative SV (millions)	SV at Zero Date
-7.0	Conceptual design	$ 50.0	$179.15	$ 0	
-5.0	Place order	100.0	248.80	0	
-4.5	Inspect and accept	100.0	227.10	100.0	$227.10
-3.0	Inspect and accept	70.0	120.96	135.0	233.28
-2.0	Inspect and accept	50.0	72.00	160.0	230.40
-1.5	Transport to site	5.0	6.57	160.0	210.24

-1.0	Assemble/install	15.0	18.00	150.0	180.00
-0.5	Debug and test	10.0	10.95	155.0	169.73
	Operational date	$400.00	$883.53	—	

[a] These dates are referenced to a date zero (operational date).

Problem: Determine P_v and P_t just prior to date -7.0 (observation date) on the time scale. At this point no money has been invested.

Solution: MARR $= 20\%$ and $t = 50\%$. Just prior to date -7.0 (observation date) no money has been invested. To proceed with an economy study we have to determine the P_v and P_t values. The logical point in time to determine these values is at the point that the facility becomes operational (zero date). Therefore, using our cash flow diagram we have to move all cash flows to zero date.

$$P_v = \text{cost to make operational} + \text{market value forgone}$$

$$+ \text{ income tax considerations}$$

In this case no investment has been made, so cost to make operational is our only concern.

$$P_v = \text{cost to make operational (millions of dollars)}$$

$$= \$50(F/P, 20\%, 7) + \$100(F/P, 20\%, 5) + \$100(F/P, 20\%, 4.5)$$

$$+ \$70(F/P, 20\%, 3) + \$50(F/P, 20\%, 2) + \$5(F/P, 20\%, 1.5)$$

$$+ \$15(F/P, 20\%, 1) + \$10(F/P, 20\%, 0.5)$$

$$= \$883,530,000$$

The government does not allow us to add interest into the base used for capital cost allowance.

$$P_t = \$50 + \$100 + \$100 + \$70 + \$50 + 5 + \$15 + \$10$$

$$= \$400,000,000$$

Problem: Assume that just prior to the commitment of $100,000,000 ($-4.5$ years $=$ observation date), associated with inspection number 1, Global Petroleums is considering a major change in design. What is the P_v value associated with this analysis? Assume that the tax benefits can be applied to other income in year 4.

Solution: We are now 4.5 years away from the plant going into operation. We have now invested $150,000,000, and if we are going to proceed after the inspection we have an immediate commitment of $100,000,000 and additional commitments as indicated on the cash flow diagram to make the plant operational. Assume that we are considering other alternatives at this time. Our procedure, once again, is to determine P_v and P_t values as a basis for our economy study.

$$P_v = \text{cost to make operational} + \text{market value forgone}$$
$$+ \text{ income tax considerations}$$

The cost to make operational requires that we move all remaining capital investments to zero date. The cost to make operational is

$$\$100(2.271) + \$70(1.728) + \$50(1.440)$$
$$+ \$5(1.314) + \$15(1.200) + \$10(1.095)$$
$$= \$455.58$$

Income tax considerations are based on any gain or loss on disposal if we decide to abandon the project. Current tax legislation will have an effect on this calculation. Our assumption is that the loss incurred can be written off in the year that it occurs, in this case year 4 on our cash flow diagram. We now move the tax saving due to this loss on disposal from year 4 to zero date. The income tax considerations are

$$\$150(F/P, 20\%, 4)0.50 = \$150(2.074)0.50$$
$$= \$155.55$$
$$P_v = \$455.58 + 0 + \$155.55$$
$$= \$611,130,000$$

9.4 SUMMARY

The outsider viewpoint can be used when a before-tax analysis is under consideration. This viewpoint is helpful in determining the cash flows associated with an existing asset under study.

On an after-tax analysis the tax implications must be considered. Equation 9.1 is used in arriving at a first cost for an existing asset under these conditions.

When commitment to an investment occurs in stages the analysis becomes somewhat more complex. We suggest that, for convenience, all cash flows associated with the capital investment be brought to an operational date. Income tax considerations associated with a loss on disposal or recaptured depreciation will come into play in the year they occur if the company is presently in business. If a new company has been formed for this particular investment, income tax considerations will be applied after the facility becomes operational.

Commitment to an investment represents a sunk cost. The percentage of this sunk cost that is recoverable on disposing of the asset will vary depending on a number of factors, such as specialization of use and technological change. These factors are critical considerations when the asset is purchased and enter into the asset value on disposal.

PROBLEMS

9.1 Assume that you purchased an automobile last week for $7,000. This week circumstances dictate that you change this car for a pickup truck. The salesperson is prepared to give you $6,000 for your car as a trade-in on the pickup, which he has quoted at $9,000. Without a trade-in the pickup can be purchased for $8,000. The net salvage value on the market for your car is only $5,000. What is the P_v value to be used in an economy study using the outsider viewpoint for (ignore any tax effects):

    a. The old car?

    b. The new truck?

9.2 Last week ABC Consultants purchased an electric typewriter for $2,000 and its salvage value today is $1,500. This week the company has decided to replace the typewriter with another model. The retail value of the newer model is $2,500, but the discount price is $2,000. The annual maintenance cost is $150 for the defender and $50 for the challenger. Assume that the economic life for both models is 5 years. The salvage value in 5 years equals 10% of the original cost for both models. Use the declining-balance method and a 30% CCA rate. $r = 0$, $t = 50\%$, and MARR $= 15\%$.

    a. What is the P_v value to be used for the defender on a before-tax basis?

    b. What is the P_v value to be used for the defender on an after-tax basis?

    c. Should the defender be replaced?

9.3 The D-J company has initiated a study to determine whether or not a currently used piece of equipment should be sold in favor of a new model. The following records and estimates have been obtained with regard to the present equipment. Use the declining-balance method and a CCA rate of 30%.

Original cost 2 years ago	$25,000
Gross resale value today	7,000
Dismantling, transportation, and selling costs	1,500
Salvage value	5,500

$r = 0$, $t = 40\%$, and MARR $= 15\%$. What is the first cost, P_v, to be used in the evaluation of costs associated with a decision to retain or sell the machine?

9.4 Reynolds Corporation is considering the replacement of a three-piece can-making system with a two-piece can-making system for the manufacture of beverage cans.

Information	Defender	Challenger
Original cost 2 years ago	$200,000	
Replacement cost today		$250,000
Remaining economic life	5 years	5 years
Salvage value 5 years from today	$25,000	$150,000
Tax rate	50%	50%
MARR	15%	15%
Operating cost per year	$50,000	$25,000

Use the fast-write-off method.

a. If the market value is equal to the book value, what is the P_v value to be used in calculating the annual equivalent revenue requirements for the defender?

b. If the market value is $100,000, what is the P_v value to be used in calculating the annual equivalent revenue requirements for the defender?

c. Assuming the same condition as in part (b), an overhaul cost of $30,000, which is capitalized, is required to make the three-piece can-making system operational. What is the P_v value to be used in calculating the annual equivalent revenue requirements for the defender?

d. If the conditions for the defender are the same as in part (b), should the defender be replaced?

9.5 P.K. Electronics has an integrated-circuit (IC) chip-producing machine that is 3 years old. New technology has made the old machine inefficient in operation. The salvage value of the old machine today is $250,000. The book value is $400,000. The new machine under consideration costs $1,000,000 and $10,000 for transportation and installation. Use the declining-balance method and a CCA rate of 20%. $t = 48\%$ and MARR $= 15\%$. Economic life for the old as well as the new machine is 6 years with salvage values of $200,000 and $350,000, respectively. What are the P_v and P_t values for the two machines?

9.6 A real estate project requires capital expenditures of $100,000, $300,000, and $150,000 for this year, next year, and the year after, respectively. The project becomes operational at the end of the third year. Assume cash disbursements at the beginning of each year and immediate tax benefits on termination. MARR = 15% and t = 40%.
  a. Find P_v and P_t as observed now, before any capital disbursements.
  b. Find P_v just prior to the last expenditure. Assume a salvage value of $250,000 at that stage. No depreciation allowance (CCA) has been taken to date.
  c. If t was 50% instead of 40%, how would P_v in part (b) change?

9.7 Over the last 3 years, one organization has made year end capital expenditures for a facility as follows:

Year	Amount
1	$ 75,000
2	180,000
3	240,000

Because of the nature of this facility, no tax benefits are applicable. Assume that the work is done by a contractor who is paid at the end of each year. A serious review of the expected cash flows from this project is now under way. To complete the facility, an additional $300,000 is required next year. The project becomes operational at the end of the fourth year. At its present stage, the development has a market value of $300,000. BTRR = 20%.
  a. What was the P_v value at the start of the project, that is, before a contractor was engaged?
  b. In view of management's decision to make a reevaluation, what is the P_v value to be used now?

9.8 The city council has completed work on a 2-kilometer stretch of road, paying its contractor $40,000. According to a new study released recently, the expected usage of the road has changed substantially from the figures when the design work was done. This means spending an additional $10,000 a year in operating costs. Alternatively, to keep the repair costs at the same level as earlier, the road has to be resurfaced. This will cost $75,000. BTRR = 10%.
  a. Find the P_v value today for each alternative.
  b. Evaluate the AER for each alternative, assuming an economic life of 25 years.
  c. What economic life in part (b) would make the two alternatives equivalent?

9.9 A manufacturer has made the following estimates with regard to construction expenditures over a 5-year period to build an additional production line. In view of the uncertainties with regard to the market behavior and the performance of the new technology line, the chance of abandoning the project at any stage before the operational date cannot be ruled out. MARR = 15% and t = 48%. Any tax savings or losses can be written off in equal amounts over 2 years.

Year	Investment Beginning of Year	Year-End Salvage Value
1	$ 500,000	$ 0
2	700,000	0
3	1,200,000	800,000
4	5,000,000	3,800,000
5	600,000	3,600,000

The operational date is at the end of year 5.
a. What is the P_v value prior to the commencement of the project? What is the P_t value?
b. If a study is initiated at the end of year 3 (just prior to investing the $5,000,000), what is the P_v value to be used for the production line under construction?

9.10 One year ago, Western Equipment Ltd. purchased a machine for $100,000 (includes installation). The market value of the machine is $35,000. Due to rising labor costs, the company has been forced to reconsider the merits of that machine in view of a less-labor-intensive design. A trade-in of the existing machine on the new design requires an additional $160,000. The machine can be purchased for $200,000 with no trade-in. Use the declining-balance method and a CCA rate of 30%. SV = 0, r = 0, t = 50%, and MARR = 20%, OC (challenger) = $10,000/year. The economic life for both machines is 8 years. Should the company trade in the old machine or sell it outright and buy the new machine? (Hint: Compare AER of the challenger using the trade-in with AER of the challenger with a direct purchase of the new machine).

9.11 The Willet Corporation purchased a duplicating machine 1 month ago for $30,000. Today there is a new model available which sells for $50,000. The old machine has a resale value of $10,000 or can be traded in on the new machine. If traded in on the new machine, a cash outlay of $35,000 will be required for the new machine. Use the declining-balance method for CCA purposes and a 30% rate. Assume that the economic life of both machines is 10 years and the salvage value in 10 years is negligible. Operating costs for the defender =

$20,000 per year. Operating costs for the challenger = $10,000 per year. $r = 0$, $t = 40\%$, and MARR = 15%.

a. Would you sell the old machine or trade it in on the new one? Why?

b. Should the new machine be purchased?

9.12 You purchased dairy equipment 1 month ago for $50,000. Today there is a new model available which has a true market value of $100,000. The old equipment has a resale value of $30,000. Assume that the economic life for both alternatives is 10 years. Use the declining-balance method and a 20% CCA rate. The new equipment will reduce operating costs by $15,000 per year. Assume zero salvage value. $r = 0$, $t = 46\%$, and MARR = 15%. Should the new equipment be purchased?

9.13 Last week the Concrete Mix Corporation purchased some new equipment for $100,000. Its salvage value today is $50,000. The salvage value in 10 years is zero. A new government contract received today means operating an additional shift on this equipment raising operating costs to $59,200 per year or replacing the equipment. Automatic equipment can be purchased for $150,000 that will handle the job and has annual operating costs of $30,000 per year. The salvage value in 10 years is zero. Use the declining-balance method and a CCA rate of 20%. $r = 0$ and $t = 50\%$. Should the equipment be replaced?

9.14 For a particular project, the tax effects of any "loss on disposal" lags the transaction by 1 year. $t = 46\%$ and MARR = 15%. Capital expenditures are at the beginning of the period and salvage values represent the salvage value just after the investment is made.

Date	Expenditure	Actual Salvage	Stage of Commitment
−3.0	$ 2,550	$ 0	Evaluation
−2.5	6,480	0	Order guarantee
−2.0	4,120	0	Transportation
−1.5	80,600	52,400	Inspection
−1.0	11,250	45,200	Installation
−0.5	8,375	40,400	Debug and training

a. What are the P_v and P_t values just prior to date −3.0?

b. Find P_v just prior to date −2.0.

c. Find P_v just after date −1.5.

9.15 Peekay Construction Company has a project under way with the following capital outlay: $500,000, $800,000, $1,200,000, and $600,000 in years 1, 2, 3, and 4, respectively. The project becomes operational at the end of the fifth year. All cash disbursements are at the end of the respective years. MARR = 15% and t = 46%.

a. Find P_v and P_t before any financial commitment on the project.

b. Find P_v just after the second year. The salvage value at this stage is $700,000. Any loss on disposal may be spread equally over 2 years.

c. If there is an additional capital expenditure of $900,000 at the end of the second year after the operational date, what is the total P_v value? [The rest of the data are as in part (a).]

9.16 Global Petroleums is doing a feasibility study for a tarsands plant. The one area of analysis still to be considered is the method of moving the raw tarsands from the mine to the refinery.

Alternative 1

   Three draglines + 3,000 feet of conveyor

Alternative 2

   Six bucket wheel loaders + 60 trucks + 500 feet of conveyor

   Draglines = $50,000,000 each

   Conveyor = $3,000/foot

   Bucket wheels = $20,000,000 each

   Trucks = $600,000 each

The plant becomes operational on January 1, 1987. Data are as follows (all dollar values in millions):

		Down Payment		Balance	
	OC	Date	Amount	Date	Amount
		Alternative 1			
Dragline 1	2.0	1/1/82	10	1/1/84	40
Dragline 2	2.0	1/1/83	10	1/1/85	40
Dragline 3	2.0	1/1/85	10	1/1/88	40
Conveyor	5.0	1/1/82	3	1/1/87	6
	11.0				
		Alternative 2			
Three bucket wheel loaders		1/1/82	15	1/1/84	45
No. 4 bucket wheel loader	0.5/unit	1/1/84	5	1/1/85	15
No. 5 bucket wheel loader		1/1/85	5	1/1/87	15

No. 6 bucket wheel					
loader		1/1/85	5	1/1/89	15
20 trucks	0.1/unit	1/1/83	3	1/1/84	9
20 trucks	0.1/unit	1/1/84	3	1/1/87	9
20 trucks	0.1/unit	1/1/87	3	1/1/89	9
500-ft conveyor	0.8	1/1/82	1	1/1/87	0.5
	9.8				

1. $i_{df} = 12\%$, $i_e = 9\%$, $i_f = 7.1\%$, $t = 40\%$, and $r = 50\%$.
2. Annual operating costs listed above are as of January 1, 1987, and are fully responsive to inflation.
3. Base calculations on a 25-year life and zero salvage value. Assume that investments made after the operational data have a salvage value equal to book value 25 years from the operational date.
4. Use the straight-line method (equation 4.2). No CCA has been taken prior to operational date.
   a. Calculate the P_v and P_t values for alternative 2 as of January 2, 1984 (observation date). Assume zero salvage value on any investment made prior to January 2, 1984. Spread the loss on disposal over the 25-year operating period.
   b. Assuming that cost is the deciding criterion, which alternative would you select prior to any investment being made?

CHAPTER TEN

Replacement Analysis

Replacement analysis is one of the most interesting and challenging topics in economic analysis. Replacement brings together all the concepts discussed previously, in addition to deterioration, economic life, sunk cost, remaining value, and obsolescence. In this chapter we present these concepts and provide examples of their application in evaluating replacement alternatives.

10.1 THE LIFE OF AN ASSET

An asset, for the purposes of this discussion, is a means of production and requires a commitment of capital which has relative merit according to its contribution to profitability. An asset has a variety of concurrent life cycles, which when defined, describe its function.

1. *Service life* is the period of production for which assets are required. Consider a forest reserve that will provide timber for 20 years; vehicles and equipment will be required for a service life of 20 years in order to harvest the timber. At an interest rate of 15%, the present equivalent of $100, 60 years from now, is 2 cents. The world went from Model

T Fords to space shuttles in 60 years, and the next 60-year planning cycle will be at least that eventful.

2. *Physical life* includes the entire life of an asset, from its manufacture to its final disposal. Those Model T Fords that have survived until now have served many functions and their physical life is far from over. However, most mechanical assets end their life as scrap metal in a much shorter time.

3. *Economic life* of an asset is that period of life from acquisition to replacement or retirement for which the cost of production, for a given service, has been minimized. The determination of economic life is the goal of this chapter. Economic life is tied to specific service. The disposal of a car, for example, from a rent-a-car company does not imply that the car would not be an economical family car in its second economic life.

10.2 THE NEED FOR REPLACEMENT

The overriding principle in determining the economic life of an asset is that its replacement should be based on economy and the profit of the whole organization. The goals of profit maximization or cost minimization lead to four general categories for replacement:

1. *Reliability or availability:* Many intended services require that an asset be available at the instant of demand or within a certain time frame. A fire truck, for example, must be ready at the sound of an alarm; if it were not, the resulting cost could be astronomical. Reliability may be assured for several years by an intensive, preventative maintenance program; however, there will come a time when the likelihood and cost of a failure exceeds the cost of the asset's replacement.

2. *Productivity:* As an asset ages, the quality and quantity of the good or service it produces deteriorates. Consider a machine that punches precise metal parts. As the machine wears with use, the precision of the parts declines, thereby increasing quality control rejections. The cost of these rejections, or of the declining capacity, may warrant the replacement of the punch. Another less obvious example might be the replacement of one machine with another which is safer to operate, thereby reducing risks and increasing productivity. A decline in productivity means less revenue.

3. *Obsolescence:* Uncertain economic futures and rapidly changing technology pose limits to the accuracy of long-term planning. This is perhaps most evident in the field of computers.

Not only has technology reduced both the size and cost of computers, but it has also increased their potential for productivity. The improvement transcends analytical devices to manufacturing tools. Computer-operated machines can now economically replace several equivalent manually operated machines and their operators.

4. *Physical impairment:* Mechanical assets wear out and fail. There are two general categories of impairment:

   a. *Deterioration:* Mechanical devices deteriorate, resulting in increasing operating costs. There is no better example than the family car, which after 5 or 6 years of service, begins accumulating higher repair bills.

   b. *Catastrophic failure:* Catastrophic failure refers to sudden and complete failure. A common example is the light bulb; however, most electronic components fall into this group.

10.3 LIFE-CYCLE COSTING

All assets go through a life cycle. The idea or need is conceived and as a result a decision is made to acquire the asset. After debugging, the asset usually goes through a stage of high productivity with minimum downtime. As the asset ages operating costs increase and its functional advantages decrease leading to disposal. There are two major stages in the life cycle of an asset: (1) the initial purchase of the asset, including the total conceptual design and research associated with the acquisition, and (2) the operational or productive stage.

Life-cycle costing considers all costs associated with the asset over its total life cycle from acquisition to disposal. The objective is to minimize these costs over the economic life of the asset.

10.3.1 Capital Cost Consideration

Initial Investment

The initial investment includes all costs associated with the purchase, transportation, and installation of the asset ready for productive use. The costs associated with research and development through to final design and production are included by the manufacturer in the purchase price of the asset. The transportation, installation, and debugging costs may be significant and are usually in addition to the purchase price. These costs, unless expensed, must be included to arrive at a total initial (installed) investment value.

Salvage Value

The selection of one asset over another is a question of economic value. One factor in determining that value is the market value on disposal. This market value for most equipment is predictable through auction summaries and previous experience with similar assets. The specific amount depends on a number of factors, such as:

1. Appearance
2. Amount of potential use remaining
3. Mechanical condition
4. Economic climate
5. The versatility of the asset
6. The cost to dismantle, remove, and sell the asset

It is a misconception to think that a piece of equipment should be retained until it is completely worn out. As we shall see later in the chapter, the life cycle of a piece of equipment is determined to minimize life-cycle costs for a particular service. It may be that one company minimizes their costs with very short life cycles and sells those assets to another company, which retains them for a much longer period.

10.3.2 Operating Cost Considerations

In the economic questions we have dealt with previously, all the costs associated with the operation of an asset were lumped together. To evaluate alternatives with an appreciation for their differences, it is important to discuss the individual cost components. The major operating cost components that should be considered are labor, energy, maintenance, downtime, and overhead.

Labor

Labor costs include:

1. Hourly wage
2. Company-provided benefits
3. Training costs
4. Supervision
5. Administration

Energy

Generally speaking, production tools require either petroleum fuels, electricity, or thermal energy to function. Energy costs play an important, even deciding role in replacement studies and other capital investment decisions. Rising prices and uncertain future energy supplies require close attention by someone well versed and current. One may choose bigger for economy of scale, smaller for energy efficiency or conversion to more economical alternate fuels. The issue of energy conservation is a critical cost component.

Maintenance

All mechanical devices are subject to wear and ultimate failure. Preventive maintenance before, or corrective maintenance after, is a significant expense. An assumption made by reliability or maintenance engineers is that the frequency of component failure follows a series of Wiebull probability density functions. This series has been referred to as the bathtub curve. Figure 10.1 illustrates the typical bathtub curve and the relative maintenance costs associated with this curve.

These costs or failure frequencies are not inevitable but may be used to promote longer life and give increased reliability or in fact the reverse. Emergency equipment, for example, may have such an intensive preventative maintenance program that failures do not occur. Equipment subject to a preventative maintenance program usually lasts much longer than equipment on a program that allows only corrective maintenance. Rental car companies, on the other hand, are not concerned with the total useful life of a car. Large-volume purchases of new cars give a tremendous price advantage, and individual selling on retirement maximizes salvage value. Under these conditions, it is advantageous to the rental company to minimize short-term ownership costs further by minimizing maintenance expense.

FIGURE 10.1 Maintenance cost prediction.

Downtime

A cost that is often overlooked because it is difficult to quantify is the cost of downtime. When a production tool is inoperative there are the costs of:

1. Lost production
2. Unproductive labor while waiting
3. Cost of short-term replacements
4. Cost of overtime

These costs are the most obvious and tangible; however, different industries have different costs more specific to their environment and operation.

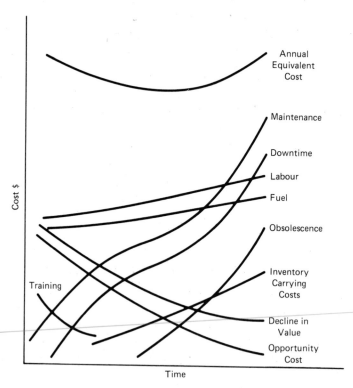

FIGURE 10.2 An asset's life-cycle costs.

Overheads

Administrative costs, maintenance of buildings and tools, or any costs not specifically directed to one function are accounted for as overhead. Budget presentations, policies, and marketing affect the allocation of overheads and as a result they are subject to change year to year. It is important to question increases or decreases in costs in order to separate the effect of changing overheads from the effect of changing direct costs. Figure 10.2 describes some of the costs to be considered in a replacement study.

10.4 SOLUTION METHODS

The solution to the replacement problem depends on such factors as the industry involved, the type of equipment, ease of data collection, the reason for replacement, the precision required, and the intended use of the solution. The goals and parameters must be clearly established before proceeding with any economy study.

The solution models presented in this chapter center on three measurements considered to be the theoretically correct criteria for measuring investment feasibility:

1. Present value
2. Annual equivalent revenue requirement
3. Rate of return

There are other methods of dealing with replacement problems. For example, large military fleets or ships might use repair expenditure limits. In essence, the values of the repair limits are dependent on the type, age, value of the asset, and probability measures such as MTBF (mean time between failure) or MTTR (mean time to repair). Product lines or very specialized equipment may rely on retirement tables based on profit, expense, and specific equipment parameters. A manager may, for ready reference, require a profitability index or payback period. The effect of technology might be considered through productivity criteria quotients or other measures of obsolescence. The examples presented in this chapter rely on estimates of salvage value and operating cost. Probability values may be applied to provide expected values or other forms of sensitivity analysis.

The analyst may even choose the technique that is easiest to understand or present to those administering the company finances. Any model may be judged according to the effectiveness with which it meets the goals of the company. Solution models make certain assumptions and are limited by specific conditions. The final judgment is whether the model is sensitive to the various ways in which these assumptions or conditions depart from the facts of the case and whether these departures are significant.

10.5 BASIC PRINCIPLES

The question of replacement involves the decision as to whether an asset should be disposed of, or retained. In other words, we must determine if the asset has reached the end of its economic life.

In the introduction to this chapter, we related economic life to that point where the cost of production for a given service was minimized. We can determine that point by calculating the annual equivalent revenue requirement of the asset on a year-by-year basis. The plot of AER versus years gives the U-shaped curve illustrated in Figure 10.3. If all other things were equal, which in practice is unlikely, and the replacement asset was identical to the existing asset, the economic solution would be to replace the asset every N^* years.

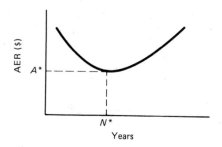

FIGURE 10.3 AER versus years of ownership.

When assessing the annual cost of ownership of a currently owned asset (the defender), it is apparent that there are three basic factors at work: (1) with each succeeding year of ownership the annual equivalent of the original cost declines; (2) with each year of use the resale value declines, although by a diminishing portion as the age increases; and (3) with each year of service, the quality of service decreases and the cost of operation increases. The result of these factors is to produce a U-shaped AER curve.

Let us first consider a simple replacement question.

Example 10.1

Problem: Determine the economic life of the investment. $P = \$1,000$, $A = \$100$, $G = \$200$, $i = 20\%$, $t = 0$.

Year		1	2	3	4	5	6	7
OC	($)	100	300	500	700	900	1100	1300
SV	($)	600	500	400	300	200	100	0

Solution

$$AER_1 = AEOC + [P(A/P, 20\%, 1) - NSV(A/F, 20\%, 1)$$
$$- t(AED)]/(1 - t)$$
$$= \$100 + [\$1,000(1.20) - \$600(1.000) - 0(0)]/1 = \$700$$
$$AER_2 = \$100 + \$200(A/G, 20\%, 2) + [\$1,000(A/P, 20\%, 2)$$
$$- \$500(A/F, 20\%, 2)] = \$618$$
$$AER_3 = \$100 + \$200(A/G, 20\%, 3) + [\$1,000(A/P, 20\%, 3)$$
$$- \$400(A/F, 20\%, 3)] = \$641$$
$$AER_4 = \$100 + \$200(A/G, 20\%, 4) + [\$1,000(A/P, 20\%, 4)$$
$$- \$300(A/F, 20\%, 4)] = \$685$$
$$AER_5 = \$100 + \$200(A/G, 20\%, 5) + [\$1,000(A/P, 20\%, 5)$$
$$- \$200(A/F, 20\%, 5)] = \$736$$
$$AER_6 = \$100 + \$200(A/G, 20\%, 6) + [\$1,000(A/P, 20\%, 6)$$
$$- \$100(A/F, 20\%, 6)] = \$786$$
$$AER_7 = \$100 + \$200(A/G, 20\%, 7) + [\$1,000(A/P, 20\%, 7)$$
$$- 0] = \$835$$

The minimum annual equivalent revenue requirement is $618 at year 2, which represents the economic life of the asset.

Example 10.2

Problem: This example is identical to Example 10.1 except that $t = 50\%$ and the depreciation allowance is calculated using the straight line method over 5 years and a zero salvage value.

Solution: $D = \$1,000/5 = \$200/$year. Spread any recapture or loss on disposal over 5 years to calculate a net salvage value.

Year	1	2	3	4	5	6	7
BV $	800	600	400	200	0	0	0
SV $	600	500	400	300	200	100	0
NSV $	660	530	400	270	140	70	0

$$NSV_1 = \$600 + (\$200/5)(t)(P/A, 20\%, 5) = \$660$$

$$AER_1 = \$100 + [\$1,000(A/P, 20\%, 1) - \$660(A/F), 20\%, 1)$$
$$- 0.5(\$200)]/0.5 = \$980$$
$$AER_2 = \$100 + \$200(A/G, 20\%, 2) + [\$1,000(A/P, 20\%, 2)$$
$$- \$530(A/F, 20\%, 2)$$
$$- 0.5(\$200)]/0.5 = \$818$$

The remaining AER values were calculated in a similar manner.

Year	1	2	3	4	5	6	7
AER($)	980	818	805	827	859	903	947

The minimum AER value is $805 in year 3, which represents the economic life of the investment. Figure 10.4 graphically illustrates the AER curves for these two examples. These curves are typical AER curves.

The annual equivalent revenue curve is generally flat in its operating range, and this tends to allow flexibility in capital planning.

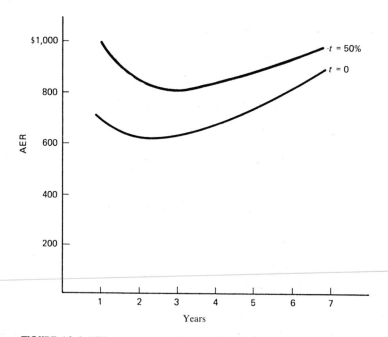

FIGURE 10.4 AER versus years of ownership.

10.6 OPTIMUM LIFE CYCLES AND REPLACEMENT OF ASSETS

10.6.1 Determining an Economic Life Cycle

In our descriptions of economic life and the replacement problems, we paraphrased the replacement question as: What is the economic life? In this section the economic life cycle is described both mathematically and graphically using the following models.

First is the AER equation:

$$AER_z = AEOC + [P(A/P, i_a, z) - NSV(A/F, i_a, z) - t(AED)]/(1 - t)$$

which calculates the revenue requirements on an annual basis. The subscript z refers to the replacement period used in the calculation. That is, AER_2 is for a 2-year period, AER_5 is for a 5-year period, and so on. The second equation is the calculation of the revenue requirement on a year-by-year basis. That is, for the next year the asset must generate sufficient revenue to cover:

1. The operating costs (OC)
2. The loss in value the asset will experience during the year
3. Interest on invested capital during the year
4. Income tax considerations

C_z is equal to the sum of these costs for year z:

$$C_z = OC_z + [P_z(1 + i_a) - NSV_z - tD_z]/(1 - t) \qquad (10.1)$$

where OC_z = operating costs in year z

D_z = depreciation charge in year z

NSV_z = net salvage value at the end of year z

P_z = cost to make operational + market value forgone +

income tax considerations at the beginning of year z

A close look at equation 10.1 reveals that it is identical to the AER equation for a 1-year period.

Example 10.3

Problem: This problem will demonstrate the method of calculating C_z and AER values on a year-by-year basis. $P = \$6,000$, $i_a = 10\%$, and $t =$

50%. Use the declining-balance method with a 30% rate

Year	Salvage Value	OC
1	$4,000	$ 500
2	3,000	500
3	2,500	500
4	2,000	1,000
5	1,750	1,500
6	1,500	2,000

Solution

$$C_z = OC_z + [P_z(1 + i_a) - NSV_z - t(D_z)]/(1 - t)$$

For the declining-balance method using the CTF, this equation simplifies to

$$C_z = OC_z + [(P_z(1 + i_a)(CTF) - (SV_z)(CTF)]/(1 - t)$$

$$CTF(P) = 1 - \{[(td)/(i + d)][(1 + i/2)/(1 + i)]\} = 0.642$$

$$CTF(SV) = 1 - (td)/(i + d) = 0.625$$

The calculation of year-by-year costs is given in the following table.

Year	$C_z = OC + [P_z (1 + i_a) (CTF) - SV_z(CTF)]/(1 - t)$	C_z
1	$C_1 = \quad$ $500 + [$6,000(1.1)0.642 - $4,000(0.625)]/0.5 = $3,975	
2	$C_2 = \quad$ $500 + [$4,000(1.1)0.625 - $3,000(0.625)]/0.5 = 2,250	
3	$C_3 = \quad$ $500 + [$3,000(1.1)0.625 - $2,500(0.625)]/0.5 = 1,500	
4	$C_4 = $1,000 + [$2,500(1.1)0.625 - $2,000(0.625)]/0.5 = 1,938	
5	$C_5 = $1,500 + [$2,000(1.1)0.625 - $1,750(0.625)]/0.5 = 2,063	
6	$C_6 = $2,000 + [$1,750(1.1)0.625 - $1,500(0.625)]/0.5 = 2,531	

$$AER = AEOC + [P(A/P, i_a, n) \, CTF - SV \, (A/F, i_a, n)$$
$$CTF]/(1 - t)$$

$$AER_1 = \$500 + [\$6,000(A/P, 10\%, 1)0.642 - \$4,000(A/F), 10\%, 1)$$
$$0.625]/0.5 = \$3,975$$

$$AER_2 = \$500 + [\$6,000(A/P, 10\%, 2)0.642 - \$3,000(A/F, 10\%, 2)$$
$$0.625]/0.5 = \$3,154$$

$$AER_3 = \$500 + [\$6,000(A/P, 10\%, 3)0.642 - \$2,500(A/F, 10\%, 3)$$
$$0.625]/0.5 = \$2,654$$

$AEOC_4 = \$500 + \$500(A/F, 10\%, 4) = \$608$

$AER_4 = \$608 + [\$6,000(A/P, 10\%, 4)0.642 - \$2,000(A/F, 10\%, 4)$

$\quad 0.625]/0.5 = \$2,500$

$AEOC_5 = \$500 + [\$500(F/P, 10\%, 1) + \$1,000](A/F, 10\%, 5)$

$\quad = \$754$

$AER_5 = \$754 + [\$6,000(A/P, 10\%, 5)0.642 - \$1,750(A/F, 10\%, 5)$

$\quad 0.625]/0.5 = \$2,428$

$AEOC_6 = \$500 + [\$500(F/P, 10\%, 2) + \$1,000(F/P, 10\%, 1) + \$1500]$

$\quad (A/F, 10\%, 6) = \$915$

$AER_6 = \$915 + [\$6,000(A/P, 10\%, 6)0.642 - \$1,500(A/F, 10\%, 6)$

$\quad 0.625]/0.5 = \$2,441$

Following is a summary of the AER and C values.

Year	OC	AEOC	AER_z	C_z
1	$ 500	$500	$3,975	$3,975
2	500	500	3,154	2,250
3	500	500	2,654	1,500
4	1,000	608	2,500	1,938
5	1,500	754	2,428*	2,063
6	2,000	915	2,441	2,531

Notice from the table or Figure 10.5 that the year-by-year, C_z, costs first decline more rapidly than the AER, and then climb more rapidly, the C_z curve crossing the AER curve at the point where the AER is a minimum. This can be deduced on intuitive grounds as follows. If the next year's costs, C_z are lower than the AER, the AER curve will continue to decline. When the C_z value is greater than the AER, the AER curve will rise. Thus the C_z curve will cross the AER curve at the minimum AER. This crossing point is the optimum economic life for the asset.

10.6.2 Replacement of Assets Using an Infinite Service Period

Most corporations over a period of time develop an identity that is associated with specific products and/or services. Therefore, the function of the assets used by a specific corporation do not change to any large degree over significantly large time spans.

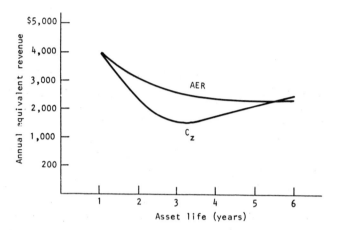

FIGURE 10.5 Optimum economic life.

With this fact in mind, it is reasonably safe to say that most replacement decisions involve the replacement of existing assets with basically an identical unit or with a unit that performs the same function but is technologically superior. The service requirement can be considered to be infinite.

The problem is how to compare an existing asset, the defender, which has already experienced the high value loss of early years, with a yet-unpurchased asset, the challenger. It is the curves of Figure 10.5 that provide the key. An asset is unlikely to become a candidate for replacement as long as its yearly costs, C_z, are declining. However, once the yearly costs start to increase, a replacement study should be conducted. When the year-by-year costs of the defender equal or exceed the minimum AER of the challenger, the defender has reached the end of its economic life and should be replaced.

If the asset performed exactly as forecast at the time of purchase, and there were no changes in either price or technology for this equipment, this policy would result in replacing the asset every N^* years. However, since the C_z value is calculated from actual operating and market data, and the AER is calculated from the manufacturer's information on the best available replacement, the calculation method provides a dynamic replacement policy.

Generally, an asset is not considered for replacement when it is on a decreasing C_z curve. The exception is caused by technological innovation and the appearance of a challenger, which promises considerable efficiencies or cost savings. In this situation it is appropriate to compare the minimum AER of the defender with that of the challenger.

The two decision rules for problems concerning assets that have an infinite service period are:

1. If the year-by-year costs (C_z values) for the defender are increasing, compare the next year's cost for the defender with the minimum AER for the challenger. If the C_z value for the defender is higher than the minimum AER for the challenger, replace the defender. Otherwise, defer the replacement question for one more year. This decision rule is applicable because if the defender is operating on the increasing portion of its year-by-year cost curve, the next year's costs are the minimum annual cost the defender has to offer.

2. If the year-by-year costs for the defender are decreasing, compare the minimum AER of the defender with the minimum AER of the challenger. Accept the alternative with the lower AER value.

Example 10.4

Problem: Global Pipelines wishes to evaluate its current policy of replacing pumps in its major pumping stations every 6 years. The replacement units are similar pumps. These pumps initially cost $80,000. Operating costs in year 1 are $40,000 and increase by 6% each year. Use the declining-balance method and a CCA rate of 30%. Salvage values are as indicated in the table below. What is the economic life of these pumps? MARR = 12% and t = 40%.

Solution: CTF(P) = 0.7296; CTF(SV) = 0.7146. Generate a table showing all costs.

$$C_z = OC_z + [P_z(1 + i_a)CTF - (SV_z)CTF]/(1 - t)$$

$$C_1 = \$40,000 + [\$80,000(1.12)0.7296 - (\$65,000)0.7146]/0.6$$

$$= \$71,571$$

$$AEOC_2 = [\$40,000(P/F, 12\%, 1) + \$42,400(P/F, 12\%, 2)]$$

$$(A/P, 12\%, 2) = \$41,133$$

$$AER_2 = \$41,133 + [\$80,000(A/P, 12\%, 2)0.7296 + \$53,000$$

$$(A/F, 12\%, 2)0.7146]/0.6 = \$68,930$$

The additional C_z and AER values in the following table were calculated in a similar manner.

Year	SV Year End	OC	C_z	AER
1	$65,000	$40,000	$71,571	$71,571
2	53,000	42,400	65,971	68,930
3	43,000	44,944	64,420	67,594
4	33,000	47,641	65,688	67,195

5	25,000	50,499	64,737	66,808
6	20,000	53,529	63,053	66,345
7	20,000	56,741	69,598	65,677
8	20,000	60,145	63,002	65,459*
9	20,000	63,754	66,611	65,537
10	20,000	67,579	70,436	65,816

Conclusion: Economic life = 8 years

Company policies should be reviewed periodically to ensure that they are based on sound economics plus good judgment. The policy to change pumps may have initially been made strictly on the basis of judgment, or on an unfortunate experience with one pump. The decision could still be valid on the basis of other factors not reflected in our economy study. For example, a strong weighting may be placed on reliability due to the high cost of shutdown. Nevertheless, all company policies should be dynamic in nature and subject to periodic review.

Example 10.5

Problem: Assume that you are considering replacement of the family car with a new model. The defender can be sold for $1,000. Its estimated operating costs for the next year are $1,500 for gas and oil plus $800 for maintenance and repairs (OC). One year from now it is estimated the car can be sold for $750. The required rate of return is 10%. The year-by-year costs for the defender are increasing. The new car, the challenger, can be purchased for $6,000 and has the following year-by-year resale values and operating costs:

Year	Resale Value at End of Year	Annual Operating Costs (OC)
1	$4,800	$ 900
2	3,800	1,100
3	3,000	1,300
4	2,500	1,500
5	2,000	1,700
6	1,500	1,900
7	1,200	2,100
8	1,000	2,300

The problem represents a replacement of the defender with a similar challenger. The service requirement is for a perpetual period of time.

Solution: Calculate next year's cost for the defender. The cost of retaining the defender in service for one more year is

$$C_z = OC_z + [P_z(1 + i_a) - SV_z - t(D_z)]/(1 - t)$$

$$= \$2,300 + [\$1,000(1.1) - \$750 - 0]/1$$

$$= \$2,650$$

Determine the minimum AER, which occurs at the economic life, for the challenger.

$$\text{AER}_1 = \$900 + \$6,000(A/P, 10\%, 1) - \$4,800(A/F, 10\%, 1)$$

$$= \$2,700$$

$$\text{AEOC}_2 = [\$900(P/F, 10\%, 1) + \$1,100(P/F, 10\%, 2)](A/P, 10\%, 2)$$

$$= \$995$$

$$\text{AER}_2 = \$995 + \$6,000(A/P, 10\%, 2) - \$3,800(A/F, 10\%, 2)$$

$$= \$2,643$$

In a similar manner the remaining AER values were calculated to complete the table.

Year	1	2	3	4	5*	6	7	8
AER ($)	2,700	2,643	2,594	2,530	2,517	2,528	2,530	2,538

The minimum AER of the challenger ($2,517) is less than next year's costs for the defender ($2,650).

Conclusion: Buy the new car.

Example 10.6

Problem: A company is considering the replacement of a forklift unit purchased 5 years ago with a similar type of forklift. The year-by-year costs for the defender are increasing and a major repair, costing $2,500, which will be capitalized, is required to make it operational. The net salvage value for the defender today is $9,000 with no overhaul. The net salvage value for the defender 1 year from today is estimated to be $6,800. Next year's operating costs are estimated to be $24,000 and the allowable capital cost allowance is $2,500. The new forklift unit has an initial cost of $40,000 and salvage values each year as indicated in the table below. Use the declining-balance method and a CCA rate of 30%. Operating costs in year 1 are estimated to be $24,000 and expect to increase by 7% per year. $i_a = 15\%$ and $t = 46\%$.

Solution

$$P_v = \text{cost to make operational} + \text{market value forgone}$$
$$+ \text{ income tax consideration}$$

$$P_v \text{ (today) } = \$2,500 + \$9,000 = \$11,500$$

$$P_v \text{ (in 1 year) } = \$6,800$$

Determine next year's cost for the defender.

$$C_z = OC + [P_z(1 + i_a) - SV_z - t(D_z)]/(1 - t)$$

$$= \$24,000 + [\$11,500(1.15) - \$6,800 - 0.46(\$2,500)]/(1 - 0.46)$$

$$= \$33,769$$

For the challenger:

$$CTF(P) = 0.7133$$

$$CTF(SV) = 0.6933$$

$$AER_1 = \$24,000 + [\$40,000(A/P, 15\%, 1)0.7133$$

$$- \$30,000(A/F, 15\%, 1)0.6933]/0.54$$

$$= \$46,247$$

$$AEOC_2 = [\$24,000(P/F, 15\%, 1) + \$25,680(P/F, 15\%, 2)](A/P, 15\%, 2)$$

$$= \$24,781$$

$$AER_2 = \$24,781 + [\$40,000(A/P, 15\%, 2)0.7133$$

$$- \$22,000(A/F, 15\%, 2)0.6933]/0.54$$

$$= \$44,146$$

The cost data for the challenger are summarized as follows.

Year	SV Year End	OC	AER
1	$30,000	$24,000	$46,247
2	22,000	25,680	44,146
3	15,000	27,478	43,154
4	10,000	29,401	42,264
5	10,000	31,459	40,947
6	10,000	33,661	40,335
7	10,000	36,018	40,119*
8	10,000	38,539	40,144
9	10,000	41,237	40,324
10	10,000	44,123	40,606

Conclusion: The minimum AER for the challenger is $40,119, which is greater than next year's cost for the defender. Therefore, retain the defender for one more year. A new economy study, based on new data, should be initiated one year from now.

10.7 REPLACEMENT ANALYSIS FOR EQUAL VERSUS UNEQUAL LIVES AND FINITE SERVICE LIVES

As discussed previously, most organizations offer services and/or products of a similar nature for sale throughout their lifetime. They may add or drop some services and/or products, but the type of business in which they are engaged usually does not differ significantly from year to year. Therefore, in the majority of replacement problems, the assumption that the service life is infinite is a realistic assumption. The analysis involves replacing the defender with either a like-for-like asset or with a technologically improved asset. However, the analysis may involve:

1. The selection of a challenger from a group of two or more challengers having different economic lives. The service to be provided continues indefinitely.
2. The provision of assets to provide a specific service for a fixed study period. A defender may or may not exist and the challengers may have different service lives.

In previous chapters we have assumed that the costs for each alternative are identical throughout each economic life cycle. Also, we have considered only alternatives such as alternative A with an economic life of 5 years and alternative B with an economic life of 10 years and a service requirement of 20 years. The lives are an easy multiple of each other and the analysis can be undertaken without any major difficulty. However, we could be faced with the situation where the economic life of alternative A is 5 years and the economic life of alternative B is 10 years and the service need is for 7 years, 13 years, or for an infinite period of time. In addition, in practice the costs are rarely the same for each economic life cycle due to such factors as inflation/recession and technology.

In an economy study that involves the evaluation of challengers with unequal lives, the first step in the analysis is to select the desired challenger from those available. This requires the generation of a table showing the annual equivalent revenue requirements for the challengers. From these data the preferred challenger can be determined. If a defender exists, the preferred challenger can be compared with the defender as previously discussed.

Example 10.7

Problem: An electronics firm requires a delivery van to fulfill the obligations of a new contract. The contract is for a term of 7 years and the company is considering three different trucks. Use the declining-balance method and a 30% rate. $i_a = 15\%$ and $t = 50\%$. First cost, salvage values, and operating costs are as indicated in the table below.

$$\text{CTF}(P) = 0.6884$$

$$\text{CTF}(\text{SV}) = 0.6667$$

Year	Challenger A, P = $8,500		Challenger B, P = $13,000		Challenger C, P = $14,000	
	SV	OC	SV	OC	SV	OC
1	$7,000	$ 2,200	$10,000	$1,500	$12,500	$ 1,500
2	5,500	2,700	8,000	2,000	10,500	1,700
3	4,200	3,200	5,500	2,500	9,000	1,900
4	3,100	3,800	4,500	3,000	6,800	2,200
5	2,200	4,500	4,000	3,500	4,500	2,600
6	1,600	5,400	3,200	4,300	2,900	3,800
7	1,000	6,300	2,700	4,800	2,000	5,000
8	800	7,300	2,400	5,400	2,000	6,500
9	600	8,500	2,100	6,000	2,000	8,500
10	500	10,000	2,000	7,000	2,000	10,000

Solution: Let us first examine the year-by-year costs and annual equivalent revenue requirements.

Year	Challenger A		Challenger B		Challenger C	
	C_z	AER	C_z	AER	C_z	AER
1	$ 7,908	$7,908	$8,750	$8,750	$ 7,000	$7,000
2	6,100	7,067	6,667	7,781	6,867	6,938
3	6,033	6,769	7,433	7,681	6,000	6,668*
4	6,107	6,637	5,433	7,231	6,933	6,721
5	6,320	6,590*	5,067	6,910	7,027	6,766
6	6,640	6,596	6,167	6,825	6,833	6,774
7	7,420	6,670	6,107	6,760	6,780	6,775
8	7,767	6,750	6,340	6,729*	6,900	6,784
9	8,927	6,880	6,880	6,738	8,900	6,910
10	10,253	7,046	7,553	6,779	10,400	7,082

There are several ownership options that would fulfill a 7-year service life. Of these options, there are four which appear to be most likely:

Option 1: Buy challenger A and keep it for 7 years.

$$PER = \$6,670(P/A, 15\%, 7)$$
$$= \$27,750$$

Option 2: Buy challenger B and keep it for 7 years.

$$PER = \$6,760(P/A, 15\%, 7)$$
$$= \$28,124$$

Option 3: Buy challenger C and replace it after 3 years and after 6 years.

$$PER = \$6,668(P/A, 15\%, 6) + \$7,000(P/F, 15\%, 7)$$
$$= \$27,866$$

Option 4: Buy challenger A and keep it for 4 years and replace it with challenger C.

$$PER = \$6,637(P/A, 15\%, 4) + \$6,668(P/A, 15\%, 3)(P/F, 15\%, 4)$$
$$= \$27,657$$

There are other options that we may consider, such as buy challenger C, keep it for 3 years, and then replace it with challenger A.

An examination of the present equivalent of 7 years' service for each option slightly favors option 4. However, it is important to note that at the end of 4 years a new economy study should be initiated. Replacement with a C unit may not be the best option at that point in time. Other, more desirable options may be available.

10.8 THE IMPACT OF A CHANGING TECHNOLOGY

A common advertising expression is "Tomorrow's Technology, Today." Technological improvements allow more economical production, open new industries, challenge management's abilities, and in general make long-range planning very difficult.

The speed of change requires that any investment decision, once made, be constantly reviewed. We can apply a rule of thumb to the frequency of replacement studies:

1. When major repairs are required
2. When a new operation is about to start
3. When new models are available
4. Once a year

10.8.1 Obsolescence

Obsolescence is the handmaiden of technological advance. In a comparison of alternatives in which the challenger is technologically improved, we may quantify the improvement as either reduced operating costs or increased quality and quantity of goods or services produced.

Example 10.8

Problem: One year ago Global Industries purchased a computer system for an automated manufacturing process for $300,000. This system can be replaced today with a more efficient system for a cash outlay of $230,000. Use the declining balance method and a 30% rate. Estimated salvage values and operating costs for each system are as follows. $i_a = 15\%$ and $t = 50\%$.

Year	Defender		Challenger	
	OC	SV	OC	SV
0		50,000		
1	$100,000	30,000	$ 50,000	180,000
2	110,000	20,000	54,000	150,000
3	125,000	10,000	59,000	125,000
4	150,000	10,000	65,000	100,000
5	190,000	10,000	80,000	75,000
6			125,000	75,000

Solution: We must determine if the defender is on the increasing or decreasing part of its cost curve.

$$\text{CTF}(P) = 0.6884$$

$$\text{CTF}(SV) = 0.6667$$

The year-by-year costs and AER values for each alternative are as follows

Year	Defender		Challenger	
	C_z	AER	C_z	AER
1	$136,667	$136,667	$174,167	$174,167
2	129,333	133,256*	130,000	153,624
3	142,333	135,870	122,333	144,613
4	152,000	139,100	123,333	140,351
5	192,000	146,946	133,333	139,311*
6			140,000	139,389

The defender is on the decreasing part of its cost curve. Therefore, the minimum AER for the defender of $133,256 is compared to the minimum AER for the challenger of $139,311. Our decision should be to keep the defender for at least one more year. Sample calculations for the defender:

$$AER_1 = \$100,000 + [\$50,000(1.15)(0.6667) - \$30,000$$

$$(0.6667)]/0.5$$

$$= \$136,667$$

$$AEOC_2 = [\$100,000(P/F, 15\%, 1) + \$110,000(P/F, 15\%, 2)]$$

$$(A/P, 15\%, 2)$$

$$= \$104,467$$

$$AER_2 = \$104,467 + [\$30,000(A/P, 15\%, 2)(0.6667)$$

$$- \$20,000(A/F, 15\%, 2)0.6667]/0.50$$

$$= \$133,256$$

10.8.2 The Value of Increased Production

We compared like for like replacement on an annual equivalent basis. In considering unequal lives and finite service we compared alternatives on a present equivalent basis. In comparing technologically different alternatives, the prospective rate of return can provide the comparative base, because generally, increases in production result in an increase in revenue. The present equivalent or annual equivalent approach may also be used if MARR is known and only an accept/reject decision is required.

Example 10.9

Problem: Edmonton Licorice produces 10,000,000 licorice sticks each year. The assembly line was purchased 5 years ago. Assume that the remaining economic life for both the challenger and defender is 5 years. Business has been good, all the output is sold, and the customers are requesting more than can be produced. Therefore, the company is considering replacing the current production line with a new, higher-capacity line which will produce 15,000,000 sticks. Use the fast-write-off method for capital cost allowance. Book value for the defender = 0.

	Defender (D)	Challenger (C)
Original cost	$450,000	$600,000
Market value today	300,000	
Estimated salvage value in 5 years	50,000	100,000
Annual operating revenue next year	240,000	290,000
Annual operating costs next year	92,000	43,700
Estimated annual increase in OC	15,000	10,000

$r = 0$ and $t = 50\%$.

Solution: Try 15%:

Recaptured CCA for the defender if sold now = $300,000

$$\text{CCA} = \$150,000 \text{ per year over 2 years.}$$

Additional taxes = $75,000/year

$$\text{PE additional taxes} = \$75,000(P/A, 15\%, 2)$$

$$= \$121,928$$

$$P_v(\text{defender}) = \$300,000 - \$121,928 = \$178,072$$

$$\text{NSV}(C - D) \text{ in 5 years} = \$50,000 - (\$25,000)(t)(P/A, 15\%, 2)$$

$$= \$50,000 - \$20,322 = \$29,678$$

$$\text{Depreciation base } (C - D) = \$600,000 - 0 = \$600,000$$

$$\text{PED} = \$150,000(P/F, 15\%, 1) + \$300,000(P/F, 15\%, 2)$$

$$+ \ 150,000(P/F, 15\%, 3)$$

$$= \$455,905$$

	Defender	Challenger	C − D
P	$ 178,072	$ 600,000	$ − 421,928
Depreciation base (C − D)	0	600,000	600,000
NSV			29,678
OR	240,000	290,000	50,000
OC	− 92,000	− 43,700	48,300
Gradient (OC)	− 15,000	− 10,000	5,000

$$\text{BTCF} = \$98,300 + \$5,000(1.7228) = \$106,914$$

$$\text{AE}(C - D) = \$106,914(0.5) - \$421,928(A/P, 15\%, 5) + \$29,678$$

$$(A/F, 15\%, 5) + 0.5(\$455,905) \ (A/P, 15\%, 5)$$

$$= \$ - 7$$

$$i_a = 15\%$$

If MARR $\leq$ 15%, accept the challenger.

10.8.3 The Impact of Technology on Assets Presently Owned

Technology has a significant impact on the timing of equipment replacement. Some industries such as computing, electronics, and the aircraft industry, are much more subject than others to technological change. However, technology affects all industries at some stage of development. As indicated in Chapter 4, functional loss in value of assets, which is technology oriented, is much more significant than loss in value due to wear and tear. The most significant cause of functional loss in value, in most industries, is due to technological change.

Example 10.10

Problem: Consider an extreme example. The week after you make a $6,000 purchase of a new car, Solar Motors introduces its new wonder car, the Sun Mobile, which sells for $10,000. Operating costs are zero and the car will run perfectly for 10 years and then dissolve into a pile of dust. The resale value declines by $1,000 per year over the 10-year life. The C_z and AER values for the defender and challenger are shown below. $i =$ 10%. The introduction of the Sun Mobile has caused the resale value of the defender to vanish. Thus the salvage value of the defender is zero.

Solution: First, it is necessary to determine if the defender is on an increasing cost curve. As indicated by the calculations, a zero salvage value for the defender immediately places this automobile on an increasing cost curve. Therefore, the analysis becomes one of comparing next year's costs for the defender with the minimum AER for the challenger.

	Defender				Challenger		
Year	NSV Year End	OC	C_z	AER	NSV Year End	OC	AER
Today					$10,000	0	
1	0	$ 900	$ 900	$ 900	9,000	0	$2,000
2	0	1,100	1,000	995	8,000	0	1,952
3	0	1,300	1,300	1,087	7,000	0	1,906
4	0	1,500	1,500	1,176	6,000	0	1,861
5	0	1,700	1,700	1,262	5,000	0	1,819
6	0	1,900		1,345	4,000	0	1,778
7	0	2,100		1,424	3,000	0	1,738
8	0	2,300		1,500	2,000	0	1,699
9	0	2,500		2,574	1,000	0	1,663
10	0	2,700		1,645		0	1,628

The minimum AER for the challenger is $1,628, which occurs for a life of 10 years. Next year's cost for the defender is $900.

Conclusion: Retain the defender for one more year.

One should not compare the year-by-year cost of the defender with the challenger's minimum AER of $1,628 and infer that the defender should be retained for 4 years.

The replacement study is for the next year only. The comparison is between the minimum AER of the challenger ($1,628) and next year's defender costs ($900). Next year another replacement study will be done based on the defender's cost forecasts tempered by a year's experience.

Example 10.11

Problem: Suppose that the market reaction to the Sun Mobile's introduction was not as drastic as indicated above and the resale value of the owned car was predicted to be:

At the End of Year	Resale Value
Now	$2,000
1	1,000
2	500
3	200
4	100
5	0

Calculations are again required to determine if the defender is on the decreasing or increasing portion of its cost curve.

Solution: Calculating C_z and AER values for the defender given the $2,000 resale value today:

Year	Salvage Value	C_z	AER
Today	$2,000		
1	1,000	$2,100	$2,100
2	500	1,700	1,910
3	200	1,650	1,831
4	100	1,620	1,786*
5	0	1,810	1,790
6	0	1,900	1,804
7	0	2,100	1,835

The defender is on the decreasing portion of its cost curve. Therefore, the AER of the defender is compared with the AER of the challenger at its economic life.

Minimum AER for the challenger = $1,628

Minimum AER for the defender   = $1,786

Conclusion: Buy the Sun Mobile.

10.9 INFLATION AND REPLACEMENT ANALYSIS

Inflation can have the effect of either retarding or accelerating replacement of an existing asset. As discussed in Chapter 7, inflation does not affect all costs to the same degree. Competition and/or technological change may result in either the initial cost or operating cost of a particular asset increasing at a significantly different rate than general price increases in the economy.

Example 10.12

Problem: Assume that an asset is purchased for $100,000. Its salvage value is estimated to be the book value at any age. Operating costs are $30,000 in year 1 and increase by 5% per year. MARR = 12% and t = 50%. For income tax purposes, the asset can be written off over 5 years using the straight-line method (equation 4.2) and an estimated salvage value of $20,000 in 5 years. What is the economic life of this investment?

Solution

$$D = (\$100,000 - \$20,000)/5 = \$16,000/\text{year}$$

Following is the cost table to determine economic life.

Year	NSV Year End	OC	AER
1	$84,000	$30,000	$70,000
2	68,000	31,500	68,897
3	52,000	33,075	67,858
4	36,000	34,729	66,886
5	20,000	36,465	65,976
6	20,000	38,288	63,155
7	20,000	40,203	61,356
8	20,000	42,213	60,190
9	20,000	44,323	59,441
10	20,000	46,540	58,979
11	20,000	48,866	58,722
12	20,000	51,310	58,614*
13	20,000	53,875	59,616
14	20,000	56,569	58,701
15	20,000	59,397	58,848

The economic life of the asset is 12 years.

Example 10.13

Problem: Asssume that conditions are identical to Example 10.12 except that operating costs increased by 8% per year instead of 5%.

Solution: The cost table to determine economic life is as follows:

Year	NSV Year End	OC	AER
1	$84,000	$30,000	$70,000
2	68,000	32,400	69,321
3	52,000	34,922	68,725
4	36,000	37,791	68,212
5	20,000	40,815	67,778
6	20,000	44,080	65,449
7	20,000	47,606	64,156
8	20,000	51,415	63,511
9	20,000	55,528	63,295*
10	20,000	59,970	63,379

The economic life of the asset is now 9 years.

Example 10.14

Problem: Assume that conditions are the same as in Example 10.12 except that MARR increases to 20% and operating costs increase by 20% each year due to inflation. What is the economic life of this investment?

Solution: The cost table to determine economic life is as follows:

Year	NSV Year End	OC	AER
1	$84,000	$ 30,000	$86,000
2	68,000	36,000	85,818*
3	52,000	43,200	85,978
4	36,000	51,840	86,474
5	20,000	62,208	87,298
6	20,000	74,650	86,830
7	20,000	89,580	87,662
8	20,000	107,495	89,349
9	20,000	128,995	91,640
10	20,000	154,793	94,381

The economic life of the asset is now 2 years.

This example points out the significant impact that an inflationary economy can have on economic life assuming no change in the initial cost of the asset but significant changes in operating costs.

Example 10.15

Problem: Assume that both first cost and operating costs increase due to inflation. Conditions are identical to Example 10.14 except that the first cost increases to $150,000 and the salvage value in year 5 is $30,000 instead of $20,000. i_a = 20%, t = 50%, and operating costs increase by 20% each year.

Solution: Following is the cost table.

Year	NSV Year End	OC	AER
1	$126,000	$ 30,000	$114,000
2	102,000	36,000	112,364
3	78,000	43,200	111,165
4	54,000	51,840	110,396
5	30,000	62,208	110,049
6	30,000	74,650	107,692
7	30,000	89,580	107,219*
8	30,000	107,495	107,963
9	30,000	128,995	109,551
10	30,000	154,793	111,756

An increase in the first cost and the expected salvage value increases the economic life from 2 years in Example 10.14 to 7 years in Example 10.15. Economic life is very sensitive to an increase in both first cost and operating costs.

In summary, when operating costs are accelerating at a faster rate than initial costs, the optimum replacement period decreases and replacement becomes more urgent. When initial costs are increasing faster than operating costs, replacement is postponed.

10.10 SUMMARY

Replacement analysis is one of the more relevant but difficult topics in economic analysis. The approach to the solution of replacement problems taken in this chapter should give the reader a good understanding of the subject. The solution of replacement problems is approached in a systematic manner using a clear format.

Formal replacement analysis has been around since the 1950s, yet even today very few organizations apply it effectively. Very often, production is the industry watchword and industry maintenance is secondary. This emphasizes the need for the analyst to develop a system for periodic review and possible replacement of existing assets.

PROBLEMS

10.1 You are considering replacement of the family car, as it has passed its economic life and is on an increasing cost curve. The market value at year end for the challenger is as follows:

Year	MV	Year	MV
1	$7,000	6	$3,500
2	6,000	7	3,000
3	5,300	8	2,600
4	4,600	9	2,400
5	4,000	10	2,100

Other data pertaining to this decision are as follows:

	Defender	Challenger
Market value today	$2,000	$8,000
Market value in 1 year	$1,500	
Operating costs for the next year	$3,000	$2,000
Operating costs will increase each year by:	10%	10%
BTRR	12%	12%

Should the new car be purchased?

10.2 The Topdog Machine Shop is considering replacement of a lathe that is presently 10 years old with a similar unit. This machine has passed its economic life and is on an increasing cost curve. The C_z cost for next year is estimated to be $60,000. The new machine has a first cost of $160,000. Operating costs are $35,000 in year 1 and are expected to increase by 10% each year. MARR = 15% and t = 50%. Use the fast-write-off method for CCA. If the service life is infinite, should the lathe be replaced?

Year	1	2	3	4	5
NSV($000)	130	120	110	100	90

10.3 Arnold Corporation is considering the replacement of their 1-year-old truck with a new technologically improved model. The information pertaining to the defender and the challenger is presented in the table below. Assume that the economic life for each alternative, starting today, is 5 years. The service life is infinite.

	Defender	Challenger
Original cost	$18,000	$25,000
Operating costs for the next year:	$ 2,000	$ 1,000
Operating costs increase each year by:	10%	8%
Salvage value today	$15,000	
Salvage value 5 years from now	0	$ 5,000

$r = 0$, $t = 50\%$, and MARR $= 15\%$. Use the declining-balance method and a CCA rate of 30%. Should the defender be replaced?

10.4 Last week Global Corporation placed an order for a new computer. An initial payment of $100,000 was necessary. The delivery date is 1 year from now. Since placing this order a new design has become available from another company which is more useful to Global. Delivery dates are identical. The original order can be canceled, but the company loses the initial payment (assume that tax effects are taken at the operational date). $125,000 has to be paid in placing the order for the challenger. Assume that the economic life for both alternatives is 5 years. The service life is infinite.

	Defender	Challenger
Total first cost	$1,000,000	$1,125,000
Remaining payment	900,000	1,000,000
Annual operating costs	250,000	300,000
Annual operating revenue	500,000	601,750
Estimated salvage value in 5 years (from date of delivery)	200,000	250,000

Use the fast-write-off method. $r = 0$ and $t = 40\%$. Under what conditions should the existing order be canceled?

10.5 Last year Caravan Travels purchased a bus for the regular tourist trip to Miami. The president now feels that a "luxury coach" will attract additional customers and that the revenue per customer will be increased by $5. The service life is infinite. The economic life for each alternative is assumed to be 5 years. Use the declining-balance method and a CCA rate of 30%.

	Regular Bus	Luxury Bus
Initial cost	$125,000	$175,000
Market value today	$105,000	
Operating cost per year	$ 90,000	$100,000
Useful life	5 years	5 years
Salvage value in 5 years	0	$ 20,000
Expected number of passengers per year	4,000	4,500
Revenue per passenger	$50	$55

$r = 0$, $t = 50\%$, and MARR $= 15\%$. Should the bus be replaced?

10.6 MTS Consultants is considering the replacement of a minicomputer that is 1 year old with a more efficient model.

	Defender	Challenger
Initial cost (new)	$35,000	$50,000
Market value today	$20,000	
Operating costs	$14,500	$10,650
Remaining economic life	5 years	5 years
Salvage value in 5 years (from today)	5,000	$10,000
Income tax rate	40%	40%

Using the fast-write-off method. In addition, the challenger can do contract work for 50 hours per month, generating a BTCF of $10 per hour. Under what conditions should the defender be replaced?

10.7 Western Electronics is considering the replacement of a computer they purchased 1 year ago with a technologically improved model. The defender is still on a decreasing cost curve. Assume the remaining economic life is 5 years for both alternatives.

	Defender	Challenger
Initial cost 1 year ago	$290,000	
Market value today	$200,000	$350,000
Salvage value in 5 years	$ 50,000	$100,000
Operating costs in year 1	$ 60,000	$ 25,000
Operating costs increase each year by:	10%	10%

Use the declining balance method and a CCA rate of 30%. $r = 0$, $t = 50\%$, and MARR $= 15\%$.
a. Should the challenger be accepted?
b. If operating costs for the challenger start at $15,000 for the first year, should it be accepted?

10.8 Ace Manufacturing is planning to diversify into the manufacture of hammers and they have asked their plant engineer to prepare a list of options in order of preference. The company expects to produce the first model of hammer for 6 years. There are two options for tooling. Which option is preferred? Use the fast-write-off method. $r = 0$, $t = 50\%$, and MARR $= 8\%$.

	Option	
	1	2
Installed cost	$600,000	$600,000
Salvage value in 6 years	200,000	300,000
Annual operating costs (OC)	300,000	250,000
Gradient increase in OC	10,000	12,000
Annual operating revenue	580,000	565,000

10.9 A large company is considering the purchase of a small car with an initial cost of $7,500 and wishes to know in advance the expected life and annual equivalent revenue requirements. A local company has a similar unit and their experience is as shown below. Use the declining-balance method and a CCA rate of 30%.

Year	Net Salvage Value	Operating Cost
1	$5,000	$1,000
2	3,000	1,500
3	2,500	2,500
4	1,500	3,500
5	1,200	4,500

$t = 50\%$ and $i_a = 12\%$. Find the economic life and the corresponding AER of the automobile.

10.10 The Federal Corporation wishes to purchase a word processor. Other offices have them and their cost records show the following:

Year	Net Salvage Value	CCA	Operating Cost
1	$8,000	$4,000	$ 600
2	4,000	4,000	1,500
3	0	4,000	2,400
4	0		3,300
5	0		4,200
6	0		5,100
7	0		6,000
8	0		7,000

The depreciation allowance is as indicated. $P = \$12,000$, $t = 50\%$, $i_a = 12\%$, and book value $=$ SV each year. What are the economic life and AER for this word processor?

10.11 A milling operation uses grinding machines called ball mills, which when new cost $40,000. Their salvage value is $30,000 in year one and is $5,000 less each year thereafter. Use the declining-balance method and a CCA rate of 30%. Operating expenses are $10,000 for the first year with a gradient cost of $3,000. $t = 50\%$ and $i_a = 15\%$. What is the expected economic life and minimum AER?

10.12 A forging company is considering the purchase of a new forging machine to replace an old forging machine currently in operation. The defender is on an increasing cost curve. For the next year, total cost (C_z) for the defender is $50,000. Data pertaining to the challenger are as follows:

Initial investment	$100,000
First-year operating cost (year end)	9,000
Operating cost increase each year by	15%

Year	1	2	3	4	5
NSV($)	75,000	50,000	30,000	10,000	0

Use the fast-write-off method to calculate the CCA. $t = 50\%$ and MARR $= 20\%$. Should the challenger be accepted?

10.13 Global Manufacturing Company requires a gang drilling machine. Two alternatives are being considered:

	Machine A	Machine B
Initial cost	$150,000	$250,000
First-year operating cost (year end)	$ 20,000	$ 15,000
Operating cost increasing at:	15%	15%

Net salvage values in ($000)

Year	1	2	3	4	5
NSV(Alt.A) ($)	100	80	70	55	40
NSV(Alt.B) ($)	200	150	125	100	80

Use the fast-write-off method. $t = 50\%$ and MARR $= 15\%$. Which machine is the suitable alternative?

10.14 A new piece of construction equipment costs $300,000. Use the declining-balance method and a CCA rate of 30%. The first-year operating costs are \$50,000 and are increasing at a rate of 15% per year. $t =$ 50% and MARR $= 12\%$. Year-by-year salvage values are as follows:

Year	Year-End Salvage Value
1	\$260,000
2	220,000
3	180,000
4	140,000
5	100,000

Determine the economic life and the corresponding annual equivalent revenue requirements to justify this investment.

10.15 Global Corporation is considering the replacement of a truck which has passed its economic life. Data for this decision are as follows:

	Defender	Challenger
Market value today	\$3,000	\$9,000
Market value in 1 year	\$2,000	
CCA for the next year	\$1,000	
Current-year operating costs (year end)	\$6,000	\$5,000
OC increases each year by:		12%

MARR $= 15\%$ and $t = 50\%$. For the challenger, year-by-year salvage values are as follows:

Year	Year-End Market Value
1	\$8,000
2	7,000
3	6,000
4	5,000
5	4,200

Use the declining-balance method and a CCA rate of 30% for the challenger. Should the new truck be purchased?

10.16 Two new machines are being considered for purchase by a corporation. The options are mutually exclusive. The economic life of the first machine is known to be 6 years, while that for the second machine is 8 years. $t = 50\%$ and MARR $= 20\%$. Data pertaining to this investment are as follows:

	Machine 1	Machine 2
Initial investment	\$100,000	\$135,000
Salvage values	\$ 30,000	\$ 55,000
Economic life	6	8
First-year-end OC	\$ 10,000	\$ 8,000
OC increasing each year by:	8%	8%

Use the declining-balance method and a CCA rate of 30%. The service requirement is infinite. Which of the two machines should be purchased?

10.17 Two machines that can perform a similar task are under consideration to replace an old piece of equipment. There is no option to retain the old machine. Data pertaining to the 2 choices are as follows:

	Machine A	Machine B
Initial cost today	$20,000	$30,000
First-year operating costs (year end)	$30,000	$29,000
Annual increase in OC	15%	15%
Expected salvage value in year 10	$ 3,000	$10,000

Year	1	2	3	4	5
SV$_A$ ($)	15,000	12,000	10,000	9,000	8,000
SV$_B$ ($)	25,000	21,000	19,000	17,000	15,000

MARR $= 20\%$ and $t = 50\%$. Assume that the operating revenues are identical and that the service requirement is infinite. Use the declining-balance method and a CCA rate of 20%.

a. If it is company policy to keep its machines for 10 years, which of the two machines should be bought?

b. Is the "10-year" policy acceptable? Show your calculations.

10.18 A production machine costs $8,000 and has operating costs of $1,200 in each of the first 2 years. Net salvage values at the end of years 1 and 2 are $5,000 and $3,000, respectively. Operating costs are expected to increase to $3,000 in the third year and the net salvage value is expected to drop to $1,000. $t = 40\%$ and MARR $= 12\%$. Use the fast-write-off method to calculate the CCA.

a. Find AER based on a 2-year life.

b. Evaluate the AER if the machine is intended to be retained through year 3.

10.19 A company is considering the replacement of a conveyor system for oil sands production. The present equipment has been in service for 2 years. The new equipment is expected to reduce operating costs. The service life for this type of equipment can be considered to be infinite for all practical purposes. Use the declining-balance method and a CCA rate of 30%. The following cash flows apply.

	Defender	Challenger
Cost 2 years ago	$200,000	
Market value today	$ 50,000	$250,000
Annual operating costs in year 1	$ 90,000	$ 70,000
Operating costs increase each year by:	25%	15%

$r = 0$, MARR $= 12\%$, and $t = 50\%$. The market value for the defender will drop by \$25,000 per year. The market value for the challenger will drop by \$50,000 per year. Should the equipment be replaced?

10.20 The following costs apply to a piece of construction equipment.

Year	Purchase Price Beginning of Year	Year-End NSV	Annual Operating Costs
1	\$115,000	\$55,000	\$ 34,000
2	55,000	25,000	52,000
3	25,000	10,000	70,000
4	10,000	10,000	88,000
5	10,000	10,000	106,000

Assume an identical service level each year. $t = 40\%$ and MARR $= 20\%$. The construction company is prepared to purchase a used piece of equipment if it is more economical than a new piece of equipment. Assuming that the company has a policy of keeping such equipment for 2 years, determine the optimal age of the equipment at purchase. Use the fast-write-off method to calculate the CCA.

10.21 A government corporation owns a 100-kilometer water pipeline. Two pumps are available for pumping the water. The first model lasts 4 years and costs \$6,000, with an annual operating cost of \$1,000. The second model lasts 8 years and costs \$20,000, with an annual operating cost of \$800. The first model requires an overhaul after 2 years which costs \$1,000. The second model requires an overhaul after 4 years which costs \$2,500. Assume zero salvage values. $t = 0\%$ and MARR $= 12\%$. Determine which of the two pumps is economically more feasible with an infinite service period requirement.

10.22 A machine was bought 4 years ago for \$50,000 and had an expected life of 10 years. The first-year operating costs (year end) were \$10,000 and are increasing at a rate of 15% per year. Use the declining-balance method and a CCA rate of 30%. MARR $= 20\%$ and $t = 50\%$. The yearly salvage values for the machine are as follows:

Year	Year-End Salvage Value
1	\$33,000
2	30,000
3	28,000
Now	25,000
5	23,000
6	22,000
7	21,000

8	20,000
9	20,000
10	20,000

Is the 10 year economic life valid or should it be changed for this type of machine?

10.23 Two machines are being considered to replace an old piece of equipment whose total cost (C_z) for the next year is $12,000. Data pertaining to the two new machines are as follows:

	Machine A	Machine B
Initial cost	$25,000	$35,000
First-year operating cost (year end)	$ 4,000	$ 3,000
Annual increase in OC	15%	15%

Year	1	2	3	4	5
SV_A ($000)	23	21	20	18	16
SV_B ($000)	33	32	31	29	27

MARR = 20% and t = 50%. Use declining-balance and a CCA rate of 30%. The service requirement is infinite. Which of the two machines, if either, should be accepted?

10.24 Mr. James is considering new insulation for the family's house at an installed cost of $900. This project will reduce the heating bill by $250 per annum. Assume a BTRR of 20%. The salvage value of the existing insulation is zero. Once installed, the salvage value of the new insulation is zero.
   a. What is the minimum required economic life to justify this investment?
   b. Repeat part (a) assuming that the savings increase at a rate of 4% per year due to rising heating costs (consider $250 as the first-year-end savings).

10.25 A certain piece of equipment costs $9,000 today. Operating costs are $1,500 for the first year and increase at a rate of 10% per year.

Year	1	2	3	4	5
SV	8,000	7,500	7,000	6,500	5,500

MARR = 15% and t = 48%. This equipment is being considered to replace an old piece of equipment whose total cost (C_z) for the next year is $4,500. Use the declining-balance method and a CCA rate of 25%. Should the challenger be accepted?

Replacement Analysis Problems
Using Microcomputers

This microcomputer program is designed to generate a table of C_z and AER values in the analysis of replacement problems using Supercalc 1.12. The program has considerable flexibility in its application and can easily be adapted to any type of microcomputer or spreadsheet program. Table 10A.1 appears on your screen when using the program.

10A.1 PROGRAM OPERATION

The program offers several options to the analyst.

1. The discount rate can be entered into cell B2 or values for r, i_d, t, i_e and i_f can be entered into cells C2 to G2 and the discount rate will be calculated.
2. If the declining-balance method and the CTF (½ year rule) is to be used, enter a 1 in cell F8 and the declining-balance rate is G8.
3. Operating costs (OC) for year 1 must be entered in cell D11 for all options. The remaining values are entered as follows:
   a. If OC is a uniform amount each year, enter 0 in cell D8.

TABLE 10A.1 Replacement analysis program for microcomputers using Supercalc 1.12.

	A	B	C	D	E	F	G	H	I	J	K
1	n	i	r	id	t	ie	if	ia			
2		.1			.5			.1			
3											
4											
5											
6							DB				
7				INFL		CTF	RATE	CTF-SV	CTF-P		
8				CODE				1	1		
9	YEAR	A/P	P/F	OC	D	NSV	P	CZ	AER	OC	NSV
10	0	1									
11	1	1.1000	.90909			0	0	0	0		
12	2	.57619	.82645	0		0	0	0	0		
13	3	.40211	.75131	0		0	0	0	0		
14	4	.31547	.68301	0		0	0	0	0		
15	5	.26380	.62092	0		0	0	0	0		
16	6	.22961	.56447	0		0	0	0	0		
17	7	.20541	.51316	0		0	0	0	0		
18	8	.18744	.46651	0		0	0	0	0		
19	9	.17364	.42410	0		0	0	0	0		
20	10	.16275	.38554	0		0	0	0	0		
21											
22											

   b. If OC increases by some percentage each year, enter $1 + \%$ in cell D8 (e.g., if OC increases by 10% each year, enter 1.1 in cell D8).
   c. Otherwise, enter values for operating costs in year 2 through year 10 in cells J12 to J20.
4. When using a method other than declining-balance and the CTF, enter depreciation allowance values in cells E11 to E20.
5. Two options are open to calculate salvage values each year:
   a. If SV = BV each year and the depreciation allowance is entered into cells E11 to E20, the program calculates the SV each year (K11 must be equal to zero when using this option).
   b. If individual salvage values (SV) are to be used, enter these values into cells K11 to K20. If you are using a method other than declining-balance and the CTF, the values entered in K11 to K20 must be *net* salvage values.

When the input data are entered as indicated, the program calculates C_z values (H11 to H20) and AER_z (I11 to I20) values.

   If you are interested only in calculating AER for a specified economic life, enter values for i, OC, P, and the depreciation allowance as indicated above. Then enter a 1 in cell K11 and the SV for year n in the appropriate cell in column K (e.g., if $n = 5$, enter SV in cell K15).

Example 10A.1

Problem: Example 10-3 will be used to demonstrate how the program works. P = \$6,000, i = 10%, and t = 50%.

Year	1	2	3	4	5	6
OC (\$)	500	500	500	1,000	1,500	2,000
SV (\$)	4,000	3,000	2,500	2,000	1,750	1,500

Use the declining-balance method and a 30% rate.

Solution: See Table 10A.2. (Enter .1 in cell B2, .5 in E2, 0 in D8, 1 in F8, .3 in G8, 6,000 in G10, 500 in D11, remaining OC values in J12–J16 and SV in K11–K16.

Problem: Assume that operating costs are \$500 in year 1 and increase by 10% each year. (Enter 1.1 in cell D8; enter 0 in cell J12.)

Solution: See Table 10A.3.

Problem: Use the fast-write-off method and operating costs the same as in part (a). (Enter 0 in cell D8, enter 0 in F8, enter 500 in cell J12, enter depreciation allowance in cells E11 to E13.)

Solution: See Table 10A.4

TABLE 10A.2

	A	B	C	D	E	F	G	H	I	J	K
1		n	i	r	id	t	ie	if	ia		
2			.1			.5			.1		
3											
4											
5											
6							DB				
7				INFL		CTF	RATE	CTF-SV	CTF-P		
8				0		1	.3	.625	.64205		
9		YEAR	A/P	P/F	OC	D	NSV	P	CZ	AER	OC	NSV
10	0		1					6000			
11	1		1.1000	.90909	500		4000	4000	3975.0	3975.0		4000
12	2		.57619	.82645	500		3000	3000	2250	3153.6	500	3000
13	3		.40211	.75131	500		2500	2500	1500	2654.0	500	2500
14	4		.31547	.68301	1000		2000	2000	1937.5	2499.6	1000	2000
15	5		.26380	.62092	1500		1750	1750	2062.5	2428.0	1500	1750
16	6		.22961	.56447	2000		1500	1500	2531.3	2441.4	2000	1500
17	7		.20541	.51316	0		0	0	2062.5	2401.5		
18	8		.18744	.46651	0		0	0	0	2191.5		
19	9		.17364	.42410	0		0	0	0	2030.1		
20	10		.16275	.38554	0		0	0	0	1902.7		
21											

TABLE 10A.3

	A	B	C	D	E	F	G	H	I	J	K
1	n	i	r	id	t		ie	if	ia		
2		.1			.5				.1		
3											
4											
5											
6							DB				
7				INFL		CTF	RATE	CTF-SV	CTF-P		
8				1.1		1	.3	.625	.64205		
9	YEAR	A/P	P/F	OC	D	NSV	P	CZ	AER	OC	NSV
10	0	1					6000				
11	1	1.1000	.90909	500		4000	4000	3975.0	3975.0		4000
12	2	.57619	.82645	550		3000	3000	2300	3177.4	0	3000
13	3	.40211	.75131	605		2500	2500	1605	2702.3	0	2500
14	4	.31547	.68301	665.5		2000	2000	1603	2465.5	0	2000
15	5	.26380	.62092	732.05		1750	1750	1294.6	2273.7	0	1750
16	6	.22961	.56447	805.26		1500	1500	1336.5	2152.2	0	1500
17	7	.20541	.51316	885.78		0	0	2948.3	2236.1		
18	8	.18744	.46651	974.36		0	0	974.36	2125.8		
19	9	.17364	.42410	1071.8		0	0	1071.8	2048.2		
20	10	.16275	.38554	1179.0		0	0	1179.0	1993.6		

TABLE 10A.4

	A	B	C	D	E	F	G	H	I	J	K
1	n	i	r	id	t		ie	if	ia		
2		.1			.5				.1		
3											
4											
5											
6							DB				
7				INFL		CTF	RATE	CTF-SV	CTF-P		
8				0		0	0	1	1		
9	YEAR	A/P	P/F	OC	D	NSV	P	CZ	AER	OC	NSV
10	0	1					6000				
11	1	1.1000	.90909	500	1500	4000	4000	4200	4200.0		4000
12	2	.57619	.82645	500	3000	3000	3000	300	2342.9	500	3000
13	3	.40211	.75131	500	1500	2500	2500	600	1816.3	500	2500
14	4	.31547	.68301	1000		2000	2000	2500	1963.6	1000	2000
15	5	.26380	.62092	1500		1750	1750	2400	2035.1	1500	1750
16	6	.22961	.56447	2000		1500	1500	2850	2140.7	2000	1500
17	7	.20541	.51316	0		0	0	3300	2262.9		
18	8	.18744	.46651	0		0	0	0	2065.0		
19	9	.17364	.42410	0		0	0	0	1913.0		
20	10	.16275	.38554	0		0	0	0	1792.9		
21											

10A.2 PROGRAM LISTING

The cell numbers containing equations and the program listing are given in Table 10A.5.

TABLE 10A.5 Program listing

H2	P = (C2*D2)*(1 − E2) + (1 − C2)*((1 + F2)*(1 + G2) − 1) + B2
H8	P = 1 − (E2*G8)/(H2 + G8)
I8	P = 1 − (E2*G8)/(H2 + G8)*(1 + H2/2)/(1 + H2)
B11	P = (H2*(1 + H2^A11)/((1 + H2)^A11 − 1)
C11	P = (1/(1 + H2)^A11)
F11	P = IF(K11 = 0,G10 − E11,K11)
G11	P = F11
H11	P = D11 + IF(F8 = 0,(G10*(1 + H2) − F11 − E11*E2)/(1 − E2),(G10*I8 *(1 + H2) − F11*H8)/(1 − E2))
I11	P = NPV(H2,D11:D11)*B11 + IF(F8 = 0,G10*B11 − F11*(B11 − H2) − E2 *NPV(H2,E11:E11)*B11,G10*I8*B11 − F11*(B11 − H2)*H8)/(1 − E2)
B12	P = (H2*(1 + H2)^A12)/((1 + H2)^A12 − 1)
C12	P = (1/(1 + H2)^A12)
D12	P = IF(J12 = 0,D11*D8,J12)
F12	P = IF(K11 = 0,G11 − E12,K12)
G12	P = F12
H12	P = D12 + IF(F8 = 0,(G11*(1 + H2) − F12 − E12*E2)/(1 − E2),(G11*H8 *(1 + H2) − F12*H8)/(1 − E2))
I12	P = NPV(H2,D11:D12)*B12 + IF(F8 = 0,G10*B12 − F12*(B12 − H2) − E2 *NPV(H2,E11:E12)*B12,G10*I8*B12 − F12*(B12 − H2)*H8)/(1 − E2)
B13	P = (H2*(1 + H2)^A13)/((1 + H2)^A13 − 1)
C13	P = (1/(1 + H2)^A13)
D13	P = IF(J12 = 0,D12*D8,J13)
F13	P = IF(K11 = 0,G12 − E13,K13)
G13	P = F13
H13	P = D13 + IF(F8 = 0,(G12*(1 + H2) − F13 − E13*E2)/(1 − E2),(G12*H8 *(1 + H2) − F13*H8)/(1 − E2))
I13	P = NPV(H2,D11:D13)*B13 + IF(F8 = 0,G10*B13 − F13*(B13 − H2) − E2 *NPV(H2,E11:E13)*B13,G10*I8*B13 − F13*(B13 − H2)*H8)/(1 − E2)
B14	P = (H2*(1 + H2)^A14)/((1 + H2)^A14 − 1)
C14	P = (1/(1 + H2)^A14)
D14	P = IF(J12 = 0,D13*D8,J14)
F14	P = IF(K11 = 0,G13 − E14,K14)
G14	P = F14
H14	P = D14 + IF(F8 = 0,(G13*(1 + H2) − F14 − E14*E2)/(1 − E2),(G13*H8 *(1 + H2) − F14*H8)/(1 − E2))
I14	P = NPV(H2,D11:D14)*B14 + IF(F8 = 0,G10*B14 − F14*(B14 − H2) − E2 *NPV(H2,E11:E14)*B14,G10*I8*B14 − F14*(B14 − H2)*H8)/(1 − E2)
B15	P = (H2*(1 + H2)^A15)/((1 + H2)^A15 − 1)
C15	P = (1/(1 + H2)^A15)
D15	P = IF(J12 = 0,D14*D8,J15)
F15	P = IF(K11 = 0,G14 − E15,K15)
G15	P = F15
H15	P = D15 + IF(F8 = 0,(G14*(1 + H2) − F15 − E15*E2)/(1 − E2),(G14*H8 *(1 + H2) − F15*H8)/(1 − E2))

TABLE 10A.3 (cont.)

I15	P = NPV(H2,D11:D15)*B15 + IF(F8 = 0,G10*B15 − F15*(B15 − H2) − E2 *NPV(H2,E11:E15)*B15,G10*I8*B15 − F15*(B15 − H2)*H8)/(1 − E2)
B16	P = (H2*(1 + H2)^A16)/((1 + H2)^A16 − 1)
C16	P = (1/(1 + H2)^A16)
D16	P = IF(J12 = 0,D15*D8,J16)
F16	P = IF(K11 = 0,G15 − E16,K16)
G16	P = F16
H16	P = D16 + IF(F8 = 0,(G15*(1 + H2) − F16 − E16*E2)/(1 − E2),(G15*H8 *(1 + H2) − F16*H8)/(1 − E2))
I16	P = NPV(H2,D11:D16)*B16 + IF(F8 = 0,G10*B16 − F16*(B16 − H2) − E2 *NPV(H2,E11:E16)*B16,G10*I8*B16 − F16*(B16 − H2)*H8)/(1 − E2)
B17	P = (H2*(1 + H2)^A17)/((1 + H2)^A17 − 1)
C17	P = (1/(1 + H2)^A17)
D17	P = IF(J12 = 0,D16*D8,J17)
F17	P = IF(K11 = 0,G16 − E17,K17)
G17	P = F17
H17	P = D17 + IF(F8 = 0,(G16*(1 + H2) − F17 − E17*E2)/(1 − E2),(G16*H8 *(1 + H2) − F17*H8)/(1 − E2))
I17	P = NPV(H2,D11:D17)*B17 + IF(F8 = 0,G10*B17 − F17*(B17 − H2) − E2 *NPV(H2,E11:E17)*B17,G10*I8*B17 − F17*(B17 − H2)*H8)/(1 − E2)
B18	P = (H2*(1 + H2)^A18)/((1 + H2)^A18 − 1)
C18	P = (1/(1 + H2)^A18)
D18	P = IF(J12 = 0,D17*D8,J18)
F18	P = IF(K11 = 0,G17 − E18,K18)
G18	P = F18
H18	P = D18 + IF(F8 = 0,(G17*(1 + H2) − F18 − E18*E2)/(1 − E2),(G17*H8 *(1 + H2) − F18*H8)/(1 − E2))
I18	P = NPV(H2,D11:D18)*B18 + IF(F8 = 0,G10*B18 − F18*(B18 − H2) − E2 *NPV(H2,E11:E18)*B18,G10*I8*B18 − F18*(B18 − H2)*H8)/(1 − E2)
B19	P = (H2*(1 + H2)^A19)/((1 + H2)^A19 − 1)
C19	P = (1/(1 + H2)^A19)
D19	P = IF(J12 = 0,D18*D8,J19)
F19	P = IF(K11 = 0,G18 − E19,K19)
G19	P = F19
H19	P = D19 + IF(F8 = 0,(G18*(1 + H2) − F19 − E19*E2)/(1 − E2),(G18*H8 *(1 + H2) − F19*H8)/(1 − E2))
I19	P = NPV(H2,D11:D19)*B19 + IF(F8 = 0,G10*B19 − F19*(B19 − H2) − E2*NPV (H2,E11:E19)*B19,G10*I8*B19 − F19*(B19 − H2)*H8)/(1 − E2)
B20	P = (H2*(1 + H2)^A20)/((1 + H2)^A20 − 1)
C20	P = (1/(1 + H2)^A20)
D20	P = IF(J12 = 0,D19*D8,J20)
F20	P = IF(K11 = 0,G19 − E20,K20)
G20	P = F20
H20	P = D20 + IF(F8 = 0,(G19*(1 + H2) − F20 − E20*E2)/(1 − E2),(G19*H8 *(1 + H2) − F20*H8)/(1 − E2))
I20	P = NPV(H2,D11:D20)*B20 + IF(FB = 0,G10*B20 − F20*(B20 − H2) − E2*NPV (H2,E11:E20)*B20,G10*I8*B20 − F20*(B20 − H2)*H8)/(1 − E2)

CHAPTER ELEVEN

Economic Analysis of Public Projects

Government funds should be invested to promote economic and social objectives in an effective manner, utilizing an efficient allocation of resources among competing programs. This objective requires a systematic approach to the problem of evaluating and selecting individual government projects.

The evaluation of projects within the public sector is conceptually more complicated than evaluating similar projects within the private sector. For example, consider the economic analysis associated with the building of a pipeline across a river to replace a present long haul by tank trucks, versus the building of a bridge across a river to replace a present ferry system.

In both cases an economic evaluation includes consideration of all capital and operating costs associated with the movement of oil by the oil company and the movement of traffic by the government. However, there is an important point in difference. The oil company bases the evaluation and final decision to invest on whether or not an adequate return may be expected through reduced operating costs and possible increased revenue through such factors as improved service. With respect to the bridge, a government agency makes the investment and pays all operating costs associated with the bridge; however, any benefits that may result by building the bridge accrue to the general public and not the government agency. It is also readily apparent that the difficulty in estimating and evaluating the outcome of each project is much less for the pipeline than it is for the bridge.

Government studies are more subject to value judgments than are studies in the private sector, and often it is not possible or realistic to quantify all cash flows associated with the study. For example, improvements in a roadway intersection through the placement of traffic lights and improved visibility in all directions will probably result in reduced accidents. This investment is certainly desirable to those who use the intersection, but is a disadvantage to such segments of the society as the auto-body shops and the legal profession.

For most of society, a reduction in accidents is desirable. Therefore, this type of value judgment is not difficult to make. However, under many circumstances the decision is not so straightforward.

Benefit-cost analysis is a method of evaluating the relative merits of alternative public investment projects to achieve efficient allocation of resources. In this method the net benefits from a project are divided by the net costs, and the resulting *benefit-cost ratio* is used as the criterion for acceptance or rejection of the project. This method is used in many parts of the world for evaluating public investments and will be fully discussed herein, but first several areas of concern will be introduced.

11.1 WHOSE POINT OF VIEW IS CRITICAL TO THE STUDY?

Economy studies within the public sector may be approached from several viewpoints. The viewpoint taken may have a significant impact on the outcome of the analysis. The viewpoint taken may be:

1. National in scope
2. That of the government agency conducting the study
3. That of a localized area (e.g., town, city, municipality)

It is essential that the analyst have clearly in mind which group is being represented before proceeding with the economy study. If the objective is to promote the general welfare of the public at large, it is necessary to consider the impact of alternative government policies on all the people, not merely on the income and expenditures of a selected group. Practically, however, without regulations the best that can be hoped for in most studies by local government is a representation of the viewpoint within a specific city or municipality. The broader question of the effect of one community's action on another usually requires regulation by government bodies that control both communities. For example, the dumping of sewage into lakes and rivers requires health regulations that protect all communities using this water source.

The national viewpoint would seem to be the correct one in all federally financed public works projects. However, even most nationally financed projects provide the major benefits to a local area making it difficult, if not impossible, to trace and evaluate quantitatively the national effects. However, the local impact is reasonably clear. Example 11.1 parallels an actual case history.

Example 11.1

Problem: This example discusses a government subsidy of a pulp mill in a depressed area. The mill is expected to generate jobs for 200 people and further stimulate the economy in the area through commercial ventures and tourist trade. The government subsidy amounts to $5,000,000. The benefits due to jobs created and improved trade in the area are estimated to be $1,000,000 per year. Six percent is considered to be a fair discount rate. The study period is 20 years. Calculate the benefit-cost ratio.

Solution

$$\text{PE of benefits} = 1,000,000 \ (P/A, 6\%, 20) = \$11,470,000$$

$$\text{Benefit-cost ratio} = 11,470,000/5,000,000 = 2.3$$

Conclusion: The benefit-cost ratio is 2.3, and hence the investment is considered to be quite beneficial. A benefit-cost ratio of 1.0 represents the accept/reject cutoff point.

The plant was established based on the study above, but pollution control equipment was not installed. Raw by-products were dumped into the river, causing major downstream water problems. Fish in the river became almost extinct. The viewpoint used as a basis for the study was much too localized. The installation of pollution control equipment after the plant was in operation made the entire project uneconomical.

This is an example of what can happen if the proper viewpoint is not established and all factors are not considered when a study is initiated.

11.2 USING THE PROPER INTEREST RATE

The appropriate interest rate to use in evaluating public investments has been a subject of discussion for many years without any reasonable consensus arising.

The rate of interest used in evaluating public projects is referred to as the social discount rate. The question that still remains is some agreement as to how to calculate this rate. Should this rate be the government borrowing rate, that is, the cost of capital to the government? This rate adjusted by $\frac{1}{2}$ to 1% for risk may represent a reasonable base.

The minimum attractive rate of return (MARR) in the private sector for any individual firm is usually somewhat higher than the cost of capital to the firm. The difference is due primarily to two factors, risk and capital rationing. Both of these factors exist in the public sector. The problem of capital rationing at various levels of government suggests that a realistic and consistent method of screening investments is highly desirable.

Governments pour millions of dollars every day into high-risk ventures. For example, what is the risk associated with a major injection of funds into a company on the verge of bankruptcy?

We may also gain some insight into what the social discount rate should be by evaluating the alternative use of government funds generated through various means of taxation. If money generated through taxes was left with the taxpayer, these funds could be applied to such items as house mortgages, furniture payments, car payments, and other investments.

The social discount rate using this approach would not be difficult to calculate. Presently, the rate would be in the order of 12% (1985), depending somewhat on the specific mix of taxpayer investments utilized.

However, it is unfortunate that the use of a social discount rate considerably lower than the rates based on methods suggested above seems to be the rule rather than the exception.

Not all legislators and social interest groups agree that benefit-cost analysis is applicable to all public projects. To support their view, they point to the difficulties of measuring, that is, assigning dollar values to some benefits. For example, calculating the dollar benefit of a public recreation area or the savings from accident reduction by constructing a divided highway are certainly open to debate. Further, they feel that the discounting process, which weights near benefits higher than distant benefits, discriminates against future generations. These are major objections and are issues in the field of welfare economics associated with studies involving government allocative processes. Discussion of these issues is beyond the scope of this text. Here we can note only that current government practice is that in projects where there are generally acceptable methods of measurement, benefit-cost analysis is applied. Where the benefits and costs are difficult to quantify, in areas such as social assistance, health care, and some areas of research, the decisions are often made by the political process of debate, opinion, and consensus.

Example 11.2

Problem: A program is under consideration to pave a certain section of a highway. The benefits and costs associated with the pavement are estimated as follows:

Capital costs per mile (upgrading, paving, etc.)	$40,000
Annual operating costs per mile	3,000
Annual benefits per mile	10,500

The estimated useful life is 10 years. Determine the benefit-cost ratios at 5%, 10%, and 15%.

Solution

Try 5%:

$$PE \text{ (capital costs)} = \$40,000$$

$$PE(BTCF) = (\$10,500 - \$3,000)(P/A, 5\%, 10) = \$57,913$$

Try 10%:

$$PE \text{ (capital costs)} = \$40,000$$

$$PE(BTCF) = (\$10,500 - \$3,000)(P/A, 10\%, 10) = \$46,085$$

Try 15%:

$$PE \text{ (capital costs)} = \$40,000$$

$$PE(BTCF) = (\$10,500 - \$3,000)(P/A, 15\%, 10) = \$37,641$$

	Benefits, B	Costs, C	B/C
5%	$57,913	$40,000	1.45
10%	46,085	40,000	1.15
15%	37,641	40,000	0.94

Conclusion: The social discount rate used to evaluate alternatives is critical to the accept/reject decision. It is essential that public investments use a social discount rate that is based on sound economic judgment.

11.3 TAX AND OTHER CONSIDERATIONS IN COMPARING THE COSTS OF PUBLIC PROJECTS

A large number of projects undertaken by governments at all levels could be undertaken at least equally well by the private sector. A few of the

numerous investments that fall into this category are utilities, power, telephone, natural gas, ambulance service, garbage pickup, postal services,
and most government corporations.

To fully evaluate investment alternatives within the public sector, all
costs, including taxes when applicable, must be considered.

Example 11.3

Problem: The city of Edmonton is considering the merits of a city-
owned versus an investor-owned garbage collection system. The cost per
unit per mile is required to make a comparison between a city-owned
collection system and a privately owned system. The analyst has considered
the following costs in her report.

Solution

1. Operating costs per unit per mile (includes such costs as maintenance, fuel, and wages)	$1.25
2. Cost per unit per mile for depreciation of buildings and equipment	0.40
3. Cost per unit per mile for the 10% interest on bonds ($r = 60\%$)	0.20
4. Cost per unit per mile for the 12% return on the 40% financed by internally generated funds	0.16
5. Cost per unit per mile for property taxes	0.15
6. Cost per unit per mile for provincial income tax	0.08
7. Cost per unit per mile for federal income tax	0.12
The cost per unit per mile	$2.36

The report was presented by the analyst to the budgeting committee.
In her presentation she pointed out the reasoning behind the costs listed.

1. Operating costs were based on an annual mileage of 25,000 miles.

2. Depreciation of buildings and equipment, plus interest charges on
   bonds and internally generated funds, represent the after-tax cash flow
   requirement. This cash flow covers repayment of the investment and
   interest on the investment.

3. Bond costs to the city this year are 9% plus an estimated 1% to include
   flotation, underwriting, and legal and administration costs.

4. The 12% on funds generated internally represents an estimate of the
   rate of return that the citizens of Edmonton may earn on these funds
   through such items as reduction in mortgage and car payments.

5. The selection of a city-owned garbage collection system versus a
   privately owned system has no effect on the property tax requirements

of the city. Therefore, if the city decides to operate its own garbage collection system, the property tax revenues that would have been collected from a privately owned concern are forgone. This loss in property taxes must then be collected from others who pay property taxes, resulting in higher rates.

6. Similar to the property tax situation, the selection of a city-owned garbage collection system versus a privately owned system has no bearing on the income tax needs of the province or federal government. Thus the selection of a city-owned garbage system means that income taxes forgone by this decision increase the tax burden to citizens throughout the province and nation. The spreading of this tax burden over a much larger population than the city may not appear to be sound economics to the city administrators. However, when one considers the many similar decisions made throughout the country, a more national viewpoint should be adopted by all municipalities, and income taxes should be added as a cost when applicable.

Tax considerations are a legitimate cost associated with many publicly financed investment decisions. To ignore these costs may result in allocating public funds to investments that can be handled much more effectively by the private sector.

11.4 BENEFIT-COST RATIOS AS A MEASURE OF PROJECT ACCEPTABILITY

Benefit-cost analysis is a method in common use throughout the world for evaluating the relative merits of alternative public investment projects to achieve effective allocation of government resources.

Within the private sector, evaluating the merits of alternative investment opportunities begins with technical feasibility. The next step involves a comparison at some MARR value of the estimated stream of benefits in the form of revenues with the estimated stream of costs over the expected economic life of each investment. In this sense, benefit-cost analysis is not a new technique. Evaluating alternative investment proposals using the benefit-cost approach as the basis for comparison requires an initial assessment of technical feasibility and engineering studies to establish the basic data. The estimated benefits and costs are then compared, usually on a present-value basis, using a predetermined discount rate.

The measure of acceptability is represented by a benefit-cost ratio of 1.0. We should have no problem in accepting a benefit-cost ratio of 1.0 as an accept/reject cutoff point. Throughout the text we have used

$$\text{PE revenues} = \text{PE costs}$$

at some predetermined interest rate, MARR/BTRR, as a measure of project acceptability within the private sector. We are now taking the ratio of these two values and referring to this ratio as a benefit-cost ratio.

$$B/C = PE \text{ (net revenues)}/PE\text{(investment costs)} = 1.0$$

The problem in using benefit-cost analysis is in identifying and quantifying what should be classified as benefits and what should be classified as costs. Quantifying all factors in terms of a dollar value is also often extremely difficult. In many cases, the outputs of public investment projects are supplied free of a direct charge to the public (e.g., parks, highways). A dollar-value estimate for these projects is usually calculated on the basis of what the consumer may be willing to pay for the service. In addition, projects financed by public funds often take a much broader perspective than projects in the private sector.

11.5 COMPARISON OF DIFFERENT BENEFIT-COST RATIOS

There are some differences of opinion as to what should be included in the numerator and what should be included in the denominator of the benefit-cost ratio. The confusion results from the fact that where public investments are concerned, the initial investment and annual operating costs are the responsibility of some department of government, whereas the net benefits (service value to the user minus any user costs) accrue to the user of the facility (the citizens of the city, area, or nation). Therefore, for many years most government agencies used what is referred to as the *conventional benefit-cost ratio:*

B/C (conventional)

$$= \frac{PE \text{ of net benefits to the user}}{PE \text{ of capital } + \text{ operating costs to the supplier}} \quad (11.1)$$

Calculations of benefit-cost ratios using the conventional method may lead to erroneous decisions.

The benefit-cost ratio, if properly applied, should lead to the same accept/reject decisions as the present equivalent, annual equivalent, or rate-of-return approaches. To maintain consistency between the various methods, it is necessary to calculate the benefit-cost ratio as follows:

1. Determine the present equivalent or annual equivalent of the net annual cash flows associated with the project. That is,

Annual operating revenues (OR) − annual operating costs (OC)
$$= BTCF$$

The annual operating revenues represent the benefits to the user. The annual operating costs represent operating costs to the user plus operating cost to the agency supplying the service. (The government agency supplying the service is actually using taxpayers' dollars. Therefore, in effect, these operating costs are a cost indirectly paid for by the user.)

2. Determine the present equivalent or annual equivalent of all capital investment costs.
3. Calculate the ratio of these two values:

$$B/C \text{ (modified)} = \frac{\text{PE of the annual BTCF values}}{\text{PE of all capital investment costs}} \qquad (11.2)$$

Note: The annual equivalent (AE) approach may be used. If income tax is a consideration, use the respective after-tax cash flow (ATCF) values.

The *modified benefit-cost ratio* treats all annual cash outflows [regardless of whether these cash outflows originate with the user (citizen) or the supplier (government agency)] as an annual operating cost. The benefits that result are usually benefits to the user. For example, highway or bridge toll fees represent a cost to the user and thereby a cost to the system under study. One may prefer to think of the user and supplier as being the same group of people. That is, the funds supplied by the government agency to cover capital and operating costs are actually supplied by the user through different forms of taxation and service charges.

The reason for using the modified benefit-cost ratio should become readily apparent when we look at a simple problem in tabular format.

Example 11.4

Problem: Three alternative proposals for a government facility are under consideration. Each proposal requires an investment of $100,000 and has an estimated life of 15 years. A before-tax rate of return value of 8% is used to compute equivalence between alternatives.

Alternative	User Annual Benefits	User Annual Costs	Supplier Annual Costs	BTCF
A	$30,000	$3,000	$12,318	$14,682
B	20,000	4,000	1,318	14,682
C	16,000	800	518	14,682

Compare these three alternatives using:

1. The conventional benefit-cost ratio
2. The modified benefit-cost ratio
3. The rate-of-return approach

Solution: Investment costs to the supplier converted to an annual equivalent BTCF.

$$\$100,000(A/P, 8\%, 15) = \$11,680$$

Alternative A

$$\text{B/C (conventional)} = \frac{\text{AE of net benefits to the user}}{\text{AE of capital + OC to supplier}}$$

$$= \$27,000/(\$11,680 + \$12,318) = 1.13$$

$$\text{B/C (modified)} = \frac{\text{AE of the annual BTCF values}}{\text{AE of all capital investments}}$$

$$= \$14,682/\$11,680 = 1.26$$

For the rate of return:

$$AE = \$14,682 - \$100,000(A/P, i, 15)$$

$$(A/P, i, 15) = 0.1468$$

$$i = 12\%$$

Alternative B

B/C (conventional) = $\$16,000/(\$11,680 + \$1,318) = 1.23$

B/C (modified) = $\$14,682/\$11,680 = 1.26$

Rate of return = 12%

Alternative C

B/C (conventional) = $\$15,200/(\$11,680 + \$518) = 1.25$

B/C (modified) = $\$14,682/\$11,680 = 1.26$

Rate of return = 12%

Summary

Alternative	Conventional B/C ratio	Modified B/C ratio	ROR (%)
A	1.13	1.26	12
B	1.23	1.26	12
C	1.25	1.26	12

Each alternative involved the same investment over the same time span with identical annual before-tax cash flows; therefore, no incremental analysis is necessary.

Conclusion: The modified benefit-cost method and the rate of return method give identical results. Using the conventional benefit-cost ratio can introduce a bias into the comparison of alternatives, thereby investing funds in undesirable alternatives. This problem applies to independent investment alternatives as well as mutually exclusive alternatives. Therefore, the conventional benefit-cost ratio is not recommended for the comparison of either mutually exclusive or independent alternatives.

11.6 APPLICATIONS OF BENEFIT-COST ANALYSIS

To conduct a benefit-cost analysis for an investment project, it is important that the analyst completes the following steps.

1. Clearly identify the problem.
2. Clearly define and set the objective(s) to be accomplished.
3. Generate alternatives that satisfy the objective(s) stated.
4. Clearly identify the constraints (e.g., technological, political, legal, social, financial) that exist within the system environment. This approach will help narrow the alternatives generated.
5. List the benefits and costs associated with each alternative. Specify each benefit and cost in monetary values. If there are factors associated with any alternative that the analyst cannot express in monetary terms, these factors should be clearly stated in the report.
6. Calculate the benefit-cost ratios and other indicators (e.g., present value, rate of return) for each alternative.
7. Prepare the final report comparing the results of the evaluation for each alternative examined.

Example 11.5

Problem: Periodic floods in the spring and drought in the summer occur in a river basin. These abnormal conditions cause economic losses and social hardships to the people located in the area. The affected area encompasses 5,000 square miles and a population of 25,000 people. The villages and towns in the area also suffer from periodic water shortages.

Objectives: The objective of the study is to manage the water and related resources of the river in the best interests of the residents of the region. Any peripheral impact outside the regions should also be considered.

Alternatives Considered

1. Damming the river to provide flood control and irrigation
2. Damming the river to provide flood control, irrigation, and a recreational area
3. Damming the river to provide flood control, a power source, and a recreational area
4. Relocating the residents most seriously affected

Constraints

1. The project must not reduce the net existing arable land area.
2. Technologically all alternatives are feasible.
3. Politically, there are pressures on both local and national governments from the area.
4. No specific limitation on funds has been set. However, care must be exercised to ensure that any funds utilized maximize the benefits to all concerned.
5. It is not socially desirable to try to relocate residents of the area.
6. Environmentally, the project should have no ill effects regarding wildlife in the area. Land slopes are conducive to irrigation with a minimum of cost.
7. The project should be targeted for completion within 3 years.

The constraints have narrowed the alternatives to two:

Alternative A: Construct a dam on the river to provide flood control and irrigation.

Alternative B: Construct a dam on the river to provide flood control, irrigation, and recreational facilities.

Assumptions

1. 10,000 acres can be irrigated once the dam is constructed, increasing the net agricultural area by 6,000 acres.
2. All investments are estimated to have zero salvage value at the end of 50 years.
3. A 50-year estimated life will be used in the calculations from the date the project becomes operational.

	Alternative	
	A	*B*
Year 1: Engineering and environmental studies	$ 500,000	$ 600,000
Year 2: Dam, irrigation, and recreational facility construction	4,000,000	4,500,000
Year 3: Dam, irrigation, and recreational facility construction	2,000,000	2,100,000
Annual operating costs to the supplier	500,000	500,000
Annual operating costs to the user for irrigation purposes	200,000	200,000
Annual operating costs to the recreational user		50,000
Additional capital costs will be required every 10 years for recreational facilities		50,000
Annual benefits to the user		
Flood damage reduction	1,000,000	1,000,000
Irrigation	700,000	700,000
Recreation		150,000

Use a social discount rate of 10%, $t = 0\%$. Evaluate these two proposals using the modified benefit-cost ratio method.

Solution: The cash flow diagram for alternative A is shown below. All values are in thousands of dollars.

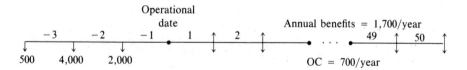

Using the modified benefit-cost ratio:

PE (at operational date) of annual before-tax cash flows

$$= (\$1,700,000 - \$700,000)(P/A, 10\%, 50) = \$9,914,800$$

PE (at operational date) of capital costs

$$= \$500,000(F/P, 10\%, 3) + \$4,000,000(F/P, 10\%, 2)$$
$$+ \$2,000,000(F/P, 10\%, 1)$$
$$= \$665,500 + \$4,840,000 + \$2,200,000$$
$$= \$7,705,500$$

$$\text{B/C (modified)} = \$9,914,800/\$7,705,500 = 1.29$$

Comparing the incremental benefit-cost ratio on the minimum investment (alternative A) over doing nothing, $\text{B/C} = 1.29 > 1.0$, so alternative A is acceptable.

Using the rate-of-return approach:

$$PE = +\$1,000,000(P/A, i, 50) - \$500,000(F/P, i, 3)$$
$$- \$4,000,000(F/P, i, 2) - \$2,000,000(F/P, i, 1)$$

Try 12%:

$$PE = \$8,304,500 - \$702,500 - \$5,016,000$$
$$- \$2,240,000$$
$$= \$346,600$$

Try 15%:

$$PE = \$-1,687,500$$
$$i = 12.5\% \text{ (which exceeds the 10\% social discount rate)}$$

Therefore alternative A is acceptable.

Problem: Use the rate-of-return method to evaluate the proposals.

Solution: The cash flow diagram below represents the incremental investment in B − A. The question to be answered is: Should the recreational facilities be installed? All cash flow values are in thousands of dollars.

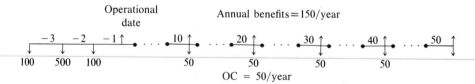

Installation of the recreational facilities at a later date cannot be accomplished with the same degree of success as installing these facilities now. If the trees are not removed now, they will cause major problems for boats, skiers, and fishermen should the facility be opened to recreation at a later date.

Using the modified benefit-cost ratio:

$$B/C \text{ (modified)} = \frac{\text{PE of the annual BTCF values}}{\text{PE of all capital costs}}$$

Bringing all cash flows to the operational date:

$$\text{PE of annual BTCF} = (\$150,000 - \$50,000)(P/A, 10\%, 50)$$
$$= \$991,480$$

$$\text{PE of capital costs} = \$100,000(F/P, 10\%, 3)$$
$$+ \$500,000(F/P, 10\%, 2) + \$100,000(F/P, 10\%, 1)$$
$$+ \$50,000(A/F, 10\%, 10)(P/A, 10\%, 40)$$
$$= \$878,782$$

$$B/C \text{ (modified)} = \$991,480/\$878,782 = 1.13$$

Using the rate-of-return approach:

$$\text{PE} = \$-100,000(F/P, i, 3) - \$500,000(F/P, i, 2)$$
$$- \$100,000(F/P, i, 1) - \$50,000(A/F, i, 10)\,(P/A, i, 40)$$
$$+ \$100,000(P/A, i, 50)$$

Try 12%:

$$\text{PE} = \$-140,500 - \$660,630 - \$112,000$$
$$- \$23,487 + \$830,450$$
$$= \$-106,167$$

Try 10%:

$$\text{PE} = -\$878,782 + \$991,480 = \$112,698$$
$$i = 10\% + 2\%(\$112,698/\$218,865)$$
$$= 11\% \text{ (which is greater than the 10\% social discount rate)}$$

Sensitivity analysis with respect to the social discount rate is normally undertaken. Use BTRR values of 5%, 10%, and 15%.

Try 5%:

$$PE \text{ of annual benefits} = \$100,000 \, (P/A, 5\%, 50) = \$1,825,590$$
$$PE \text{ of capital costs} = \$840,007$$
$$B/C \text{ (modified)} = \$1,825,590/\$840,007 = 2.17$$

Try 10%:

$$B/C \text{ (modified)} = \$991,480/\$878,782 = 1.13$$

Try 15%:

$$PE \text{ of annual benefits} = \$666,050$$
$$PE \text{ of capital costs} = \$944,456$$
$$B/C \text{ (modified)} = (\$666,050/\$944,456) = 0.71$$

In summary:

	Alternative B − A	
5%	10%	15%
B/C = 2.17	B/C = 1.13	B/C = 0.71

Sensitivity analysis could also be done with respect to the future cash flows associated with the recreational facility. For example, the growth in use of the facility may be assumed to parallel population growth in the area with a gradual increase in tourist trade.

11.6.1 Evaluation of Projects Having Several Objectives

With respect to projects that have several objectives, it is important to treat the incremental costs and benefits associated with any particular project separately whenever possible. For example, a flood control project should not be required to support recreational facilities partially, nor should the flood control project be used as a means to "carry" any other uneconomic project.

Many government projects fall into the category of multiobjectives. A feasibility study for a dam project involving one or more dams may have been initiated for flood control purposes. However, in the analysis, hydroelectric power, recreation, irrigation, and fish and wildlife as possible beneficial objectives can and probably should be explored. Too often, the approach is either one of suggesting that additional benefits for other purposes accrue at no additional cost without a proper analysis, or a makeshift effort is made to include other uses after the fact.

Example 11.6

Problem: An incremental analysis using benefit-cost ratios to evaluate a project having several objectives. The social discount rate used to calculate annual equivalent costs is 10%. Salvage values for each alternative are assumed to be negligible. Determine which alternatives should be selected.

Alternative	Investment Required	Estimated Life (years)	Net Annual Benefits BTCF	Annual Investment Costs P(A/P, 10%, 30)	Modified Benefit-Cost Ratios
A	$1,000,000	30	$350,000	$106,080	3.3
B	1,500,000	30	420,000	159,120	2.6
C	1,800,000	30	460,000	190,944	2.4
D	2,000,000	30	475,000	212,200	2.2

Solution

Method 1: Determine the annual equivalent of benefits minus costs for each alternative:

Alternative		AE
Do nothing	=	0
A	$350,000 − $106,080 =	$243,920
B	=	260,880
C	=	269,056
D	=	262,800

Select alternative C. Alternative C maximizes the annual equivalent value and therefore minimizes the combined costs to the user and supplier of the facility.

Method 2: An alternative method of analysis utilizing modified benefit-cost ratios may be done as follows:

a. Compare the incremental benefit-cost ratio on the minimum investment (alternative A) over doing nothing. B/C = 3.3 > 1.0, so alternative A is better than doing nothing.

b. Compare the incremental benefit-cost ratio on the next increment of investment in alternative B over alternative A.

$$B/C = \$70,000/\$53,040 = 1.3$$

1.3 > 1.0, so accept alternative B.

c. Compare the incremental benefit-cost ratio on the next increment
of investment in alternative C over alternative B.

$$B/C = \$40,000/\$31,824 = 1.3$$

1.3 > 1.0, so accept alternative C.

d. Compare the incremental benefit-cost ratio on the next increment
of investment in alternative D over alternative C.

$$B/C = \$15,000/\$21,256 = 0.7$$

0.7 < 1.0, so alternative D is not acceptable.

Alternative C is the preferred choice in the set of alternatives.

11.6.2 Benefit-Cost Analysis and Capital Rationing

Capital rationing is a fact of life within the public sector, and as a
result many worthwhile projects must be postponed or canceled because
of a shortage of funds.

To eliminate projects in a capital-rationing situation, certain government
agencies use a benefit-cost ratio greater than 1. Raising the benefit-cost
ratio as a means of eliminating certain projects may result in errors in
assigning priorities to investment alternatives. For example, a benefit-cost
ratio of 1.5 on one independent project A may correspond to a rate of
return of 15%, whereas a benefit-cost ratio of 3.0 on another independent
project B may correspond to a rate of return of 10%. The benefit-cost ratios
of 1.5 and 3.0, respectively, for the two independent projects indicate that
both projects meet the accept/reject measure, which is a benefit-cost ratio
of 1.0. It does not imply that project B is superior to project A.

When there is a shortage of funds (capital rationing) the social discount
rate should be revised upward to reflect the opportunity cost of capital.
Raising the social discount rate will allow the accept/reject measure, a
benefit-cost ratio of 1.0 to perform its intended function. That is, the desirable
set of independent projects will be selected to match funds available.

Example 11.7

Problem: Several independent projects are under study in a capital-
rationing situation. All projects are considered to be equivalent in risk, and
capital rationing is considered to apply indefinitely. Salvage values are
expected to be negligible with respect to all projects. Which of these independent projects should be accepted?

Project	Investment Required	Est. Life (years)	Annual User Benefits	Annual User Costs	Annual Supplier Costs	BTCF
1	$ 500,000	5	$300,000	$ 50,000	$100,000	$150,000
2	1,000,000	10	500,000	100,000	220,000	180,000
3	1,200,000	15	600,000	200,000	260,000	140,000

Solution: Calculate the benefit-cost ratios for each project using interest rates of 5%, 10%, and 15%. The calculations for 5% are as follows:

Project 1

B/C (modified) = $150,000/[$500,000(A/P, 5%, 5)] = 1.3

Project 2

B/C (modified) = $180,000/[$1,000,000(A/P, 5%, 10)] = 1.4

Project 3

B/C (modified) = $140,000/[$1,200,000(A/P, 5%, 15)] = 1.2

In a similar manner the remaining benefit-cost ratios in the table were calculated.

Project	B/C Ratio (5%)	B/C Ratio (10%)	B/C Ratio (15%)
1	1.3	1.1	1.0
2	1.4	1.1	0.9
3	1.2	0.9	0.7

Project acceptability represents a benefit-cost ratio ≥ 1.0. The projects that should be selected based on funds available can now be listed.

Funds Available	Social Discount Rate (%)	Projects Accepted
$ 0	Over 15	0
500,000	15	1
1,500,000	10	1, 2
2,700,000	5	1, 2, 3

The fact that project 2 has the highest benefit-cost ratio at a discount rate of 5% does not mean that this project should be given first priority. All three projects exceed the accept criteria, that is, a benefit-cost ratio ≥ 1.0. Therefore, one cannot say that any one of the three projects should be given a priority rating. If funds are not available for all three projects, the opportunity cost of capital is greater than 5%.

An alternative approach to the problem is to calculate the rate of return on each independent project and ladder the projects in descending order of rate of return. The cutoff point can then be determined based on funds available.

Project 1

$$AE = \$150,000 - \$500,000(A/P, i, 5)$$

$$i = 15\%$$

Project 2

$$AE = \$180,000 - \$1,000,000$$

$$(A/P, i, 10)(A/P, i, 10) = 0.180$$

$$(A/P, 12\%, 10) = 0.1770$$

$$(A/P, 15\%, 10) = 0.1993$$

$$i = 12.4\%$$

Project 3

$$AE = + \$140,000 - \$1,200,000$$

$$(A/P, i, 15)(A/P, i, 15) = 0.1167$$

$$i = 8\%$$

Cutoff Rate of Return (%)	Projects Accepted	Funds Required
Over 15	0	0
12.4–15	1	500,000
8.0–12.4	1, 2	1,500,000
8.0 or less	1, 2, 3	2,700,000

11.7 SUMMARY

The evaluation of projects financed by public funds is often more complex than evaluating projects within the private sector. Projects in the public sector are often more diverse in scope than projects in the private sector, and it is sometimes difficult to define the specific groups of citizens who benefit and those groups that the project may affect adversely.

Benefit-cost analysis is used by many government agencies as a method of project analysis and project comparison. The method, when properly applied through the use of the modified benefit-cost ratio, gives results consistent with the rate-of-return, present equivalent, or annual equivalent methods. However, the benefit-cost ratio is subject to misapplication and should be used with caution. A check by another method, such as rate of return, represents a valuable safeguard.

PROBLEMS

11.1 Discuss the principle "Those who benefit should pay" with respect to a city transit system that continually operates at a loss.

11.2 The social discount rate used to evaluate publicly funded projects has a major bearing on whether or not the project is approved. Discuss alternative methods of arriving at the proper social discount rate that should be used.

11.3 Discuss the use of benefit-cost ratios for ranking projects in a capital rationing situation.

11.4 How would you determine the monetary benefits of:
a. A city park?
b. A new fire station?

11.5 The following estimates apply to a public investment. Initial investment = $1,000,000, user benefits = $460,000 per year, user costs = $50,000 per year, supplier costs = $250,000 per year, social discount rate = 8%, $t = 0$, $n = 10$ years, and SV = 0.
a. Calculate the modified benefit-cost ratio.
b. Calculate the conventional benefit-cost ratio.
c. Calculate the rate of return.

11.6 Assume that the city in which you live is considering improving the fire protection it gives its citizens. The plan is to add additional fire stations and equipment. The initial cost is estimated to be $25,000,000 with an estimated salvage value of $5,000,000 in 15 years. User benefits through increased protection and lower insurance rates are estimated to be $4,000,000 per year. Supplier operating costs are estimated to be $1,000,000 per year. The social discount rate = 8%, $t = 0$, and $n = 15$ years.
a. Calculate the conventional benefit-cost ratio.
b. Calculate the modified benefit-cost ratio.
c. Calculate the rate of return.

11.7 A program is under consideration to pave a certain section of highway. The following estimates pertaining to this project have been developed (costs per mile):

Capital cost	$42,000
Annual operating costs to the supplier	3,000
Annual benefits to the user	10,000

The estimated useful life is 10 years. Determine the modified benefit-cost ratio using a social discount ratio of 5%, 10%, and 15%.

11.8 The widening of a highway from two-lane, two-way traffic to four-lane with two lanes in each direction is proposed between two cities 200 kilometers (125 miles) apart. The proposal includes a toll system to pay for the highway.

Construction costs per kilometer	$ 100,000
Increased annual operating costs	600,000
Annual user benefits mainly due to:	
Reduced accidents	1,000,000
Reduction in travel time of	
20 minutes per trip	4,000,000

Social discount rate = 8%, $t = 0$, $n = 10$ years, and SV = 0. Use the modified benefit-cost ratio. What toll fee must be charged to warrant this investment if on the average 4,000 vehicles use the highway each day (365 days per year)?

11.9 A new bridge is proposed across the river in a major city which will lower the traffic congestion on existing bridges. As a result much valuable time will be saved by the users and a reduction in accidents is expected.

Initial cost of the project	$50,000,000
Increased annual operating costs	2,000,000
User benefits mainly due to:	
Reduction of travel time based on a time saving of 10 minutes per vehicle valued at $15.00 per hour and a traffic flow of 10,000 vehicles per day for 250 days per year	6,250,000
One less accident per day at an average cost of $1,000	250,000

Social discount rate = 8%, $t = 0$, $n = 25$ years, and SV = 0. Should the bridge be built? Use the modified B/C ratio.

11.10 Two mutually exclusive proposals for a government facility are under consideration, each with a useful life of 10 years and zero salvage value. Neither alternative has to be accepted.

	A	B
Initial cost	$200,000	$300,000
Annual operating costs	50,000	25,000
Annual revenues	80,000	80,000

Social discount rate = 10% and $t = 0$.
a. Compare these proposals using the modified benefit-cost ratio.
b. Compare these proposals using rate of return.

11.11 Two sources of water supply are under consideration for a small city.

Alternative A: Water wells expected to be sufficient for the next 10 years, and then a pipeline will have to be run to a nearby river. Some water rationing is expected to occur with the use of wells.

Alternative B: Run a pipeline to the river now.

	A	B
Initial cost		
Water wells now	$ 100,000	
Pipeline in 10 years	5,000,000	
Pipeline now		$4,000,000
Annual operating costs for the		
first 10 years	450,000	
Annual operating costs after		
10 years	250,000	
Annual operating costs with		
pipeline now		250,000
Annual user benefits over the		
first 10 years		100,000

Social discount rate is 10%, $t = 0$, $n = 30$ years, and SV = 0. Compare these alternatives: Either alternative A or alternative B has to be accepted.
a. Using the modified benefit-cost method.
b. Using the rate-of-return method.

11.12 Three alternative proposals for a government facility are under consideration. Each proposal requires an investment of $100,000 and has an estimated life of 5 years. All investments are assumed to have zero salvage value. A social discount rate of 10% is used to compute equivalence between alternatives.

Alternative	User Annual Benefits	User Annual Costs	Supplier Annual Costs	BTCF
A	$47,000	$3,000	$14,168	$29,832
B	35,000	0	5,168	29,832
C	26,000	0	−3,832[a]	29,832

[a] Indicates a negative cost to operate the facility.

Compare these three alternatives using:

a. The conventional benefit-cost ratio.

b. The modified benefit-cost ratio.

c. The rate-of-return approach.

11.13 A power dam is under consideration. In addition to using the dam for hydroelectric power, recreational facilities are under consideration. It will take 3 years to place the power project on stream. Use a life of 50 years from the operational date and an estimated salvage value of zero. The social discount rate = 10% and $t = 0$.

Alternative A: power dam only

Alternative B: power dam plus recreation facilities

	Alternative	
	A	B
Year 1: Engineering and environmental studies[a]	$ 1,000,000	$ 1,200,000
Year 2: Construction[a]	50,000,000	55,000,000
Year 3: Construction[a]	100,000,000	110,000,000
Additional capital costs will be required every 10 years for recreational facilities (year 10, 20, 30, 40)		500,000
Annual operating costs to the supplier	2,000,000	1,500,000
Annual charges to the user		1,000,000
Annual benefits to the user	20,000,000	23,000,000

[a] Beginning-of-year cash flows.

Evaluate these two alternatives using:

a. The modified benefit-cost ratio.

b. The present equivalent approach.

11.14 The following mutually exclusive alternatives represent a project that has several objectives. Salvage values for each alternative are assumed to be negligible. Use a social discount rate of 8%.

Alternative	Investment Required	Estimated Life	Net Annual Benefits BTCF	Annual Investment Costs P(A/P, 8%, 25)	Modified Benefit-Cost Ratios
A	$1,000,000	25	$400,000	$110,170	3.6
B	1,400,000	25	450,000	154,238	2.9
C	1,900,000	25	530,000	209,323	2.5
D	2,100,000	25	550,000	231,357	2.4

Determine which of the alternatives should be selected.

a. Use the benefit-cost analysis and an incremental approach.

b. Use the annual equivalent approach.

11.15 Several independent projects are under study in a capital-rationing situation. All projects are considered to be equivalent in risk, and capital rationing is considered to apply indefinitely. Salvage values are expected to be negligible with respect to all projects.

Project	Investment Required	Number of Years	Annual User Benefits	Annual User Costs	Annual Supplier Costs	BTCF
1	$1,000,000	5	$600,000	$100,000	$200,000	$300,000
2	1,000,000	10	500,000	75,000	250,000	175,000
3	1,000,000	15	550,000	150,000	275,000	125,000

Which of these independent projects should be accepted if:

1. $1,000,000 is available to invest?
2. $2,000,000 is available to invest?
3. $3,000,000 is available to invest?
   a. Use benefit-cost analysis and interest rates of 5%, 10% and 15%.
   b. Use rate-of-return analysis.

CHAPTER TWELVE

Capital Investment Decisions Involving Risk and Uncertainty

Forecasting and predictions are an integral part of economic analysis. The analyst deals with the future, and the forecasted results are only as accurate as the analyst's ability to foresee future events and project current trends into the future. Therefore, any economic analysis is at best a prediction. There is always a certain amount of uncertainty that surrounds the final results. Sometimes, this uncertainty is insignificant in relation to the overall decision problem. At other times it is the dominant element within the decision.

Statistics represent a useful tool to the decision maker in measuring in quantitative terms the amount of risk and uncertainty associated with any investment decision.

12.1 STATISTICAL ANALYSIS

Before proceeding with a specific discussion of risk and uncertainty, it is desirable to present some of the basic data associated with a statistical analysis. There are two basic purposes of statistics as used by the analyst or manager. They are:

1. To help predict the future
2. To help influence the future

In the business world, a major portion of management's time is devoted to analyzing such statistics as company financial statements and government publications on consumer spending and inflation projections. These statistics give management an indication of what to expect of the future. With this type of information, management is able to take whatever precautionary measures are necessary to see that the organization meets its projected objectives.

The organization must revolve around the total systems concept. Market forecasts are made so that manufacturing can provide the necessary equipment, materials, and personnel to meet sales demands. The accounting department predicts the inflow and outflow of money so that all accounts such as payroll and accounts payable can be satisfactorily handled. Manufacturing predicts the number and type of personnel that are required in order that the personnel department can satisfactorily supply the labor demand. Each individual, department, and division of an organization must function as an integral part of the total organization if a company is to be successful.

The past data on which a statistical analysis is made tells the decision maker what will happen if nothing changes. Statistics also tell the decision makers where to place their specific efforts to influence what will happen in the future.

To make a statistical analysis, past data are gathered. From this sample of data, decisions are made as to what is happening with respect to some particular function (e.g., operating revenues or operating costs). Therefore, the decision maker should be able to assess the value of the information on which the prediction is made and should be able to state the accuracy of the prediction made.

The intent of this section is to give sufficient background to answer three basic questions:

1. What information is required as input to make a prediction?
2. How should the prediction be made?
3. How accurate is the prediction that is made?

To answer these questions, one must have:

1. A knowledge of the distribution from which the sample is obtained, to determine the following two measures
2. A measure of central tendency
3. A measure of variability

12.1.1 Empirical Frequency Distributions

The normal method for summarizing or describing a set of data is the construction of a frequency table or frequency distribution. That is, we divide the overall range of the values in our sample data into a number of classes and count the number of observations that fall into each of these classes. Data in this form allow one readily to see the overall pattern of the distribution of the data.

Example 12.1

Problem: To illustrate the construction of a frequency distribution, we will use the operating costs per mile for a fleet of eighty 1-ton trucks.

70	66	67	67	71	70	75	70
72	69	69	70	70	68	70	68
64	73	72	72	69	67	68	70
76	71	70	71	67	71	64	68
70	72	71	70	66	75	67	70
69	76	73	69	75	72	71	72
68	75	69	67	69	70	73	73
74	65	71	72	71	72	72	71
69	64	68	70	74	69	69	70
71	66	70	68	70	71	70	68

Solution: Before constructing a frequency distribution we must determine how many classes to use. An empirical relationship that seems to hold and may be used as a guide to the number of classes is given by Sturges (see Appendix D, Ref. 46).

$$k = 1 + 3.3 \log n$$

where k = number of classes
n = total number of numerical
values in the data
$\log n$ = log of n to the base 10

In the example given:

$$k = 1 + 3.3 \log 80$$
$$= 1 + 3.3(1.9)$$
$$= 1 + 6.27$$
$$= 7.27$$

On the basis of this calculation we would use seven classes. As an additional guide, one would normally not have less than 5 or more than 15 classes.

Using seven classes, we must now specify the class intervals and the class boundaries. In the table the range of values is

$$R = 76 - 64 = 12$$

This means our class intervals must cover the range from 64 to 76. Therefore, we can use seven classes, each of which is two units wide. We could, of course, use some other combination. Equal class intervals are not essential but will be found to be desirable in terms of ease of calculation. The limits of the class intervals should be chosen so that there is no ambiguity in assigning observed values to the classes. The latter requirement is most easily satisfied by:

1. Selecting class intervals that carry one more decimal place than the original data
2. Proper use of inequality and equality signs

We shall adopt the first of these procedures. Using class intervals of width 2 cents we arrange the data in the following form:

Cost per Mile (cents)	Tally	Frequency	Relative Frequency (%)
63.6–65.5	IIII	4	5.0
65.6–67.5	ﷺ IIII	9	11.3
67.6–69.5	ﷺ ﷺ ﷺ III	18	22.5
69.6–71.5	ﷺ ﷺ ﷺ ﷺ ﷺ III	28	35.0
71.6–73.5	ﷺ ﷺ III	13	16.2
73.6–75.5	ﷺ I	6	7.5
75.6–77.5	II	2	2.5
		80	100.0

Note that in this table we no longer know the individual values of the operating cost, only in which class interval they fall. But the loss in accuracy is balanced to some extent by the gain in conciseness.

The data in the table can be used to draw a frequency histogram or polygon. The column headed "relative frequency" tells us what proportion of the total observations falls in each class. The values are found by dividing each class frequency by the total frequency. These data can be used to construct a cumulative frequency distribution.

Figure 12.1 shows these data as a frequency histogram. A frequency histogram is formed by erecting rectangles over the class intervals. The areas of the rectangles are proportional to the class frequencies. One advantage

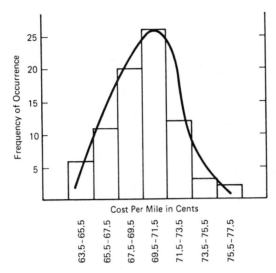

FIGURE 12.1 Histogram of operating costs per mile per unit.

of equal class intervals is that areas of the rectangles will be proportional to the heights of the rectangles. We can therefore draw the rectangles so that cost is proportional to frequency.

The frequency histogram, as well as the frequency distribution, gives us not only an estimate of the average value but also an idea of the amount of variability present in the data.

12.2 DESCRIPTIVE MEASURES NECESSARY TO MAKE A PREDICTION

We have looked briefly at tabular and graphical forms of presenting and summarizing quantitative data. These techniques are of value in displaying the outstanding features of the distribution of the data; however, the application of statistical methods, for the most part, requires a more concise description. This can be achieved by performing certain arithmetical operations on the data to obtain values for one or more descriptive measures of the data. The basic descriptive measures are:

1. The measure of central tendency or location
2. The measure of variation or scatter

Our prime objective in using statistics is to infer from a sample the characteristics of the universe or total population. The characteristics calculated from the sample chosen from a population are called statistics.

12.2.1 Measures of Central Tendency

The measures of central tendency of most interest to us in making predictions are:

1. The arithmetic mean
2. The median

The arithmetic mean is by far the more important of these two values.

Arithmetic Mean

The arithmetic mean is the most common representative value of central tendency. The *sample mean,* denoted by $\overline{X}$, is defined as the arithmetic average of all the values in the sample. The formula for calculating the sample mean is

$$\overline{X} = \frac{\Sigma X_i}{n} \tag{12.1}$$

where $\overline{X}$ = arithmetic mean
X_i = ith value observed
n = number of observations in the sample

It should be noted that the arithmetic mean is affected by every item in the sample and is greatly affected by extreme values. Two interesting properties of the arithmetic mean are:

1. The sum of the deviations from the arithmetic mean is 0.
2. The sum of the squares of the deviations from the arithmetic mean is less than the sum of the squares of the deviations from any other value.

The arithmetic mean has both advantages and disadvantages. Its advantages are:

1. It is the most commonly used average.
2. It is easy to compute.
3. It is easily understood.
4. It lends itself to algebraic manipulation.

The one major disadvantage of the arithmetic mean is that it is unduly affected by extreme values and may therefore be far from representative of the sample.

If our data are in the form of a frequency distribution, the sample mean is calculated by the formula

$$\overline{X} = \frac{\Sigma M_i f_i}{\Sigma f_i} = \frac{\Sigma M_i f_i}{n} \qquad (12.2)$$

where M_i is the midpoint of the ith interval and f_i is the frequency of the ith interval.

Median

A representative value frequently employed as an aid in describing a set of data is the median. The *sample median*, denoted by m, is the middle value of an *odd* number of observations in an array. If the sample is composed of an *even* number of observations, the median is the average value of the two middle observations.

Perhaps the most common use of the median is to replace the arithmetic mean under conditions where extreme values may tend to bias the calculation of the value to represent central tendency.

12.2.2 Measures of Variation

The representative values discussed above more specifically define the central value. It must be clear, though, that they are not sufficient by themselves to describe most populations or samples adequately. For example, if we consider two sets of data which have the same mean and the same median, but which differ significantly in the amount of variation present in each set of data, it seems reasonable that some measure of variation, or dispersion, among the individual values is desirable. Several such measures have been devised, the most common of which are:

1. The standard deviation
2. The variance
3. The range

Standard Deviation and Variance

The best known and most widely used measure of variability is the standard deviation. Of almost equal importance is the square of the standard deviation, known as the *sample variance* and denoted by S^2. The variance and standard deviation may be calculated as follows:

$$S^2 = \frac{\sum_{i=1}^{n} X_i^2 - \left(\sum_{i=1}^{n} X_i \right)^2 \Big/ n}{n - 1} \qquad (12.3)$$

$$S = \sqrt{S^2} \tag{12.4}$$

The sample standard deviation is then defined as the positive square root of the variance.

Range

The simplest measure of variation is the range. If we denote the smallest sample value by X_{min} and the largest sample value by X_{max}, the *sample range* is given by

$$R = X_{max} - X_{min} \tag{12.5}$$

The sample range, although easy to obtain, is often determined inefficient because it ignores all the information available from the intermediate sample values. However, for small samples (sample size less than 10), the efficiency, relative to other measures of variation, is quite high. Thus we find the sample range enjoying a favorable reception and wide use, because of its ease in computation.

Coefficient of Variation

To portray a valid comparison of variation among large values and the variation among small values, such as the variation among salaries of industrial executives and the variation among the wages of day laborers, the variation is expressed as a fraction of the mean, and frequently as a percentage. This measure of relative variation is called the *coefficient of variation* and is defined as

$$CV = \frac{S}{\overline{X}} \tag{12.6}$$

12.3 DISTRIBUTIONS

Most industrial data tend to group around some central value with some variability on either side of this central value. When these distributions follow definite patterns, they are given specific names. The three most useful distributions to the industrial decision maker are referred to as:

1. The normal
2. The poisson
3. The binomial

Of these three distributions the normal distribution finds the most wide application in solving industrial problems and will receive major emphasis here.

If one considers a normal distribution with mean μ and standard deviation σ, 68.27% of the area under the curve is included within the limits $\pm 1\sigma$; 95.45% is included within the limits $\pm 2\sigma$, and 99.73% is included within the limits $\pm 3\sigma$ (see Figure 12.2). The significance of Figure 12.2 is that one can now make a probability statement about values that one presumes come from the universe or population from which the sample distribution is drawn.

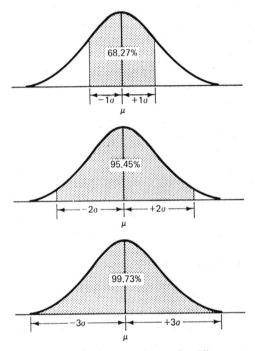

FIGURE 12.2 Areas under the normal curve for different sigma limits.

Let us assume that the sample data in Table 12.1 are representative of the population from which they were drawn. That is, the sample was randomly and independently selected.

Random selection is making sure that each item in the population nas an equally likely opportunity of being selected. Independent selection ensures that the samples selected are independent of each other; that is, the choice of one item does not affect the probability that the others will be chosen.

If one further assumes that the sample shown in Example 12.1 has been drawn from a normal population, what predictions can be made about that population?

1. The estimated mean equals

$$\overline{X} = \sum X_i/n = 70$$

2. The standard deviation equals

$$S_x^2 = \frac{\sum_{i=1}^{n} X_i^2 - \left(\sum_{i=1}^{n} X_i\right)^2 \Big/ n}{n - 1} = 7.29$$

$$S_x = \sqrt{S_x^2} = 2.7$$

Let us now present this on a standard normal distribution (a normal distribution with a mean of 0 and a standard deviation of 1) and make some predictions with respect to these data. Z values are calculated according to the formula

$$Z = \frac{\text{desired value} - \text{mean}}{\text{standard deviation}}$$

To calculate the probability of the occurrence of values below 75 cents; calculate

$$Z = \frac{75 - 70}{2.7} = 1.85$$

From the table of the cumulative normal distribution in Appendix F, a Z value of 1.85 corresponds to an area of 0.9678. Thus it can be inferred that 96.78% of the values are below 75 and 3.22% of the values are above 75.

Similarly, we can calculate that values smaller than 67.3 will occur about 15.9% of the time, values between 70 and 72.7 will occur about 34.1% of the time, and values larger than 75.4 will occur about 2.3% of the time (see Figure 12.3).

12.4 HOW VALID ARE THE PREDICTIONS?

A most important question is how accurate is the information produced by statistical analysis? If randomness and independence were indeed achieved, relatively small samples can be used to predict the population behavior.

To determine the accuracy of the data, let us look again at the sample

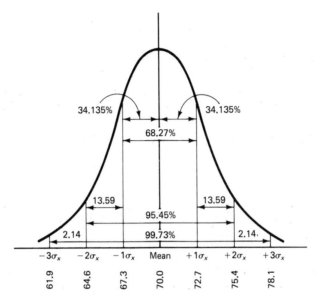

FIGURE 12.3 Relationship between standard deviation and frequency.

shown in Example 12.1. Let us assume further that eight random and independent samples of n equal to 80 were drawn. We find the means of the samples to be, say, 70.2, 69.7, 70.0, 69.6, 70.2, 70.5, 69.9, and 70.0 cents, respectively.

Using the following results from statistical theory, one can make some interesting observations.

1. The *central limit theorem* states that whatever the shape of the frequency distribution of the original population of X's, the frequency distribution of $\overline{X}$'s in repeated random samples of size n tends to become normal as n increases. The individual sample size used depends on how close to normality the original population happens to be. For most industrial applications sample sizes would not have to exceed 10. In many cases sample sizes of 4 or 5 are sufficient.
2. In samples of reasonable size, $\overline{X}$ is approximately normally distributed about $\overline{\overline{X}}$, with a standard error of $S_x/\sqrt{n}$.

Example 12.2

Problem: Using the mean values for operating costs per mile calculated above:

$$70.2, \quad 69.7, \quad 70.0, \quad 69.6, \quad 70.2, \quad 70.5, \quad 69.9, \quad 70.0$$

and assuming that sufficient samples have been drawn and that the sample size is sufficiently large to approximate the normal distribution, calculate $\overline{\overline{X}}$, $S_{\overline{x}}$.

Solution

$$\overline{\overline{X}} = \frac{\Sigma \overline{X}}{n} = \frac{560.1}{8} = 70.01 \text{ cents per mile}$$

$$S_{\overline{x}} = \frac{S_x}{\sqrt{n}} = \frac{2.7}{8.94} = 0.30 \text{ cent per mile (sample size: } n = 80)$$

(1) $\overline{X} = 70$ $S_x = 2.7$ cents per mile
(2) $\overline{\overline{X}} = 70.01$ $S_{\overline{x}} = 0.30$ cent per mile

This brief introduction to statistics should be sufficient to allow us to proceed with our discussion of risk and uncertainty.

12.5 RISK AND UNCERTAINTY

There is some confusion and ambiguity surrounding the terms risk and uncertainty. They are sometimes used in talking about the same things; other times they are used in talking about very different aspects of the same problem. Some texts do not distinguish between the two; others, for example, Canada and White (see Appendix D, Ref. 13) suggest that the element or analysis involves risk if the probabilities of the alternative possible outcomes are known, whereas a problem is characterized by uncertainty if the frequency distribution of the possible outcomes are not known. A more common use of the term *risk* connotes the prospect of unfavorable outcomes. It is usually used in situations that entail some possibility of harm either to one's person or to one's investment. For example, oil exploration companies speak of risking their capital in the search for oil, investors refer to risking their money by investment in a new venture, and drivers consider risking their lives on the highway. The term *uncertainty* implies simply that all the relevant facts are not known. These are sometimes not known because they cannot be, for example, in attempting to predict the future, or they could be unknown because it is simply too expensive or too difficult to determine them completely. Most business decisions are made under uncertainty. The uncertainty may create an element of risk, that is, a prospect of an unfavorable outcome, or it may simply create a situation where there exists a spectrum of favorable outcomes. It is the uncertainty that exists in the decision that we must address.

The following meanings will be used in this text.

Risk: The possibility that the venture has an unfavorable outcome

Uncertainty: the fact that the outcome of the venture is not, or cannot, be known

12.6 CAUSES OF UNCERTAINTY

If it were possible to encapsulate the uncertainty contained within a specific analysis, the problem would be greatly simplified. However, it is a rare situation where the uncertainty arises from one factor. Some of the obvious causes of uncertainty within an economic analysis are as follows.

1. *A misinterpretation or misunderstanding of the causative factors that worked in the past:* Most projections and analyses are based on an extrapolation of past data into the future. Usually, this is done on the basis of some assumed causative factors. For example, increased disposable income is taken as a positive sign by automobile manufacturers that people will be buying new cars. Similarly, projections for increased sales of new homes could be taken by appliance manufacturers as indicative that they will sell more appliances. If, however, the positive factor assumption is invalid, the original base series may increase, whereas the corresponding complementary product which is the one of concern will not.

2. *Changing external circumstances that invalidate past experience:* The classic example of this sort of situation is the OPEC oil cartel of 1973. Analyses made prior to that time projected low-cost energy into the foreseeable future. Analyses made after the cartel tended to project high-cost energy, together with a very high rate of increase in cost.

3. *Technological change and obsolescence:* There was a time when the most stable and conservative investment one could make would be to buy railway or street car company stock. These investments were seen as basic industries that would always be required and would continue into the future. What the analysts did not see at that time was the change in traveling patterns that came about with the advent of the automobile and the airplane.

4. *Excessive optimism:* It is not uncommon for project sponsors to become overenthusiastic and excessively optimistic about the prospects of their particular project. As a result, revenues tend to be optimistically estimated, and possible adverse cost effects minimized or ignored.

5. *Flexibility:* There is no question that the relative recoverability of investment commitments is a major function in uncertainty. A car rental agency usually has little difficulty in disposing of its automobiles; thus an error in the number and style may have some adverse con-

sequences, but it is not likely to be serious. On the other hand, a company producing a very special purpose facility or purchasing a highly specialized piece of machinery is very vulnerable to any changes that might occur.

6. *Government action:* As was suggested when the topics of income tax were introduced, the government is a major partner in almost every business and investment concern. However, some types of business and investments are highly susceptible to changes in government policy. By changing such factors as the depreciation rules, tax write-offs, investment subsidies, and licensing arrangements, the government can make some investments very profitable and others very unprofitable. An example of the extreme effect of government action is the current nationalistic activities of a number of governments. Any major trans-national company must expect to have some of its facilities and assets nationalized or expropriated over their lifetime. Furthermore, the company may or may not receive adequate compensation for the takeover.

7. *Errors:* As must be obvious from the other chapters in this text, the application of economic principles to investment evaluations are not always clear and straightforward. Thus there always exists the possibility of mistreatment or inappropriate allocation of some of the data. Furthermore, there will always exist the possibility of errors. This possibility is especially relevant now that we are moving into an era of complex computer models, where it is not always possible to check each operational step.

12.7 RISK AVERSION

The usual business reaction to uncertainty is to avoid it. Business managers will tend to select the projects that offer the smallest level of uncertainty. If this is not possible, they will move to other devices—either that of insurance or enlisting a number of partners and thus spreading the results of uncertainty. Many business ventures are entered into jointly and thus share the risk involved.

That business seeks to minimize the uncertainty and avoid risk should seem neither unnatural nor surprising. After all, most large corporations or utilities are not managed by their owners. The owners are often a widely disparate group of shareholders, including pension funds, insurance companies, mutual funds, and individual investors. These owners reflect their wishes through the vehicle of the stock market, buying the shares of companies that please them and selling the shares of companies that displease them. Generally, a steady, consistent growth of profits and markets, together with a conservative management image, is taken as being indicative of sound management practice. Excessive (i.e., predominately different from average) profits or losses, except where they can be explained by revolutionary developments (e.g., Xerox) or market aberrations (e.g., the OPEC cartel), are generally viewed as an indication of unstable management. Firms of this type tend to attract speculators rather than investors, which in turn contributes to an unstable stock price.

So generally, managers do not exhibit risk-seeking, adventurous behavior, or seek the optimum result from decision situations. Instead, they practice what Simon has described as *satisficing* behavior. That is, they seek satisfactory profits, satisfactory market penetration, and satisfactory growth.

Simon (see Appendix D, Ref. 44), in his development of the theory of human choice, distinguished between administrative man and economic man. Economic man is assumed (1) to be completely informed and (2) to maximize something. For administrative man Simon rejected these assumptions and proposed the *principle of bounded rationality*.

> The capacity of the human mind for formulating and solving complex problems is very small compared with the size of the problems whose solution is required for objectively rational behavior in the real world—or even for a reasonable approximation to such objective rationality. [Thus the] key to the simplification of the choice process . . . is the replacement of the goal of maximizing with the goal of satisficing, of finding a course of action that is good enough.*

The risk-averse behavior in a decision situation will usually manifest itself in one of four ways. One way is to choose alternatives that maintain flexibility. Consider the following alternatives to handling a major increase in output.

Alternative A: Meet the increase by working the personnel overtime.

Alternative B: Hire additional workers and purchase extra machinery for them.

Alternative C: The new output level is now high enough to justify the purchase of an expensive special-purpose, highly automated machine which was uneconomic at the previous output.

The correct alternative depends on whether the increased demand is sustained or temporary. If it is sustained, alternative C is the best choice. If it is temporary, alternative C is the worst choice, leaving one with high payments, high fixed costs, and overcapacity.

---

* H. A. Simon, "A Behavioral Model of Rational Choice," *Quarterly Journal of Economics,* Vol. 69, pp. 99–118, 1955.

If one chooses alternative A, and the increase is sustained, it is always possible to shift to alternative B or C. Furthermore, the accounting records do not measure potential profit not made. Most industries are not so competitive that miscues of this type will have any noticeable short-run effect. So alternative A, the temporary expedient, which maintains flexibility and masks indecision, is the desirable one.

The second way is to avoid alternatives that have potentially unfavorable outcomes. This behavior is briefly discussed in the section on utility theory (Section 12.10.1). Here it will only be noted that business managers tend to avoid decisions that have outcomes, however remote, that would have a major adverse effect on their careers or that of their company's.

The third and fourth behaviors are often confused. One is procrastination, the other is to seek more information. They are confused because they interact. Sometimes a delay produces more information and thus facilitates the decision. Sometimes a search for new or additional information results in nothing more than a delay. We cannot endorse procrastination as a decision technique.

12.8 SENSITIVITY ANALYSIS

In the previous chapters we have used single-valued estimates for all parameters, such as first cost, salvage value, economic life, operating costs, and operating revenues.

In effect, we have assumed that the parameter values were known with certainty. This does not mean that the analyst has not explored other possible outcomes. In some cases these single-valued estimates may implicitly consider risk through the use of conservative values for such parameters as MARR.

In a large number of evaluations, single-valued estimates accomplish our objective quite satisfactorily. However, under certain conditions we may want to check the effect of varying one or more of the investment parameters. This procedure is referred to as sensitivity analysis. In an investment analysis we are dealing with cash flows that occur in the future. We may be reasonably sure of the possible values a parameter can take on, but uncertain of their chance of occurrence. For example, we may have projected values each year for operating revenues associated with a specific investment but feel that these revenues may vary by as much as plus or minus 15%. We can hold all other parameters constant and check the impact on the rate of return, present equivalent, or annual equivalent values of varying operating revenues by plus or minus 15%. If a variation of ± 15% does not have a significant impact on our calculated results, we can say that the analysis is relatively insensitive over the range in values concerning operating revenues. If the investment is quite sensitive to changes in operating revenues, the analyst may, for example, decide that a more in-depth market study is required.

Each parameter value in the study can be varied independently or simultaneously to answer the question "What if"? That is, what happens if I vary the value of each parameter independently, holding all other values constant? What happens if I vary several values simultaneously?

Example 12.3

Problem: Sensitivity analysis

Assume that we have projected the following values for each parameter regarding a specific investment alternative. P = $150,000, SV = $50,000, n = 10, OC = $40,000 per year, t = 50%, and i_a = 15%. Use the straight line method for depreciation allowance (equation 4.2)

We would like to check the sensitivity of revenue requirements to independent variations in each variable. Calculate revenue requirements (AER) using the projected values.

Solution

$$AER = \$40,000 + [\$150,000(A/P, 15\%, 10)$$

$$- \$50,000(A/F, 15\%, 10) - 0.5(\$10,000)]/0.5$$

$$= \$84,850$$

Problem: Determine the sensitivity of AER to changes in first cost, salvage value, operating cost, and economic life.

Solution

P ($)	100,000	125,000	150,000	175,000	200,000
AER ($)	69,925	77,388	84,850	92,313	99,775
SV ($)	20,000	30,000	40,000	50,000	60,000
AER ($)	84,805	84,820	84,835	84,850	84,865
AEOC ($)	20,000	30,000	40,000	50,000	60,000
AER ($)	64,850	74,850	84,850	94,850	104,850
n (years)	5	10	15	20	25
AER ($)	94,664	84,850	82,537	81,952	81,940

The tabular format and graphic analyses (Figure 12.4) indicate that in this particular investment revenue requirements are:

1. Very sensitive to changes in the initial investment
2. Relatively insensitive to changes in salvage value
3. Very sensitive to changes in operating cost
4. Relatively insensitive to changes in estimated life beyond 10 years

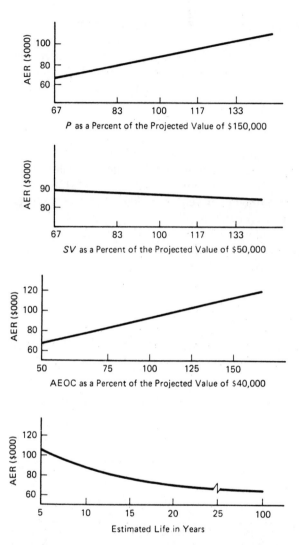

FIGURE 12.4 Sensitivity analysis of four investment parameters.

The example also illustrates a difficulty with sensitivity analysis. That is, with only four parameters and a limited analysis, we obtained 20 different results. However, the analysis gives us no guidance as to which, of the 20 different combinations of input values, is the more likely. This form of analysis is good when the sensitivity of only one or two parameters are being evaluated. With more than two parameters more powerful techniques such as Monte Carlo simulation are required.

12.9 MONTE CARLO SIMULATION

When more than one variable is varied simultaneously to display the risk associated with a given investment, the Monte Carlo method may be used. The *Monte Carlo method* consists of using random sampling of the project inputs to generate a frequency distribution of the possible project outcomes. The method and size of sampling is done according to statistical procedures, so the resulting distribution has a predetermined level of accuracy. Thus it is possible to use the distribution of outcomes to make a probability statement about the project. For example, the results of a simulation of a given project might have indicated that there is a 50% chance the project will yield a 14% rate of return, and only a 2% chance that the project will result in a loss (rate of return less than zero).

Basically, the approach in applying probability theory to investment decisions involves:

1. Assigning numerical values to the range of possible outcomes for the parameters considered in evaluating the investment (e.g., years or dollars).
2. Assigning a definite numerical weight (probability) to every possible outcome. The weights for all possible outcomes should total to 1.00.
3. Use the Monte Carlo method to develop a series of numerical weights and associated profitability distribution for the investment.

The Monte Carlo technique is best explained by an example. Suppose, for a certain product, that the marketing department gave the following estimate of sales:

Probability of sales exceeding 1,000 is 90%
Probability of sales exceeding 1,500 is 70%
Probability of sales exceeding 2,000 is 50%
Probability of sales exceeding 2,500 is 20%
Probability of sales exceeding 3,000 is   0%

The purchasing department provides the following estimate of raw material costs:

Probability that raw material will cost
  less than or equal to $5.00/unit = 10%
Probability that raw material will cost
  less than or equal to $5.10/unit = 25%

Probability that raw material will cost
  less than or equal to $5.20/unit =   50%
Probability that raw material will cost
  less than or equal to $5.30/unit =   95%
Probability that raw material will cost
  less than or equal to $5.40/unit =   100%

The labor cost per unit is $3.00, and there is a fixed cost of $3,600. The units sell for $10.00 per item.

Calculating the extremes, it can be seen that if sales are 1,000 units and material costs $5.40 per unit:

$$\text{Profit} = 1,000 \times \$10.00 - \$3,600 - 1,000 \times \$3.00 - 1,000 \times \$5.40$$

$$= \$-2000$$

and if sales are 3000 and material $5.00:

$$\text{Profit} = 3,000 \times \$10.00 - \$3,600 - 3,000 \times \$3.00 - 3,000 \times \$5.00$$

$$= \$2,400$$

Neither of these extremes are very likely, so to provide a basis for deciding, we will use the Monte Carlo technique to develop a frequency distribution of outcomes.

First, the sales and materials forecasts are plotted in Figure 12.5. We then generate a series of random numbers between 0 and 1. Computers and many calculators have random-number-generating programs. Alternatively, tables of random numbers like Table F.2 in Appendix F are generally available. The first two numbers of the fifth column of Table F.2 are .56644 and .36551.

These numbers are used to generate the sample points. The first number applies to the sales graph [Figure 12.5(a)] and a probability of 0.56644 corresponds to sales of 1,840 units. The second number, 0.36551, applies to the cost graph, which gives a cost of $5.20 per unit. The first sample point is therefore

$$P = 1,840(\$10.00 - \$3.00 - \$5.15) - \$3,600$$

$$= \$3,404 - \$3,600$$

$$= \$-196$$

The process is then repeated with the next two random numbers, and in this way sufficient sample points are generated to provide a frequency

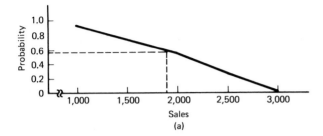

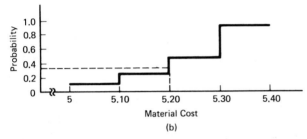

FIGURE 12.5 (a) Sales versus probability of occurrence; (b) material cost versus probability of occurrence.

distribution of the outcome. The mean and standard deviation of the distribution can be calculated, and probability estimates can be made of the product's profitability.

In a similar manner a computer program can be designed to generate a rate of return distribution, a present equivalent, or an annual equivalent distribution. The following estimates are required:

1. The probabilities and associated years for the life of the investment
2. The probabilities and associated cash values for the investment cost
3. The probabilities and associated net after-tax cash flow values for each year of the investment life

12.9.1 Steps in the Analysis

The entire analysis can be computer programmed. For example, for a present-value analysis, to determine one net after-tax present-worth value of the profitability distribution, the analysis proceeds in the following manner.

1. The computer generates a random number. This random number is then associated with a probability and the corresponding year of retirement for the investment life to fix the life of the investment in the first cycle.

2. A second random number is then generated. This random number is associated with a probability and the corresponding dollar value for the investment cost to fix investment cost in the first cycle.

3. The net after-tax cash flow value is then determined by generating a random number and associating this random number with a probability and the corresponding net after-tax cash flow value for year 1. This procedure is continued for each year of the investment life.

4. The net after-tax cash flow value determined for each year is then multiplied by $1/(1 + i)^n$ to convert the value to a present-worth value (n equals the year in question). The cost of the investment is then deducted from the value to give us one point on the present-value distribution. The procedure should be repeated as many times as necessary to develop a valid distribution from the original data. Generally, 200 runs will be sufficient. The profitability values are then placed in a specific number of class intervals. A representative present-worth value and corresponding probability is developed for each interval. The program can be designed to output these values in tabular form and as a histogram.

5. A comprehensive report, including the range, variance, and expected values for each investment, can be generated.

12.9.2 Development of Appropriate Estimates

The development of a frequency distribution for each variable is perhaps the most important task in the entire analysis. If the range of values for each variable is not formulated with care by individuals thoroughly familiar with the variables concerned, the value of the analysis will be limited.

Management should make a significant investment in time and money in the development of values for each variable. Use should be made of information available from previous forecasts, market surveys, government statistics, and the knowledge of individuals within the company.

The ranges used for each variable should be directly related to the degree of confidence that the estimator has in the estimate. The less confident the estimator is in predicting values for a specific variable, the greater should be the range of values used.

The analysis must be structured in a manner that accounts for the dependency among certain variables. For example, in the development of the after-tax cash flow values each year, the simulation may select a maximum value for operating revenues and a minimum value for operating costs. In practice, these two conditions are unlikely to occur simultaneously. The analyst who is familiar with the variables involved in the development of predicted after-tax cash flows should be in the best position to retain the necessary independence. Therefore, in order to avoid errors the method, as outlined above, uses net after-tax cash flow values with their associated probabilities rather than the individual distributions involved.

Example 12.4

Problem: Sensitivity analysis of multivalued parameters using Monte Carlo simulation.

The probabilities and associated values are recorded for each parameter. $i = 10\%$ and $t = 50\%$. Use the SL method (equation 4.2).

n (years)	1	2	3	4	5
Prob.	0.05	0.20	0.45	0.25	0.05
Cum. prob.	0.05	0.25	0.70	0.95	1.00
P ($)	850	900	1000	1100	1200
Prob.	0.05	0.20	0.35	0.25	0.15
Cum. prob.	0.05	0.25	0.60	0.85	1.00
ATCF ($)	300	400	500	600	700
Prob.	0.10	0.25	0.30	0.25	0.10
Cum. prob.	0.10	0.35	0.65	0.90	1.00

A computer program can then be used to generate the solution to the problem.

Solution

PE of ATCF	Prob.	Cum. Prob.
$-649	0.01	0.01
-560	0.04	0.05
-471	0.07	0.12
-382	0.07	0.19
-293	0.08	0.27
-204	0.13	0.40
-115	0.10	0.50
-26	0.11	0.61
63	0.08	0.69
152	0.13	0.82
242	0.06	0.88
331	0.06	0.94
420	0.04	0.98
509	0.02	1.00

The cumulative frequency diagram (Figure 12.6) indicates that there is only a 37% chance of a PE ≥ 0.

The results indicate that there is a reasonably high risk that the investment will not meet the MARR requirement of 10%.

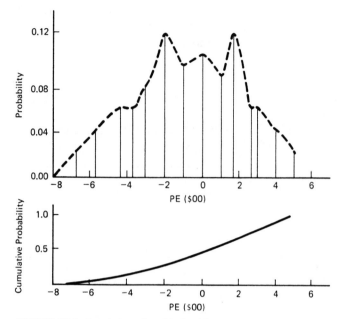

FIGURE 12.6 Simulation of multivalued input parameters.

12.10 EXPECTED VALUE ESTIMATES

Expected value may also be used as an appropriate measure of acceptability. Expected value is the average obtained when all possible values are weighted according to their respective probabilities.

It is easiest to explain expected value in terms of gambling. For example, if one were to draw a card from a deck of playing cards and receive \$10.00 if an ace were drawn, or pay \$1.00 if any other card were drawn, the expected value of the gamble is

$$E(V) = \text{(probability of ace)(payoff for ace)}$$

$$+ \text{(probability of not an ace)(payoff for not an ace)}$$

$$= \left(\frac{4}{52}\right)(\$10.00) + \left(\frac{48}{52}\right)(\$-1.00)$$

$$= -\$0.15$$

The expected value is less than zero, and if we were to play the game a large number of times, we could expect long-run losses equal to -0.15 times the number of plays.

The same type of calculation can be applied to investment opportunities. Using the values, payouts, and probabilities calculated in Example 12.4:

$$
\begin{aligned}
E(PE) = {} & (\$-649)(0.01) + (\$-560)(0.04) + (\$-471)(0.07) \\
& + (\$-382)(0.07) + (\$-293)(0.08) + (\$-204)(0.13) \\
& + (\$-115)(0.10) + (\$-26)(0.11) + (\$63)(0.08) \\
& + (\$152)(0.13) + (\$242)(0.06) + (\$331)(0.06) \\
& + (\$420)(0.04) + (\$509)(0.02) \\
= {} & \$-66.76
\end{aligned}
$$

The expected value of the present equivalent, $E(PE)$, is negative, indicating that the investment is not acceptable.

12.10.1 Utility of Money Concepts

Suppose that one of the following opportunities were available to a person:

1. Accept a gift of $1,000,000 with no strings attached.
2. Let the flip of a fair coin decide whether one would receive nothing or $5,000,000.

Which alternative would be accepted? The certain receipt of $1,000,000 or the 0.50 chance of winning $5,000,000. When confronted with these two alternatives, many persons would choose the certain $1,000,000 even though the expected value of the coin flip is equal to $2,500,000. The expected value of the outcome for each alternative is

$$
(\$1,000,000)1.0 = \$1,000,000
$$

$$
(\$5,000,000)0.50 = \$2,500,000
$$

If a person were faced with a second set of alternative choices as follows:

1. Accept a gift of $5 with no strings attached.
2. Let the flip of a fair coin decide whether one would receive nothing or $25.

which alternative would be accepted? The certain receipt of $5 or the 0.50 chance of winning $25.

When confronted with these two alternatives, many would choose the gamble rather than the sure bet. From these two examples three important conclusions can be drawn.

1. Expected monetary value is not always a valid guide in the solution of practical business problems.
2. The amount of money involved in the decision has a direct bearing on whether the expected value may be used as a valid guide in the solution of practical business problems.
3. Some persons have a greater aversion to risk than others.

This raises the question: If expected monetary value cannot always be used as a valid guide to action in the solution of practical problems, what can be used as a valid guide under these circumstances? Utility theory is one approach to the solution of the problem. The following discussion is based on this theory.

According to the theory of utility, each person has a measurable preference among alternative choices involving risk. This measurable preference is referred to as his or her utility. By answering a series of questions, it is possible to develop for each person a relationship between dollars and utility known as his or her utility function. Using utility theory as a basis, a person who wishes to act in a logical manner in evaluating decisions involving risk will choose the alternative that maximizes his or her expected utility. Therefore, once the manager's utility function is known, and probabilities are assigned to the variables in each investment decision, the acceptability of the alternative investments under analysis can be projected.

The techniques for deriving a personal utility function are beyond the scope of this text and the interested reader is referred to the writings of Schlaifer (Ref. 42) and Swalm (Ref. 47), listed in Appendix D. For our purposes it is sufficient to note that in those situations where expected monetary value is not a valid guide, it is possible to calculate expected utility. Expected utility, since it incorporates personal preference and risk aversion, can be used as a guide for decision making.

12.11 SUMMARY

Risk and uncertainty are a fact of life where capital investments are concerned. The effort that should be applied to evaluate the risk associated with a specific investment is very dependent on the size of the investment and the possible impact of the investment results on the organization.

With access to computers, the capital budgeting system within any organization should be capable of presenting to the management information that represents a reasonable projection of the risks associated with any investment.

The application of expected value to investment decisions should be very helpful to the analyst in providing the manager with recommendations. However, the method does have some limitations. If the manager cannot accept the expected-value calculations, utility theory may be used. This is difficult because both the managers and the corporation's preferences change over time and may certainly vary depending on the magnitude of the cash flows involved.

These limitations can be overcome, to a large degree, if the people involved are prepared to contribute the significant time and effort required to develop a workable system.

PROBLEMS

12.1 The following table shows a frequency distribution of the weekly wages in dollars of 66 employees at the JD Company.

250	253	305	365	390	270
260	267	357	368	395	276
265	275	365	385	380	281
259	278	354	376	385	285
280	306	385	364	375	275
275	294	395	386	397	288
293	309	365	395	395	284
310	280	380	398	376	250
317	305	358	376	375	277
301	315	367	374	387	288
295	258	360	385	375	275

a. Construct a frequency distribution for the data above. (Use seven class intervals.)
b. Construct a cumulative frequency distribution for the data above.
c. Determine the percentage of employees earning less than $300 per week.
d. Determine the percentage of employees earning less than $300 per week but more than $275 per week.

12.2 Monthly maintenance costs have been recorded as follows:

$ 250	$2000	$ 90	$1000
100	200	150	200
75	275	150	100
125	150	200	90
1200	140	150	200

On the bases of the data above:
a. What is the average repair cost?
b. What is the median repair cost?
c. Which value is more representative of central tendency?
d. Find the expected value of the maintenance costs.
e. Determine the range for maintenance costs.
f. What is the probability that the maintenance costs for any one vehicle in the group will not exceed $150?
g. Draw a graph of maintenance cost (horizontal axis) versus probability of maintenance costs not exceeding $_ (vertical axis).

12.3 Define:
a. Risk.
b. Uncertainty.

12.4 The annual operating costs for a new pumping unit are somewhat uncertain, but experience with other pumps indicates that these costs are most likely to be about $20,000. However, there is a 20% chance that they could be less than $15,000 and a 20% chance that they could exceed $25,000. Using the normal distribution, determine:
a. The standard deviation.
b. The probability that the expense will be more than $23,000.
c. The probability that the expense will be less than $16,000.

12.5 The HM Corporation has recently installed power from a central distribution facility. Prior to this point, they have had their own power-generating facilities. The question is whether or not to retain the old facilities for standby purposes. The total AER to retain the facilities is estimated at $30,000. If the central power distribution is disrupted, the hourly cost is expected to be $2,000, due to lost production. If a power failure occurs, repairs average 6 hours. Should the standby unit be retained if power failure follows the frequency distribution below?

Number of Failures per Year	Probability
0	0.05
1	0.10
2	0.30
4	0.50
5	0.05

12.6 A new milling machine is under consideration. The decision has been narrowed to two machines.

	X	Y
Installed cost	$250,000	$400,000
Annual operating costs	50,000	60,000
Estimated salvage value in 10 years	60,000	120,000

$i_a = 15\%$, $t = 48\%$, and $r = 0$. Use the SL method (equation 4.2). Either machine can handle projected revenues for the first 2 years. However, if machine X is selected now, there is a 40% chance that an additional machine X will be required in 2 years. Use a 10-year study period and assume that the second machine X can be sold for book value 10 years from now. (Depreciation = $19,000 per year.) Which machine should be purchased?

12.7 The NG Company is considering the following investment:

Installed cost	$200,000
Estimated economic life	5 years
Estimated salvage value	$100,000
Annual operating costs	$ 50,000

$i_a = 12\%$ and $t = 48\%$. Use the straight-line (SL) method for depreciation allowance (equation 4.2).
a. Calculate the annual revenue requirements.
b. Check the sensitivity of revenue requirements to changes in:
  (1) First cost: $200,000 to $300,000 in $25,000 increments.
  (2) Salvage value: $100,000 to $180,000 in $20,000 increments.
  (3) Operating costs: $50,000 to $90,000 in $10,000 increments.
  (4) Economic life: 5 to 25 years in 5-year increments.
  (5) Rate of return for 5%, 10%, 12%, 15%, and 20%. Develop a summary table similar to Example 12.3.

12.8 An investment alternative is being considered. The following estimates have been developed:

Installed cost	$80,000
Estimated economic life	5 years
Estimated salvage value	$40,000
Annual operating costs	$30,000
Annual operating revenues	$62,000

$t = 50\%$ and $r = 0$. Use the SL method (equation 4.2).
a. Calculate the rate of return.
b. How sensitive is the rate of return to variations in the revenue? (Vary by $\pm 20\%$)

12.9 Assume that you have the opportunity to start a small retail business selling aquatic recreational equipment. You have completed a feasibility study and developed the following estimates.

1. Annual direct labor costs (year 1)	$20,750
2. Annual rental costs	$6,000
3. Remodeling, furniture, and fixtures (capitalize) (fully depreciated over 2 years)	$5,000
4. Heat, light, water, telephone, and office supplies and janitor services	$2,500
5. Average inventory (working capital)	$30,000
6. Bank interest (i_d)	12%
7. Estimated life	3 years
8. Income tax rate	30%
9. Assume that you borrow $15,000 from the bank and that you invest $20,000 of your own money.	

a. What is the rate of return on the investment if sales are as follows:

Year	Sales	Direct Labor
1	$110,000	20,750
2	120,000	25,000
3	140,000	25,000

All operating costs except labor remain constant. The selling price averages 160% of the purchase price on all merchandise. Assume that the salvage value of inventory at the end of 3 years is $25,000. There is no additional salvage value from assets.

b. Check the sensitivity of the rate of return to a 5% increase in operating costs each year.

12.10 Assume that the following estimates have been made for operating costs with respect to a specific investment.

Operating Costs ($)	40,000	60,000	80,000	100,000
Probability	0.10	0.30	0.50	0.10

a. Find the expected value for operating costs.
b. Find the expected value using the random number table (use 50 values).

12.11 Past experience regarding insurance claims for break and entry into private residences are as indicated below.

Probability That a Policy Holder Will Have a Claim during the Year	Range of Claim	Average Settlement
0.97	$ 0–$ 50	$ 0
0.02	51– 300	150
0.005	301– 500	400
0.003	501– 1,500	1,000
0.002	1,501– 10,000	6,000

Find the annual premium that should be charged to a policyholder if 25% of the premium cost is required to service the claims?

12.12 A service station operator is equipped to service two vehicles simultaneously. However, he is somewhat undecided as to whether or not he should have one service or two service workers. Existing conditions are as follows. As a policy he wishes to offer service such that the average customer waiting time for service does not exceed 5 minutes.

Amount Charged		Gross Profit
Grease job	= $6.00 material	$5.00
Oil change	= $10.00 material	4.00
Filter (oil)	= $8.00	4.00
Filter (air)	= $8.00	4.00
Direct labor cost	= $8.00/hour/worker	

Overhead costs do not change with one or two service workers. All service jobs require a grease job and an oil change. Seventy-five percent of all service jobs require an oil filter; 20% of all service jobs require an air filter.

Time between Successive Arrivals (minutes)	Frequency	Service Time (minutes)	Frequency
0	5	0	0
30	15	15	5
45	20	20	10
60	35	25	15
75	20	30	20
90	5	35	20
	100	40	25
		45	5
			100

Simulate conditions for one and for two service workers using a sample of 10 arrivals in each case. (Note: In actual practice one would probably need a minimum sample of 200.)

a. What is the average customer waiting time for service?

b. What percentage of the time are the service workers idle?

c. Which alternative is the most profitable?

12.13 The probabilities and associated parameter values for a specific investment are listed below. $i = 12\%$.

n (years)	1	2	3	4	5
Prob.	0.05	0.20	0.45	0.25	0.05
P ($)	8,000	9,000	10,000	11,000	12,000
Prob.	0.05	0.20	0.35	0.25	0.15
ATCF ($)	2,500	4,000	4,500	5,000	5,500
Prob.	0.10	0.25	0.30	0.25	0.10

a. Using the following random numbers, generate 20 values for the present equivalent of the after-tax cash flows.

Simulation	RN (years)	RN (P)	RN (ATCF)	Simulation	RN (years)	RN (P)	RN (ATCF)
1	93	6	2	11	48	2	2
2	53	30	42	12	57	42	6
3	4	98	28	13	90	28	35
4	7	94	36	14	61	39	75
5	55	17	71	15	73	71	51
6	6	42	77	16	18	77	57
7	30	72	99	17	17	99	54
8	98	27	59	18	65	59	51
9	94	45	90	19	30	90	77
10	17	26	27	20	79	27	12

Note: In practice one would probably need a sample of 200 or more simulations.

b. What is the probability that the rate of return will be 12% or higher? Use the following class intervals for the present equivalent values.

Class Interval		
−9000	to	−7501
−7500		−6001
−6000		−4501
−4500		−3001
−3000		−1501
−1500		−01
1500		2999
3000		4499
4500		5999
6000		7499

    c. What is the expected present equivalent value?

12.14 Suppose that the following opportunities were available to you:
    a. (1) A gift of $500,000 with no strings attached.
       (2) The possibility of winning $2,000,000 on the flip of a fair coin. Which alternative would you choose? Why?
    b. (1) A gift of $25 with no strings attached.
       (2) The possibility of winning $100 on the flip of a fair coin. Which alternative would you choose? Why?
    c. Calculate the expected value for parts (a) and (b).

CHAPTER THIRTEEN

Cost Estimating

Cost estimating is concerned with the development of reasonably accurate costs to satisfy a specific purpose(s). The previous chapters have concentrated on the methodology associated with an economy study. The cost data have been treated as a "given," without questioning their source or validity. We have neatly fit our study of costs and benefits to an analysis involving one of the equations in Chapter 6. For example,

$$AE = (AEOR - AEOC)(1 - t) - P(A/P, i_a, n)$$
$$+ NSV(A/F, i_a, n) + t(AED)$$

To arrive at an annual equivalent, all the values in this expression are determined from an estimate. The validity of the analysis is very dependent on the accuracy of the estimated values. Perhaps the most difficult part of any economy study is the development of accurate estimates. This chapter discusses some of the estimating methods used in practice that should be very helpful to the analyst.

13.1 WHY ESTIMATE?

Estimating is an everyday function; for example, we estimate the amount of money we need to go shopping, the weather forecast is an estimate, and we ask for an estimate before we have the car fixed. In the questions and examples in this text, we use estimates to measure costs and compare alternatives.

Cost estimates are used to prepare bids and to verify that quotations given are fair. The selection of one design over another usually rests on an estimate of cost. Estimates are used in comparing manufacturing processes. A cost or time estimate may serve as an initial job standard. Many businesses depend on the ability of an estimator. Accurate and timely estimates allow accurate and timely business decisions. Good decisions generate profit and errors can be disastrous.

Design teams solve problems using a systematic approach, which is often referred to as systems design. Good design identifies cost-effective solutions to problems, using estimates. Figure 13.1 represents a typical approach to systems design.

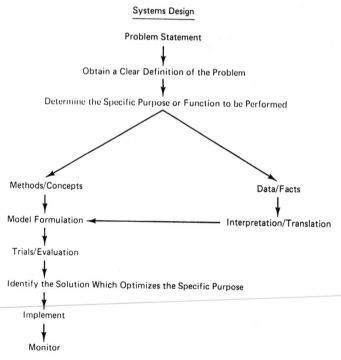

FIGURE 13.1 Typical systems design process.

13.2 TYPES AND SOURCES OF DATA

Most costs are derived from three types of data: (1) historical, (2) measured, and (3) policy. As the term implies, *historical data* refer to previous experience, either our own or someone elses. The data concern processes, money, materials, labor, services, and other expenses. Historical data tell us how we did it before and how well it turned out. Important types of historical information are standards and standard costs.

Measured data refer to the design of a specific system to measure actual performance: for example, the establishment of time standards in the manufacture of a product through such methods as stopwatch time study and work sampling.

Policy data are usually data that are fixed and accepted as factual and beyond the control of the estimator. The origins of policy data are varied and may be internal or external to the organization. These data include union–management wage settlements, budgets, government legislation with respect to such factors as taxation, environmental controls, fire protection, and fringe benefits.

13.2.1 Internal Sources of Data

Accounting records represent the major source of cost information internal to the organization; however, the goals of the accountant are fundamentally different from those of the cost estimator. This difference requires a record system that satisfies the specific needs of estimating.

Estimates produced today must be documented and filed in a format that makes them readily available and useful for future estimating. Contractors use the following procedures to ensure that estimating information is retained.

1. Every conversation and phone call pertaining to a specific estimate should be recorded and filed. This documentation is essential not only for future estimating but for the specific cost estimate under study.
2. The first step is the recording of all drawings, specifications, and tendering documents. List the documents as received, the date received, the set or section number, and the number of pages in each section. Determining that a page is missing in the initial stages can save costly last-minute problems.
3. Drawings should be filed in a systematic manner listing the customer's project number, the number of each drawing, the date of each drawing, and any revisions so noted on each drawing; also, a record of any adenda issued during the period of tender. The importance of recording all data pertaining to each tender submitted cannot be overemphasized.

The purchasing section and the personnel section are also excellent sources of information for both current costs and for comparisons. Finally, the line departments, "the doers," can provide an enormous wealth of information, as to both processes and costs. As a general rule, in researching internal services one should touch all the appropriate bases.

13.2.2 External Sources of Data

Governments are a good source of information for cost estimators. Virtually all government agencies publish information for the benefit of the public. A very broad range of analyzed data is available, such as the Consumer Price Index and inflation indices for almost every major business field.

There are many businesses, trade associations, and publications that provide information valuable to cost estimators. The information takes the form of indices, cost factors, and measures of productivity.

13.3 COST FACTORS

Every dollar generated by a firm has an end use. The cash flow diagram shown in Figure 2.1 is, in general terms, the use made of each dollar.

All firms share one common objective: to maximize profit. Profit is demonstrated on an accounting document, the income statement. The long-term benefits that a firm enjoys are demonstrated on another accounting document, the balance sheet. The final proof of any benefit/cost analysis is shown on the accounting profit or loss statement.

The traditional accounting of costs follows a set of rules referred to as GAAP (generally accepted accounting principles). There are three principal rules which explain the accounting approach to value: assets and costs should be valued objectively, verifiably, and consistently. These accounting principles are the rationale for accounting issues such as historic cost and timing of recognizing revenue and expenses. However, even accountants, who pride themselves on the principle of verifiability, must confess to using estimates.

Analysts, in performing the economic analyses presented in the previous chapters, apply a relative monetary value to time, in the form of an interest rate. That interest rate is an arrived-at rate, which represents the opportunity cost or cutoff cost of capital. It is an estimate.

If a firm were to purchase a storage shed for $10,000 from a cash account, both an accounting and an economic analysis would say that the cost was $10,000. However, if the building was to last 10 years and the question was raised at the end of the first year, "What did it cost?", two different answers would result. An accountant might say that the first year's cost was one-tenth of the purchase price. The engineer would apply a current interest rate and might offer two answers. If the interest rate were 10%, either the first-year cost was $1,000 in repayment plus $1,000 toward return on invested capital or that an annual equivalent cost was $10,000 × 0.1628 or $1,628.

What then are costs? Costs are, in the final analysis, all the accounts found on the income statement and balance sheet. The principal cost factors are labor, materials, administration, and the environment. These will now be discussed.

13.3.1 Labor Costs

Labor costs generally refer to the wages paid to those people directly concerned with a project or the manufacturing of a product. The costs consist of an hourly rate or salary agreement plus a number of paid benefits, such as unemployment insurance, holidays, medical plans, and workers' compensation.

The rate of pay is generally given for most cost-estimating problems; however, it does not mean that the overall labor cost is fixed. Reduction in cost comes from an improvement in productivity: either doing more work with the same number of people or doing the same work with fewer people. From an estimator's point of view, work must be done "smarter," not "harder," although the latter helps.

The estimator can help control labor costs by realistic time planning. For example, overtime can be a significant cost, especially at time and a half, double, or even triple time.

It is a common practice to divide labor into two categories: direct and indirect. Direct labor is normally considered as that time spent working directly on a project or product line. Indirect labor represents the time necessary for cleanup, maintenance, and other support functions.

13.3.2 Material Costs

There are several factors beyond the control of the cost estimator which set material costs. Some of these are:

1. A minimum strength of material
2. A minimum quality required
3. Volume and size
4. Equipment processes

There is, in any material estimate, an allowance for scrap. Scrap is to be expected in the form of cutouts, ends, shavings, mistakes, and overestimates. An estimator may, in determining an allowance for waste, apply a percentage value based on experience. This is a prudent approach. However, it does not necessarily imply that the percentage could not be reduced over time.

Some ways that material cost could be controlled or reduced are:

1. Purchasing raw materials in a form that minimizes scrap
2. The setup of machines to produce components within desired tolerances
3. Improved quality control both in receipt of material and through the labor process
4. Proper training

13.3.3 The Cost of Administration

Administration exerts the greatest influence on the overall profitability of a firm. Included in administration are estimators, planners, engineers, line managers, accountants, support staff, and the owners or steward. The cost of administration includes all their salaries and benefits and also the costs that result from their decision making.

For convenience, we summarize all these costs as being administration costs, including such items as the cost of information systems, financial charges, and so on. We distinguished between material and labor costs not due to their unique importance but rather to demonstrate that costs can be reduced. Energy conservation programs, training programs, computer-programmed machines, and suggestion award programs are only a few of the ways that costs can be controlled. The cost estimator may not necessarily assign a monitary value to any of these improvements; however, the long-term benefits result in lower cost estimates.

13.3.4 The Environment

The cost estimator works in the environment of the firm. The firm, in turn, is part of the aggregate economy. An environment exerts influences, which must be recognized. Figure 13.2 pictorially demonstrates some of those influences.

13.3.5 Budgeting

A budget is the formal strategy that has been set to meet a firm's objectives. It is impossible to operate a business successfully without budgeting.

In the final analysis, every cost estimate becomes part of a budget.

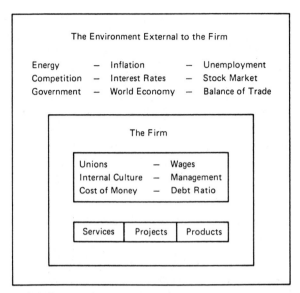

FIGURE 13.2 Micro and macro influences on costs.

These estimates include budgeting for such factors as labor, material, and capital requirements.

Budgeting will be discussed in detail in Chapter 14. However, it is briefly discussed here because of its importance to cost estimating.

The budgeting process begins with the objectives such as to build a building, to provide rental cars, or to produce widgets at the most competitive prices or within the required time frame. The second step is the collection of all the pertinent information from previous activities: costs, revenues, and performance indicators. The third step is to project or estimate the revenues and costs associated with meeting the goal. The fourth step is to reduce the costs by rethinking, taking a fresh approach, or doing things in a new way. Having set the budget, it must be monitored, preferably by comparing the performance of the budget to the performance of the steps taken to meet the objective.

A comprehensive budget system involves three types of budgets:

1. An operating budget
2. A cash budget
3. A capital budget

Operating Budget

The operating budget is the financial plan for the operation of a firm. It relates primarily to the income statement. There are many approaches to budgeting for operations. Planned program budgeting is a system used primarily by a firm with a stable and ongoing enterprise. The budget process assumes that you need all that you have and it demands that additional requirements be justified. Service-level budgeting begins with a desired budget and asks for the effects of reducing it: what things would not be done or what the effects are on quality. Zero-base budgeting implies that you begin with nothing and requires that all funds be justified.

An important aspect of budgeting, other than the process or method, is that someone is responsible for ensuring that a budget be maintained and monitored. Responsibility can be assigned by cost center, process, job, or program. In its final form an operating budget could appear as in Figure 13.3.

	Account/Process Job 1/Program	Account/Process Job 2/Program	Total by Cause
Labor	$100,000	$125,000	$225,000
Overtime	30,000	15,000	45,000
Material	450,000	375,000	825,000
Space Rent	380,000	380,000	760,000
Space Maintenance	25,000	25,000	50,000
Power	12,000	18,000	30,000
Utilities	2,500	2,500	5,000
Equipment Depreciation	2,500	2,000	4,500
Equipment Maintenance	500	1,500	2,000
Indirect Material	1,500	1,500	3,000
Indirect Labor	500	500	1,000
Total by Account Process/Program/Job	$1,004,500	$946,000	$1,950,500

FIGURE 13.3 Typical operating budget presentation.

Cash Budget

A cash budget is the planned flow of cash. It is derived from estimated cash inflows or sources and estimated cash outflows or uses. It relates to the statement of changes in working capital or financial position, in that it is the estimated statement. Cash is often the most important asset a firm has and the plan of its flow can either ensure a firm's success or its demise. It is not enough to say that cash-in will be greater than disbursements; the sequence of inflows and outflows are critical.

Sources of cash are cash sales, collection on accounts, loans, sale of assets, and sale of equity. Uses of cash typically include payrolls, materials, repayment of debt, taxes, dividends, and purchase of assets.

Capital Budget

A capital budget is a firm's plan for purchasing long-term assets and it refers to the balance sheet. The budget describes how much a firm will spend on capital assets and the timing of these expenditures. The sources of information leading to investment decisions are estimates.

13.4 STATISTICAL APPLICATIONS IN COST ESTIMATING

Statistical analysis represents a very useful tool to the cost estimator. Three specific types of forecasts will be discussed herein.

1. When information is not available or if the forecast concerns something new, *human judgment* must be relied on. Forecasting based on experience and judgment is the process of turning an opinion into a forecast.
2. A forecast can be based on *historical data*. It is unreasonable to assume that the future is identical to the past. However, one can assume that in the short term, future conditions will be extensions of the past.
3. Often, the variable that is the object of our forecast has a relationship to other variables which can be observed. Using historical data it is possible to forecast through this *causal relationship*.

13.4.1 Human Judgment

Often, "guesstimates" are not good enough, even though a lack of information or experience or the nature of the forecast precludes a formal mathematical study. In these circumstances we look to experts who can provide at least an "informed guesstimate."

A panel of experts, moderated by the cost estimator, is an opportunity to brainstorm solutions to problems. There are many approaches taken to this team approach to problem solving. Theories of groups and individuals within groups are important parts of many aspects of management. The value of the group to the cost estimator depends on his group skills, the group itself, and the format of the discussion.

To reduce the influence of such factors as social pressure, majority view, or dominant personalities, one could choose the *Delphi technique*. With this process, a group of experts are asked for their forecasts in an environment in which all know the problem, have access to all the information, yet have limited discussion with others in the group. For example, a questionnaire serves as a medium of communication for this method.

13.4.2 Historical Data

There are several methods available whereby historical data may be utilized in developing cost estimates. Only brief mention can be made of some of these methods in our discussion. Our objective is to develop an awareness in the interested reader that techniques are available whereby historical data may be analyzed. Useful estimates may be made regarding such factors as labor costs, material costs, and administrative costs by using the following forecasting methods:

1. Forecasting by moving average
2. Forecasting by exponential smoothing
3. Forecasting using the cumulative sum technique

13.4.3 Causal Relationship

Often, two items are causally related to each other. Sometimes the relationship can be fairly direct, such as the total cost of a building and the gross floor area; and sometimes it can be time lagged, such as the sales of new machines and the sales of spare parts. The mathematical method of determining the functional relationship between two variables is called regression analysis. Linear regression is discussed briefly and an example is given to indicate how it may be used by the estimator.

Least-Squares Method of Regression

The method of *least squares* presumes a linear relationship between two variables, the equation of a line being $y = a + bx$. The value of x is considered as an event and the value y as an observation subject to error. The line is given a "best fit" when the sum of the squares of the errors is minimized. The magnitude of the error is the distance from the line to the observation. To fit the equation $y = a + bx$ (y is the fitted values of y) to a number, n, of paired data points (x_i, y_i),

Slope

$$b = \frac{n \Sigma xy - (\Sigma x)(\Sigma y)}{n(\Sigma x^2) - (\Sigma x)^2} \tag{13.1}$$

y *Intercept*

$$a = \frac{\Sigma y \Sigma x^2 - (\Sigma x)(\Sigma xy)}{n \Sigma x^2 - (\Sigma x)^2} = \frac{\Sigma y - b \Sigma x}{n} \tag{13.2}$$

Example 13.1

Problem: L.J. Construction is building a 1.7-million square foot office tower. The cost estimator wishes to estimate the amount of overhead to include in the estimate. The records for their most recent projects are:

Millions of Square Feet of Building	Overhead (millions of dollars)
1.0	3.1
1.1	2.6
1.2	3.4
1.3	3.1
1.4	4.0
1.5	4.1
1.6	4.1

Solution

x	y	x^2	xy
1.0	3.1	1.00	3.10
1.1	2.6	1.21	2.86
1.2	3.4	1.44	4.08
1.3	3.1	1.69	4.03
1.4	4.0	1.96	5.60
1.5	4.1	2.25	6.15
1.6	4.7	2.56	7.52
9.1	25.0	12.11	33.34

$$b = \frac{7(33.34) - 227.5}{7(12.11) - 82.81} = \frac{5.88}{1.96} = 3.0$$

$$a = \frac{25(12.11) - 9.1(33.34)}{7(12.11) - 82.81} = \frac{-0.644}{1.96} = -0.33$$

$$y = a + bx = -0.33 + 3(1.7) = 4.77$$

The amount of overhead to include is $4.77 million.

Multiple Linear Regression

The value of a variable such as overhead depends on the influence of more than one factor. These may be introduced to weigh their effects. If we were to consider in Example 13.1 that overhead depends on size and volume of material, we would add a new term to our expression:

$$y = a + b_1 x_1 + b_2 x_2$$

The equation of the line can be determined by substituting values into the following equations and solving for a, b_1, and b_2.

$$\Sigma y = na + b_1 \Sigma x_1 + b_2 \Sigma x_2$$

$$\Sigma x_1 y = a \Sigma x_1 + b_1 \Sigma x_1^2 + b_2 \Sigma x_1 x_2$$

$$\Sigma x_2 y = a \Sigma x_2 + b_1 \Sigma x_1 x_2 + b_2 \Sigma x_2^2$$

Example 13.2

Problem: The following table (all values in millions) includes the value of material for the buildings constructed in Example 13.1. Estimate the amount of overhead for a building of 1.7 million square feet and $10 million in material costs.

Area, x_1 (ft^2)	Material, x_2	Overhead, y
1.0	$7.2	$3.1
1.1	7.8	2.6
1.2	7.7	3.4
1.3	8.4	3.1
1.4	8.7	4.0
1.5	9.2	4.1
1.6	9.8	4.1

Solution: Construct the following table.

x_1	x_2	y	$x_1 y$	$x_2 y$	$x_1 x_2$	x_1^2	x_2^2
1.0	7.2	3.1	3.10	22.32	7.20	1.00	51.84
1.1	7.8	2.6	2.86	20.28	8.58	1.21	60.84
1.2	7.7	3.4	4.80	26.18	9.24	1.44	59.29
1.3	8.4	3.1	4.03	26.04	10.92	1.69	70.56
1.4	8.7	4.0	5.60	34.80	12.18	1.96	75.69
1.5	9.2	4.1	6.15	37.72	13.80	2.25	84.64
1.6	9.8	4.1	7.52	40.18	15.68	2.56	96.04
9.1	58.8	25.0	33.34	207.52	77.60	12.11	498.90

$$25 = 7a + 9.1b_1 + 58.8b_2$$

$$33.34 = 9.1a + 12.11b_1 + 77.6b_2$$

$$207.52 = 58.8a + 77.6b_1 + 498.9b_2$$

Solving these equations simultaneously for a, b_1, and b_2, we get

$$a = -0.84 \qquad b_1 = -0.212 \qquad b_2 = 0.55$$

$$y = a + b_1 x_1 + b_2 x_2$$

$$= -0.84 - 0.212(1.7) + 0.55(10) = \$4.303 \text{ million}$$

Correlation

A correlation coefficient is a measure of the extent to which two variables are associated. Perhaps the most useful coefficient "r" is referred to as the sample product moment correlation coefficient. A linear relationship is calculated as follows for two variables x and y.

$$r = \frac{n \sum xy - (\sum x)(\sum y)}{\{[n \sum x^2 - (\sum x)^2][n \sum y^2 - (\sum y)^2]\}^{1/2}} \tag{13.3}$$

The product is 0 for no correlation and 1 for perfect correlation.

If we were to determine the relationship between x and y for Example 13.1,

$$r = \frac{7(33.34) - 9.1(25)}{[(7(12.11) - 9.1^2)(7(92.44) - 25^2)]^{1/2}} = 0.894$$

The value of r shows a high correlation. The positive sign indicates the slope of the relationship. A negative sign would have indicated that as x decreases, y increases.

Model Fitting

Analysts must deal with other than linear relationships much of the time. Mathematical expressions can be determined, as well as measures of dependence. In this section we briefly discussed linear relationships because many others can be "made linear" by making assumptions using a logarithmic relationship or other techniques. This does not mean to suggest that linear regression is best, nor does it preclude using other models. Discussions of model fitting, as well as more complete discussions of relationships between variables, can be found in cost-estimating and statistics books.

13.5 APPLYING A SYSTEMATIC APPROACH TO ESTIMATING

Estimating is the technical process of predicting the costs of supplying a service and/or product. For example, an estimate is prepared prior to constructing a project and thus is, at best, a close approximation of actual cost.

Estimates must be prepared in a way that is explicit and consistent and which takes account of the specific methods to be used and all circumstances that may affect the execution of work. A sound estimate can be achieved only when each operation is analyzed into its elements and the cost estimated methodically on the basis of factual information.

The estimate must be developed in a systematic manner. Whether you operate a one-person shop or you are a large contractor, there are certain basic principles that must be adopted to maintain a healthy and prosperous business. Of prime importance is the necessity to be certain that you are bidding on the same job that the client has requested of you. That is, be sure that you and the client fully understand the scope of the work associated with the cost estimate required.

For many small contracts (using contract in the broad sense as an obligation to fill an order for a product or service), you may only receive a verbal description of the work to be done, quite often from a person who does not understand the total job requirements (e.g., government regulations regarding safety, material specifications, etc.). To gain an interpretation of the client's request, without offending the client, is a real test of your professional ability with respect to salesmanship, trade knowledge, and estimating experience.

The precise approach to preparing a detailed estimate depends on the size of the bid, the number of staff available to do the take-offs necessary, and the complexity of the project. Serious consideration must be given to the time allowed to prepare a tender. When the documents are first received, the estimator in charge of the job should scan through the bid documents (drawings and specifications) to obtain an approximate number of labor hours to prepare the complete tender. It is highly advisable at this point to develop a critical path schedule for the tender submission.

Several estimators may work together as a team on a large estimate. This approach requires that the plans and specifications be broken down into workable packages. This breakdown requires considerable expertise on the part of the chief estimator to coordinate the abilities of the persons involved to produce a competitive and feasible bid on schedule.

Do not overextend your capabilities. If you foresee timing problems on submitting a tender, contact your client at the earliest possible date and negotiate a time extension, or possibly, forgo the bid. In summary:

1. Carefully choose the jobs you want to tender.
2. When you receive the bid documents, be absolutely certain that you have a complete set of documents.
3. Review the method of tendering requirements.

4. Determine the total estimating time you will require to produce the tender. After dividing the document into workable packages, develop a critical path for tendering.

5. Once a decision is made to submit a tender, complete a bid specification review sheet. This sheet is developed by reading the specifications thoroughly and listing all items specifying type. List all subtrades to be used and notify these subtrades.

13.6 DEPTH OF ANALYSIS

Everyone would like the ability to predict the future perfectly. Cost estimators work toward that goal with methods that depend on the enterprise: construction, manufacturing, or service. Estimating itself has a cost: the estimator, the collection of information, and the application of solution methods. These costs are subject to a great many influences. However, two can be noted as common: time and accuracy. Their relative effects on the cost of the estimate are shown in Figure 13.4. Perfect estimates are usually not realistic; in fact, most feasibility studies are less than complete in detail. The estimator must match the method he or she chooses to the objective of the estimate and weigh the benefit, compared to the cost. Three levels of detail and accuracy will be discussed, however, in reality, there is a sliding scale. These levels are:

1. Order-of-magnitude estimates
2. Semidetailed estimates
3. Detailed estimates

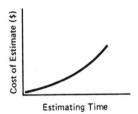 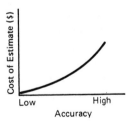

FIGURE 13.4 Factors affecting the cost of an estimate.

13.6.1 Order-of-Magnitude Estimates

An *order-of-magnitude estimate* is the lowest order of estimate in terms of accuracy and of effort and cost involved. It is a value that is expected to vary from the actual cost by 15 to 50%. The magnitude of this large variance depends on the skill and experience of the estimator and the accuracy of the information at the estimator's disposal. This broad-gauge estimate can be used for:

1. Initial screening of a large number of alternatives
2. Aid in budget preparation
3. To decide whether to continue design effort

Decisions based on this first estimate may lead to financial commitment or to valuable alternatives being eliminated. Mistakes can be costly. The order-of-magnitude estimate can be very useful, but its usefulness is related directly to the skill of the estimator. The following methods are used in the development of these estimates.

Conference Method

The *conference method* provides an estimate based on experience. Very simply, all the concerned departments within a firm meet to discuss a proposed project, product, or service. The format of the meeting is left to the estimator. All aspects and elements of cost and price can be discussed— however, usually without many facts. The conference method would not be a reliable way to create a bid, but may be an acceptable way to assess one.

Unordered Ranking and Graphical Methods

These methods represent two simple techniques that use limited amounts of readily available information. An unordered ranking is simply a list of data gathered for a unit comparison. An example would be a list of alternatives for construction material priced as a common unit.

There are a variety of graphical techniques used in estimating, such as exclusion charts and band charts. Figure 13.5 demonstrates a band chart. It is a plot of all points including a representative mean. These methods at least force identification of the variables that are significant to the value of the estimate.

Comparison Methods

An estimate may be determined by a direct comparison, either to a recent similar design of known cost or to a simpler design problem for which an estimate is determined more easily. This comparison can be expressed mathematically as follows:

$$C_1/C_2 = (S_1/S_2)^x \tag{13.4}$$

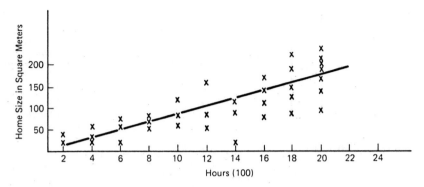

FIGURE 13.5 Labor hours to frame a standard three-bedroom home.

where C_1 = cost of the design to be estimated
C_2 = cost of the design standard
S_1 = size of the design to be estimated
S_2 = size of the design standard
x = a cost-capacity factor determined historically or taken from published tables (see Table 13.1)

The method can be applied in four simple steps:

1. Start with a recent project of known cost similar in nature to the one being estimated.
2. Convert the cost of the previous project to current dollars by an appropriate index such as the Consumer Price Index, specific to the particular industry.
3. Define the relative size of the two projects in terms of capacity or size.
4. Knowing the size and cost of the first project and the size of the second project, the cost of the second project can be estimated.

Table 13.1 lists some typical estimating factors.

Expressions of comparisons are not restricted to a cost capacity ratio. Published sources give a wide range of data, from cost per square foot in construction to other relative weights of material, labor, and direct and indirect costs. Included as well are standard time and methods data. In using any comparison factor, the estimator must know what is included in order to determine the limits of the estimate.

Some sources of published indices are:

1. Boeckh's Index of Construction Costs (reported annually in the *Engineering News-Record*)

TABLE 13.1 Typical cost-capacity factors for different types of facilities.

Type of Facility	Cost Capacity	Units
Acetylene plant	0.73	Tons/day
Aluminum plant	0.76	Tons/day
Ammonia plant	0.72	Tons/day
Steam boiler plant	0.75	Pounds/hour
Cement plant	0.86	Tons/day
Chlorine plant	0.62	Tons/day
Electric generating plant (nuclear)	0.68	Megawatts
Electric generating plant (steam)	0.79	Megawatts
Industrial building	0.67	Square feet
Municipal incinerator	0.80	Tons/day
Oxygen plant	0.72	Tons/day
Public housing project	0.75	Number of rooms
Refrigeration system (mechanical)	0.70	Tons
Sewage treatment plant (primary only)	0.68	Gallon/day
Sewage treatment plant (primary and secondary)	0.75	Gallons/day
Storage tank	0.63	Gallons
Sulfuric acid plant	0.67	Tons/day
Utility distribution main (gas and water)	0.91	Pipe diameter
Utility distribution main (gas and water)	0.82	Length installed

Source: Reprinted by permission of John Wiley & Sons, Inc., from *Cost Engineering Analysis—A Guide to the Economic Evaluation of Engineering Projects*, by William R. Park, © 1973, p. 137.

2. Cost and Production Handbook
3. Engineering News-Record Construction Cost and Building Cost Indices
4. Marshall and Stevens Installed Equipment Cost Index
5. The Nelson Refinery Construction Cost Index
6. Chemical Engineering Plant Construction Cost Index
7. Government Departmental Indices (e.g., commerce, public works, etc.)
8. Manufacturers' associations

Example 13.3

Problem: Global Industries has asked you to develop an order-of-magnitude estimate for a new manufacturing facility. (Assume that the published Engineering News-Record Index = 1,000.) The building will be 100,000 square feet. Data regarding a similar building with 50,000 square feet of floor space which was constructed 3 years ago (assume that the Engineering News-Record Index = 800) indicates a cost of $24 per square foot. Total cost = (50,000)$24 = $1,200,000. Determine the construction cost per square foot.

Solution: Current cost of a 50,000-ft^2 building:

$$\$1,200,000(1000/800) = \$1,500,000$$

Current cost of a 100,000-ft^2 building:

$$\$1,500,000 \, (100/50)^{0.67} = \$2,386,609$$

$$\text{Cost per square foot} = \$23.87$$

Problem: Determine the estimating error introduced if the actual cost-capacity factor = 0.8.

Solution: Using a cost estimating factor of 0.8:

$$\$1,500,000(100/50)^{0.8} = \$2,611,652$$

$$\text{Cost per square foot} = \$26.12$$

Estimating error = $225,043, an error of approximately 9% in the estimate.

Factor Method

The *factor method* represents an additional refinement over the unit cost method. The factor method essentially determines the estimate by summing the product of several quantities. For example:

$$C_T = \left(C_E + \sum_{i=1}^{n} F_i C_E \right)(F_x + 1) \tag{13.5}$$

where C_T = total estimated cost of a specific major item
$\quad C_E$ = a specific major equipment cost
$\quad F_i$ = factor for estimating the cost of a building,
$\qquad$ instrumentation, installation, etc., associated with C_E
$\quad F_x$ = factor for estimating indirect expenses, such as
$\qquad$ engineering, administration, and profit

The unit cost estimating method uses a single factor for calculating overall costs. The factor method achieves improved accuracy through the adoption of separate factors for specific cost items. For example, a unit cost estimate for an office building may be arrived at by using a unit factor of X dollars per square foot. The factor method can improve the accuracy of this estimate by using individual costs per unit of area for such factors as heating, lighting, and air conditioning.

Example 13.4

Problem: A mountain of napthalene cyanide, a major base of paint and ceramic products, has been discovered. Studies and assays have shown that there is enough good-quality ore to meet world demands for 50 years. Napthalene cyanide is sold by the ton of processed ore at a price determined by quality and type of packaging. The firm that acquired the mineral rights now wishes to evaluate the feasibility of designing, constructing, and operating a mill on-site which would process the ore into finished product. Several proposals were entertained. The most likely proposal, which would process 700,000 tons of ore annually into 325,000 tons of product, is evaluated according to a preliminary flowsheet design.

Solution: The firm has some experience in mine and mill operations. Their previous experience is with a much smaller mine; however, the process was very similar.

Given this preliminary information, the firm chooses to proceed with a more-inclusive estimate. The first step begins with the design, represented by the flowsheet in Figure 13.6.

The smaller mill processed 425,000 tons of ore annually and cost $4,500,000 to build 4 years ago. The Engineering News-Record Index was 725 and is now 970. A cost capacity relationship was used as a preliminary estimate. Using a factor of 0.65,

$$C = \$4,500,000(970/725)(700/425)^{0.65}$$

$$= \$8,327,357 = \$8,327,000$$

Annual operating expense is estimated to be 25% of the capital cost.

$$AEOC = \$2,082,000$$

Excluded from this estimate are the costs to mine and transport the ore from the mine site to the mill, which are estimated to be about $3.15 a ton. The firm has already paid $10,000,000 for the rights to mine. Depletion and depreciation allowance are assumed to be based on the straight-line method (equation 4.2). The term of the study is for 25 years because it is felt that a substitute for napthalene cyanide will be found by that time.

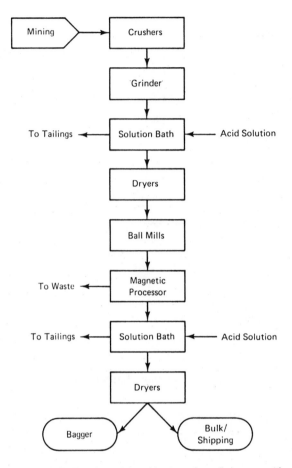

FIGURE 13.6 Flowsheet for production of napthalene cyanide.

The price per processed ton of product is currently $47.38 a ton and is seen to be stable. Assume a 50% tax rate and zero salvage value:

$$AE = (AEOR - AEOC)(1 - t) - P(A/P, i_a, n) + t(AED)$$

A trial-and-error approach indicates a projected rate of return of about 32%.

The next step to improve accuracy involves including equipment, which entails an initial equipment list with costs. The following table shows this equipment list, together with the number required and their total cost. The next table also sets factors as percents, for all other costs, including construction costs. These factors may be found in either internal or external sources, and would be based on experience.

Equipment	Quantity	Cost/Delivered and Installed
Ball mills	6	$ 168,000
Crushers	2	164,000
Grinder	1	114,000
Heaters	4	40,000
Dryers	4	64,000
Magnetic processer	2	420,000
Conveyors	30	116,000
Bagger	1	135,000
Hoppers	5	140,000
Water facilities	1	382,000
Pumps	50	210,000
	C_E = cost of major equipment =	$1,953,000

Factor, F_i (%)		Indirect Cost, F_x (%)	
Building	110	Contractor's profit	22
Electrical	26	Overheads	23
Plumbing	22	Contingency	2
Plant items	23		F_x = 47%
Painting	07		
Insulation	14		
Storage building	12		
F_i = 214%			

Apply the equation introduced in this chapter:

$$C_T = \left(C_E + \sum_{i=1}^{n} F_i C_E \right)(F_x + 1)$$

$$= [\$1,953,000 + 2.14(\$1,953,000](1.47)$$

$$= \$9,014,657 \cong \$9,000,000$$

This estimate indicates a projected rate of return of 31%.

13.6.2 Semidetailed Estimates

Semidetailed, conceptual, or *budget estimates* should be accurate to within about 10% of the actual project cost. This level of accuracy is usually adequate for making decisions regarding project feasibility, that is, sufficiently accurate for the owner to decide if the project is economically feasible. The engineer's estimate on new construction work might be considered a semidetailed estimate in most cases. The accuracy of the semidetailed estimate depends on the amount and quality of information available at the time the estimate is made.

More information is required for making semidetailed estimates than for making order-of-magnitude estimates. Instead of using mathematical relationships between historical costs and estimated costs on proposed work, the new project must be considered on its own. Actual quotations should be obtained on major equipment and material items, and some design data are necessary to obtain an accuracy of ± 10%–15% on take-offs.

The semidetailed estimates can be used to advantage by the analyst in several ways. Since the accuracy of this estimate is generally sufficient to warrant the authorization of the project by an owner, the semidetailed estimate may be all that is needed, as long as the owner is aware of the accuracy limitations. This type of estimate is also useful in providing a check on detailed estimates obtained through more refined methods.

An example of a semidetailed estimate, with reference to a service industry, is presented below.

Example 13.5

Problem: Gord's car rental agency rents two sizes of cars: big and small. He charges by the day and by the kilometer, and in return provides everything, including gas. Gord has received numerous requests from his regular customers for a preliminary estimate of charges in 1984.

Solution: Gord gathered the following information:

	1981 ↓	1982 ↓	1983 ↓	1984 (Est.)
No. of cars	97	112	110	99
Avg. value ($)	4,100	4,400	4,700	4,900
Avg. days rented	250	210	238	238
Avg. annual km	24,500	22,500	22,000	22,000
Avg. rate/km ($)	0.18	0.20	0.24	0.27
Avg. rate/day ($)	12.00	13.50	15.00	17.00
Revenue ($)	718,770	821,520	973,500	966,500
Expenses ($)				
Rent	50,000	55,000	61,000	67,000
Salary	210,000	236,000	265,000	270,000
Gas	58,000	72,000	86,000	86,000
Maint.	180,000	201,000	225,000	234,000
Depr.	119,310	147,000	155,000	134,000
Interest	24,000	31,000	30,000	24,000
BTCF	77,770	79,520	151,500	151,500
Taxes	32,000	32,500	70,000	70,000
Net income ($)	45,770	47,020	81,500	81,500

Gord's experience suggested that in 1984, business will decline about 10%. He feels that the best way to scale down his operations is to reduce the number of vehicles he has and aim for the same utilization as this year.

Gord predicted next year's costs by applying an inflation factor for price and a decline in value to those expenses he could reduce. He then set a financial goal to maintain the same net income. The last step was to set a rate that would match expenses and meet his goal. That rate may have to be lowered due to competition, reducing net income.

This estimate is preliminary. Final rate setting would be done by breaking down revenues and expenses in detail. It is important to match revenues and expenses in that Gord may find that his big cars are used proportionately less yet cost proportionately more. This same approach could be used by any service industry. It requires an in-depth knowledge of the service, not only to identify costs, but more important, to know how these costs relate to profit.

13.6.3 Detailed Estimates

Detailed estimates are used as a basis for making bids. These estimates should be accurate within ±5%, having been prepared from complete engineering specifications, drawings, and site surveys. They are, however, time consuming and costly to prepare and should be used only when necessary. In many cases, on a construction project, the contractor is the only one capable of making a good detailed estimate.

Considerable information is required for a detailed estimate, often more than is available at the time the engineer's estimate must be made. In preparing the estimate, the specifications and plans must be studied carefully, quantity take-offs made, prices obtained, labor availability and wage rates checked, subcontractors' estimates requested, and schedules developed. There are, in every detailed estimate, many opportunities to make mistakes, usually resulting in estimates that are too low. The most common errors leading to low estimates include the omission of items, under measurement of quantities, and underestimation of labor requirements.

13.7 PROJECT COST ESTIMATING

An assessment of the total cost of an engineering project usually consists of the following items:

1. Land and easements
2. Legal expense
3. Bond expense
4. Cost of construction
5. Engineering expense
6. Interest during construction
7. Contingencies

The methods for determining item 4, the cost of the construction, are of specific interest to our discussion. A project cost estimate has three major categories: direct costs, indirect costs, and profit.

Profit is determined by management and the market and has been fully discussed previously.

Indirect costs, sometimes called head office overheads, include such items as office expenses, promotional (estimating) costs, service personnel, and so on. These costs are determined, or estimated, by the cost accountant and generally apportioned to the different projects that the company is doing on a prorata basis.

Direct costs are the labor, equipment, and material that contribute directly to the construction of the facility. Also included in direct costs for a project are project overhead items, such as supervisors, site offices, signs, and so on.

13.7.1 Developing a Detailed Estimate

A clear understanding of the scope of the project is an essential first step in arriving at a detailed cost estimate. As stated previously, this requires mutual understanding with the client as to what is to be accomplished, deadlines to be met, and the total resource requirement to produce the tender.

The estimation process, which contains the following steps, can then be accomplished in a systematic manner.

1. Break down the project into specific job packages or operations.
2. Do the necessary material take-offs from the plans and specifications.
3. Determine the costs for labor and equipment.
4. Determine the production rates for labor and equipment.
5. Determine the costs of materials.
6. Synthesize the rates, costs, and material quantities to arrive at an estimated cost for the specific job.
7. Summarize the costs of the specific jobs to arrive at the cost of the project.

13.7.2 Breaking Down the Project

In preparing an estimate, the estimator will first break down the project into as many jobs or operations as required. The operations should appear in the estimate in the order that they will be performed. For example, on a construction site the first operation may be clearing the site; followed by erection of site offices; then excavation, pilings, and so on. Experienced estimators usually rely on checklists and forms.

The first step in the estimating of an operation is the quantity take-off. This will involve all materials placed on the project, plus earth excavation and fill. Materials for each operation should be listed separately, in the correct quantities, according to their classification and units.

13.7.3 Completing a Take-off from the Plans and Specifications

After completing your review of the tender documents and breaking the job into specific job packages, you are ready to proceed with individual take-offs. A *take-off* is the act of systematically measuring and counting the various quantities of equipment and materials to complete each work package (subsystem).

Every contractor realizes the necessity for a form on which the estimator can record the measured unit quantities. There is no unique set of forms to use for any specific type of contract. Each contractor designs forms to suit their specific purposes, and these forms tend to change over time. But it is a fact that a form that is well laid out, designed to suit the nature of your work, and agreeable to the estimating staff is an effective component to the overall estimating system.

Table 13.2 represents a sample of the detail required in a typical checklist covering the concrete work for a building estimate.

This table is developed from a series of forms that list the detailed data associated with each item on Table 13.2. For example, Figure 13.7 represents a form that may be used to develop the required information for item 400 on Table 13.2. Other items, such as item 900, may require several forms to obtain all the detailed information for a final cost. For example, a form similar to Table 13.3 may be used to determine labor costs.

TABLE 13.2 A building estimate checklist and summary for the concrete work.

Item	Operation	Amount ($)
100	Site clearing	
200	Temporary construction, sheds, site enclosure, etc.	
300	Excavation, grading, backfill, special fill	
400	Foundation support, piles, caissons, cribbing	
500	Shoring, sheeting: temporary and permanent	
600	Underpinning: temporary and permanent	
700	Water lines, gas, drains, sewers, conduct	
800	Paving, curbs, sidewalks, drives	
900	Concrete requirements	
1000	Concrete surfacing and cement work	
1100	Concrete forms, wood, metal	
1200	Reinforcing rods and mesh; metal inserts	
1300	Concrete blocks	

Item No.	Description	Calculations	No. Units	Unit	Unit Cost ($)	Mat'l. Cost ($)	Equip. Cost ($)	Labor Cost ($)	Total Cost ($)
400	Furnish and drive 100 treated timber piles. Drive piles to full penetration in normal soil length of piles, 50 ft. size of piles 16" butt and 6" tip diameter								
10	Materials-Piles-add 5% for breakage	105 piles × 50ft	5,250	lin. ft	3.00	15,750.00			15,750.00
20	Equipment	100 piles 2.5							
	Pile Driving unit	piles/hour	40	hr	75.00		3,000.00		3,000.00
	Travel time		8	hr	75.00		600.00		600.00
	Rig up + tear out		6	hr	75.00		450.00		450.00
30	Labor (including fringe benefits)								
	Foreman	8	8	hr	30.00			240.00	240.00
	Operator	1. man 40+8+6	54	hr	25.00			1,350.00	1,350.00
	Helpers	2. man 40+8+6	108	hr	20.00			2,160.00	2,160.00
40	Subtotal direct cost					15,750.00	4,050.00	3,750.00	23,550.00
	Overhead	15% × $23,550.00							3,532.50
60	Subtotal costs								27,082.50
70	Profit (before-tax)	20% × 27,082.50							5,416.50
80	Subtotal cost								32,499.00
85	Performance bond	1% × $32,499							324.99
90	Total cost of bid								32,823.99
95	Cost per linear foot	$32,823.99/5,000							6.56

FIGURE 13.7 Form used to estimate construction costs.

TABLE 13.3 Labor hours required to place 1 cubic yard of ready-mixed concrete.

Type of Structure	Method of Handling	Unskilled Labor	Supervisor	Crane Operator	Skilled Labor
Large foundation	Discharge directly from truck (chute)	0.5	0.05		0.07
Large pier	Crane and bucket	0.5	0.05	0.3	0.2
Slab at ground level	Crane and bucket	1.5	0.06	0.2	0.1
Slab above ground level	Crane, bucket hopper, buggies	1.5	0.10	0.3	0.2
Foundation wall	Hand buggies	1.0	0.10		0.1

These costs may then be transferred to a form similar to Figure 13.7 and from this form to Table 13.2.

The unit costs for labor are usually based on labor hours. For each class of labor a cost per hour is developed which includes wages, pension, unemployment insurance, and so on. The cost of equipment can be developed only by analysis of the specific situation. If the equipment is to be on the job for almost the full duration of the project, for example, a tower crane on a high-rise building, it can be treated as a lump-sum cost based on the estimated duration. If, however, it is brought onto the project for a specific operation, for example, a backhoe for excavation, it is the production rate and hourly (or weekly) costs that are relevant. Also, equipment is either rented or purchased. If it is rented, the rental cost is the unit cost. If the equipment is purchased, the unit cost must be developed from consideration of depreciation, maintenance and repairs, investment, and fuel and lubrication.

Production rates are the number of units of work produced by a worker or a machine in a specified time, usually an hour or a day. To be realistic, they should usually include an allowance for the fact that a worker does not usually work 60 minutes during an hour. Published tables of production rates can be found in most estimating texts and handbooks. In addition, most estimators develop their own rates from previous jobs. However, the rate must be varied to fit the circumstances that apply to the particular operation being estimated. It is in the assigning of production rates that the estimator's skill and experience are apparent.

Costs of materials are usually provided by suppliers. The normal practice is to ask for firm quotations.

The synthesis of rates, costs, and materials to produce the estimate is a fairly mechanical procedure that is facilitated by the use of prepared forms. Normally, after arriving at the total cost, the estimator will compare this figure with past jobs or with bids submitted by competitors. If the figures are radically out of line, the estimator will recheck rates, costs, and quantities to determine whether there is an error or why this job is different.

The summation consists of adding together the component parts to arrive at a total cost. This total is also checked to see if a gross error has been made.

Project overheads are also considered as a direct cost. These include such items as permits, cold-weather expense, and cleanup. The total overheads can be added as a lump sum or, more commonly, distributed over all the operations.

13.8 PRODUCT COST ESTIMATING

The purpose of product cost estimating is to arrive at a selling price for the product. The estimate may be used as a basis to develop a tender or a price list to a distributor, wholesaler, or end user. In any event the selling price of the product is composed of the following basic costs:

Selling price (SP) = research and development costs (RDC)

+ manufacturing costs (MC)

+ administration costs (AC)

+ selling costs (SC) $\qquad$ (13.6)

+ income tax (IT) + profit (P)

SP = RDC + MC + AC + SC + IT + P

MC = direct material (DM) + direct labor (DL)
+ overhead (OH)

Figure 13.8 shows these costs in graphic form.

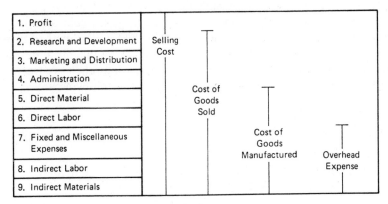

FIGURE 13.8 Representative cost and price structure.

1. *Profit:* after-tax profit requirement
2. *Research and development:* includes all costs associated with research through to the finalization of blueprints
3. *Marketing and distribution costs:* includes such items as salaries and commissions for sales personnel, advertising, entertainment, and travel expense
4. *Administration costs:* includes such items as salaries and bonuses of management personnel, wages of clerical and accounting staff, office rent, supplies, and telephone
5. *Direct material costs:* materials directly consumed in the manufacture of a product (including scrap)
6. *Direct labor costs:* costs associated with specific machine operators
7. *Fixed and miscellaneous expenses:* includes such costs as rent, taxes, depreciation, maintenance, heat, light, water, and tools
8. *Indirect labor costs:* includes such costs as supervision, inspection, factory clerical staff, and maintenance personnel
9. *Indirect material costs:* includes such items as factory supplies and lubricants

The development of an estimate for a product requires that the estimator gather the necessary blueprints and as a first step proceed with a work study analysis.

13.8.1 Work Study

Work study is composed of two basic components: methods design and work measurement. The purpose of *methods design* is to find the best method of doing a job. The purpose of *work measurement* is to determine how long a specific method will take.

Methods design and work measurement are inseparable. The standard time, the time established by management as the allowable time for the performance of an operation, cannot be established until a method of performance is chosen. The method chosen should be the most desirable method that is available at this point in time. The comprehensive treatment of work study requires consideration of at least the following areas:

1. Methods design
2. Work measurement by:
   a. Time study
   b. Standard data, developed through time study and/or by use of predetermined times
   c. Predetermined time systems

d. Work sampling

e. Historical data

f. Estimating

g. Negotiation (unions)

Methods Design

Methods design involves establishment of the method of performing a specific function and the standardization of all aspects of the job. A large number of analytical tools are at the estimator's disposal to develop the desired method for any given job. How successfully the estimator makes use of these tools depends on the questioning attitude of the person.

The models of product flow serve as a guide to develop the method to be used. A material take-off list is formed from the blueprints, and the models of product flow are then generated. Once again, the assumption is made that the blueprints and material specifications are acceptable. In practice, a good estimator will continually question the basic blueprints and material specifications. Often, major savings can be made through minor changes in basic design.

Material List: A systematic study of the blueprints and material specifications for the product should be made by the estimator before proceeding to outline the materials list.

The original blueprints should be designed with full economic consideration by the design engineers regarding such factor as:

1. *Material selection:* A methods study should consider the possibility of using different types of material (e.g., steel versus plastic) plus the availability of the materials.

2. *Manufacturing processes:* The type of material and the methods available from a manufacturing standpoint should be fully considered.

The form of the materials list varies from organization to organization. Figure 13.9 outlines a comprehensive materials list showing all recommended data that should be included.

Process Charts: Using the materials take-off list, process charts can be developed. These charts basically consist of:

1. The operation process chart

2. The flow process chart

3. The routing sheet or operation sheet

The symbols normally used in charting are the symbols published by the ASME:

Subject:

Drawing No. 16725   Sheet 1 of 1   Date _____   Charted By: _____

Part No.	Description	Material Specifications	Raw Material Size (Inches)	Raw Material Weight (lbs.)	Finished Product Size (Inches)	Finished Product Weight (lbs.)	Scrap (lbs.)	Quantity	Total Weight (lbs.)	Per Unit ($)	Total ($)	Remarks
1	Mid-section	ASTM-A366 MSG 16 or Equivalent	48 × 48	40	48 × 48	40.00	0	1	40	4.80	4.80	Based on 12c/lb.
2	Heads	ASTM-A366 MSG 16 or Equivalent	24 × 24	10	17.8 dia.	8.75	1.25	2	17.5	1.05	2.10	Based on 12c/lb.
3	Tank Feet	Hot Rolled Carbon Steel or Equivalent	3/4 × 1/8	1.25	3/4 × 1/8	1.25	0	1	1.25	0.15	0.15	Based on 12c/lb.
4	Wrought Iron Screwed Coupling		Nom 1-1/4 1-7/8 OD 2-1/16 Long	2.7 lbs/ft	Same as Stock	0.46	0	3	1.38	0.41	1.23	8.5% Tax Included. 10,000 PCS/Order
5	Wrought Iron Screwed Coupling		Non 3/8 7/8 OD X 1-3/16 Long	0.57 lbs/ft	Same as Stock	0.06	0	1	0.06	0.16	0.16	8.5% Tax Included. 10,000 PCS/Order
				54.52		50.52			60.19		8.44*	

*Material Cost (With Scrap) $8.74

FIGURE 13.9 Material list for a residential water system pressure tank.

Activity	Symbol
Operation	◯
Transportation	⇨
Delay (e.g., in process storage)	D
Storage	▽
Inspection	▢

1. *The operation process chart.* The *operation process chart* is a graphic representation of the points at which materials are introduced into the process and of the sequence of inspections and all operations except those involved in materials handling. It may also include information considered desirable for analysis, such as time required and location.

The operation process chart is useful in layout work because it encourages an overall, yet systematic view of the manufacturing process. Figure 13.10 is a typical example of an operation process chart.

2. *The flow process chart.* The *flow process chart* is similar in concept to the operation process chart except that it adds more detail and has a slightly different field of application. The flow process chart adds transportation and storage activities to the information already recorded in an operation process chart. Thus, whereas the operation process chart shows only the productive activities, the flow process chart focuses on nonproductive activities as well. These nonproductive activities are very critical to the overall effectiveness of the system. These nonproductive activities require labor and equipment, and thus capital, for such factors as material handling and inventory storage (work-in-process, raw materials, and finished products). These activities, therefore, may be the major deciding criteria in the development of the final layout.

In analyzing the flow process chart, it is most important that the estimator question:

1. The purpose of each operation
2. The location of each operation
3. The timing of each operation
4. The people involved in each operation
5. The procedure followed in each operation

Figure 13.11 shows a portion of the flow process chart for the manufacture of the residential water system pressure tank. The flow process chart may be used for either materials or people.

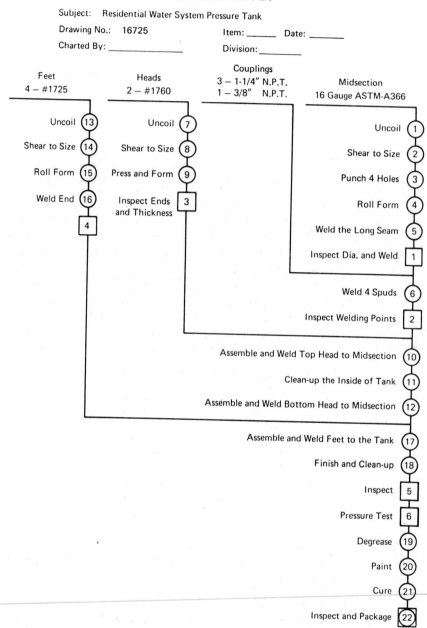

FIGURE 13.10 Operation process chart for a residential water system pressure tank.

Flow Process Chart

	Present		Proposed		Difference	
	No.	Time	No.	Time	No.	Time
Operations						
Transportations						
Inspections						
Delays						
Storages						
Distance Travelled						

Job: _____

Man Or Material: _____

Chart Begins: _____

Chart Ends: _____

Charted By: _____   Date: 6/12/85

Details of (Present) Method	Operation	Transportation	Inspection	Delay	Storage	Distance (ft.)	Quantity	Time (min.)	Eliminate	Combine	Sequence	Place	Person	Improve	Notes
1. Material Racked in Uncoiler								3.0							Coil
2. Uncoil Material and Feed Thru Straightening Rolls								0.02							Ft./Min.
3. Shear Material to Size								0.01							
4. Punch 4 Holes								0.01							Do simultaneously
5. Roll Form Into Required Diameter								0.5							
6. Weld Seam to Form Tank Body								5.0							M.I.G. Welder
7. Inspect Diameter and Weld								5.0							
8. Work-in-process (W-I-P) storage															
9. Jig for Coupling Welding								4.0						X	Improve Jig
10. Weld in Couplings								15.0							
11. Inspect Coupling Welds								5.0						X	Improve
12. Transport to Head Assembly and Welding															

FIGURE 13.11 Flow process chart for a residential water system pressure tank.

3. *Routing sheet (operation sheet)*. The following minimum information is included on a *routing sheet:*

1. Part: name and number
2. Part drawing
3. Operation: name and number
4. Machine–tool name
5. Tools, jigs, and fixtures
6. Standard time

The actual appearance of the routing sheet differs from industry to industry and from company to company. However, basically the process planning determines the sequence of operations to take place on a given part. This sequence of operations is specified on the routing sheet and is of utmost interest to the person doing the layout. In the product-type layout the sequence of operations practically determine the layout.

The routing sheet has many names, which vary from industry to industry. Some of the more common names that apply to this particular form are:

1. Operation sheet
2. Process sheet
3. Production routing sheet
4. Planning sheet

Figure 13.12 represents a typical routing sheet.

PRODUCTION ROUTING

Subject $\underline{24-B}$ $\underline{\text{Tank Midsection}}$ Drawing No. _____
      (No.)   (Name)

Raw material: $\underline{\text{16 gauge cold roll steel (coiled)}}$

Comments: _____

Operation No.	Operation	Machine	Tools and Fixtures	Standard Time (Min./pc)
1	Uncoil Sheet Material	Rowe-Type 1-H		0.6
2	Straighten Material	Rowe Y-50		
3	Shear to Size	Rowe S-50		0.5
4	Punch 4 Holes	Cincinnati Series 4	Special Jig	1.1
5	Roll Form	Keetona SWG Model KIPCD		5.5
6	Weld Seam	Lindle — MIG Type 252/23	Special Fixture	15.2
7	Inspect Dia. and Weld			5.3
8	Move to Work Centre Two For Fittings Welding		Pallets	6.2

FIGURE 13.12 Typical routing for a 24-B tank midsection.

Work Measurement

Establishing the job method for a particular operation and the task of establishing a time standard for the operation are performed concurrently. The *labor standard* determines the standard number of minutes that a qualified, properly trained, and experienced person should take to perform a specific task or operation when working at a normal pace.

Considerable experience is required on behalf of the estimator in arriving at this standard time. For example, in performing a time study, a judgment must be made by the time-study observer as to the speed exhibited by the operator during the study, and the selected time is adjusted by this rating factor so that a qualified operator working at a normal pace can easily do the work in the specified time. This adjusted time is called the *normal time* (NT). To this normal time are added allowances for personal time, fatigue, and delay, the result being the *standard time* (ST) for the task. Figure 13.12 represents a typical routing sheet.

$$\text{NT} = (\text{observed time})(\text{rating factor})$$

$$\text{ST} = \text{NT}[1.0/(1.0 - \text{allowance \%})] \tag{13.7}$$

On completion of the methods study the estimator will have a direct material cost and established standard times for each operation of components to be manufactured in-house.

This information may be summarized by work center and operation, or it may be summarized by cost center. The method used will depend on the company and the estimator.

13.8.2 Direct Material Costs

The major component of the direct material cost can be derived by completing a material take-off from the product blueprint(s). To this material cost one must add scrap material costs, which are represented by two components: (1) scrap due to waste material, such as the slugs that are discarded from punching a hole in the components being manufactured; and (2) components that have to be scrapped due to such factors as workmanship and machine misalignment.

These costs can be significant and thereby stress the importance of:

1. Purchasing raw materials in a form that minimizes scrap (e.g., coiled steel material versus sheet steel)
2. Machine setup to produce components within the desired tolerances
3. Proper training programs
4. Communications throughout the organization

Direct material costs represent the major cost component in most manufacturing operations. For example, in a metal fabrication shop this cost will probably represent 40 to 50% of the manufacturing cost.

13.8.3 Direct Labor Costs

Direct labor costs are those labor costs incurred by such persons as machine operators in the manufacture of a product. This cost is usually estimated for a specific product or component from a combination of historical records, measured data, and policy decisions.

Several factors that enter into the labor estimate are:

1. The fringe benefits associated with labor costs are significant. Fringe benefits will normally amount to 25 to 35% of the labor rate or salary.
2. Large variances exist between workers with respect to production output. Surveys indicate that if the sample is sufficiently large (say 100 workers), the workers on the top end of the curve will have an output double that of the workers on the lower end of the curve. This factor stresses the need for careful planning in estimating production time to complete a particular operation.

3. Environmental working conditions affect production output. If significant changes in the environment exist, these changes must be considered.

4. Policy changes, such as the implementation or removal of a wage-incentive plan, may result in major changes in production output. Direct labor costs represent a significant portion of the job cost. For example, in a job shop the direct labor cost will normally amount to 20 to 30% of the manufacturing cost. Naturally, this cost depends on how labor intensive the operation is.

13.8.4 Overhead Costs

Overhead costs are those costs that cannot be associated with particular operations or products and therefore must be prorated among all the products by some method. There are many methods whereby overhead costs may be allocated within the framework of any specific company structure. The estimator should adopt a system that gives the desired accuracy with a minimum of effort on a continuing basis.

The exact nature of the overhead rate may differ from one company to the next, and it is not uncommon to find various types of overhead rates within one firm. Some of the more important decisions that need to be considered in applying overhead rates are:

1. Whether the rate includes fixed costs
2. The base used to apply overhead, such as direct labor dollars, direct labor hours, and machine hours
3. The activity level assumed as a basis for fixed costs
4. The scope of the application of the rate (to the whole plant or to a cost center)
5. Whether the rate applied to all designs (such as product lines) or to specific design(s)

Some of the methods used to distribute factory overhead costs to job costs are the following:

1. Overhead costs applied on the basis of direct labor cost or time
2. Overhead costs applied on the basis of direct labor and direct material costs
3. Overhead applied on the basis of conversion costs
4. Overhead costs applied on the basis of direct material costs
5. Overhead costs applied on the unit of product basis
6. Overhead applied on a machine-hour basis
7. Overhead applied on the basis of a fixed and variable cost classification

For example, the overhead rate on the basis of direct labor costs would be calculated as follows:

$$\text{Rate} = \frac{\text{actual factory overhead}}{\text{actual direct labor hours or direct labor costs}} \qquad (13.8)$$

An actual overhead rate has the merit of distributing among the jobs the incurred factory overhead. Although it is possible to determine an overhead rate following the end of the period and to use this rate to apply actual costs for that period, historical overhead rates are at a disadvantage. Historical rates delay cost calculations until the end of the month and often later and fluctuate because of seasonal and cyclical influences acting on the actual overhead costs and on the actual volume of activity for which overhead costs are spread.

In most situations overhead rates are determined from data developed from an operating budget, although overhead rates can be computed without using a budget. For the latter situation the company does not compile an operating rate; rather, the estimate is made by studying the behavior of recent cost experiences, and then various adjustments are made. Regardless of the method used, allocation of overhead rates on a reliable basis is critical, as illustrated by the following example.

Example 13.6

Problem: The Argo Manufacturing Co. has determined the following breakdown as a basis for allocating costs. Use straight-line depreciation (equation 4.2). The salvage value of machinery is negligible.

Dept.	Floor Space Used by Dept. (ft²)	Horsepower of Machinery	Direct Labor Personnel	Machinery First Cost	Life (years)
A	10,000	100	50	$100,000	10
B	8,000	50	30	80,000	8
C	5,000	20	20	60,000	6
D	2,000	30	10	180,000	6
	25,000	200	110	$420,000	

Factory expense items are as follows:

Depreciation of building (allocate by floor area)	$100,000
Heat, light, and fire insurance (allocate by floor area)	10,000
General (allocate by number of workers)	25,000
Power	5,000
Superintendent's salary (allocate by number of workers)	20,000
Wages (including all fringe benefits)	

Departments A and B average $16,000/year/worker
Departments C and D average $20,000/year/worker

Find the overhead rate (ratio of dollars of overhead to dollars of direct labor cost) for the total factory, and find the overhead rate in department B by departmentalizing costs.

Solution

Item	Basis for Allocation	Total	Department			
			A	B	C	D
Depreciation						
Equipment	SL	$ 60,000	$10,000	$10,000	$10,000	$30,000
Building	Floor area	100,000	40,000	32,000	20,000	8,000
Heat, light, and						
fire ins.	Floor area	10,000	4,000	3,200	2,000	800
General	Employees	25,000	11,364	6,818	4,525	2,262
Power	HP	5,000	2,500	1,250	500	750
Supt.'s						
salary	Employees	20,000	9,091	5,455	3,654	1,827
		$220,000	$76,955	$58,723	$40,679	$43,639
Wages ($)		1,880,000	800,000	480,000	400,000	200,000
Overhead rates (%)		11.7	9.6	12.2	10.2	21.8

Problem: Find the factory cost for a product manufactured entirely in department D, assuming direct labor costs of $2.00 per unit and direct material costs of $3.00 per unit.

Solution

$$MC = DL + DM + OH$$

$$= \$2.00 + \$3.00 + (\$2.00)21.8\% = \$5.44$$

13.8.5 Marketing and Distribution Costs

There are significant variations from organization to organization in the total marketing structure. Marketing strategies of companies with closely allied product lines may vary considerably. These differences, which include methods of reimbursing sales personnel, advertising programs, extent of the marketing area, and internal training programs, exist for a multitude of reasons. However, perhaps the most significant factor to be gained through discussion of these differences is the need for a systematic approach within each organization in planning marketing strategies for both the short and long ranges.

Marketing expenses as a percentage of the total selling price for a product vary considerably depending on the type of organization under study. However, as an example, the typical small fully integrated manufacturing company would spend from 10 to 20% of the selling price on marketing its product. Variations exist for many reasons, one of the more significant being company age. A new company would expect to have higher advertising costs than an older, established firm, in order to broaden the scope of its marketing area and the makeup of its customer population.

13.8.6 Administration Costs

The traditional approach has been to view administration costs as a distinct cost separate from marketing and manufacturing. However, some companies consider administration as a direct support function to manufacturing and marketing and include these costs in manufacturing and marketing overhead. The approach used by any individual organization, once again, should be viewed within the framework of the total system, considering all factors both internal and external to the organization. For example, how does each component affect the accounting function and development of financial ratios for comparison with companies in a similar business?

13.8.7 Research, Engineering, and Development Costs

These costs are often associated directly with manufacturing expenses. Whether they are recorded independently or as manufacturing overhead, it is important from a control standpoint to be able to review these costs periodically. In many organizations these costs are not identifiable and are often overlooked in product pricing. This situation is very likely to occur within small manufacturing firms that are faced with the continual need to improve product design to remain competitive with large national or international corporations.

Without a satisfactory method of identifying these costs, the company may experience major problems because either too little or too much effort is being directed toward research, engineering, and development costs. Either approach can be detrimental.

13.9 SUMMARY

Cost estimating is an essential and difficult component of the analyst's function. The results of any economy study are only as valid as the input data.

Order-of-magnitude estimates utilizing published data and/or data internal to the firm may be used to narrow the alternatives under consideration. However, for a proper application of the data, experience on the part of the analyst is essential. Significant errors can be introduced through such factors as errors in cost capacity factors and conversion of published data to account for local differences.

Semidetailed or budget estimates accurate to within $\pm 10\%$ are normally quite satisfactory for making decisions regarding project feasibility (economy studies).

Fixed-price bids for construction projects and manufactured job lots require detailed estimates with an accuracy within $\pm 5\%$. In-house records, specific to the operations within an individual firm, are an essential component for estimates requiring this level of detail. The analyst should be thoroughly familiar with the capabilities within the company. This can be accomplished only through a continual update of cost data and a close liaison with management at all levels.

The accuracy of an estimate has a direct relationship to the cost of achieving it. We have discussed the value of time and detail to accuracy and we have discussed several methods, presented in their simplest form. There is much that can be considered; even the errors of the estimator can be included. However, stochastic models in great detail, fed by voluminous history, are expensive. Computers assist but may or may not be cost-effective. If cost estimating addresses the question of worth, the value of the estimate must also be considered.

The subject of cost estimating has merely been introduced. An estimator's success depends on knowledge of the theory and also a thorough practical knowledge of the industry and the firm. Success comes from this knowledge and experience, from learning the do's and don'ts of estimating, from understanding the reasons for high or low estimates, and from realizing that the estimate is merely an estimate, however, hopefully close to the actual value.

PROBLEMS

13.1 Project cost estimating is usually undertaken within three levels of detail and accuracy.
   a. List the three levels of detail and accuracy.
   b. What level of accuracy should the estimator be targeting for with respect to data used in an economy study?

13.2 Published indices are available from many different sources and in many different forms.
   a. What precautions should the estimator take in using indices?
   b. At what level of estimating accuracy do you think indices are valid if used with discretion?

13.3 L.J. Construction is building a convention center in a northern Alberta community. It is to be 2.3 million square feet. L.J. wishes to estimate the amount of overhead to include in its estimate. The records for their most recent projects are:

Size (millions of square feet) (x)	Overhead (millions of dollars) (y)
1.6	4.5
1.5	4.3
1.9	5.5
2.4	7.0
2.4	7.1
2.2	5.7

a. Use linear regression to estimate the overhead.
b. What is the correlation coefficient?

13.4 The federal government wishes to decrease the number of unemployed by creating tax incentives for manufacturing. They have had some experience over the last few years. How much money would they need to input to create 17,000 jobs?

Value of Incentives (billions of dollars) (x)	Jobs Created (y)
1.0	10,000
1.4	12,000
2.0	19,000
2.5	25,000
3.0	27,000

a. Use linear regression to determine the money required to produce 17,000 jobs.

13.5 Black Industries is planning to construct a new 500,000-ft^2 warehouse in 2 years' time. A similar type of building (200,000 ft^2) was constructed 2 years ago at a cost of $20.00 per square foot. The building index 2 years ago was 120, today it is 150, and in 2 years it is expected to be 180. Use a cost capacity factor of 0.7 and determine a construction cost per square foot for the warehouse in 2 years' time.

13.6 Prepare a cost estimate for a pipeline pumping station to be budgeted for construction in 1985. A pumping station of identical capacity was constructed in 1982. Costs pertaining to this project were:

| | 1982 | Index | |
Component	Costs	1982	1984
Site preparation	$ 10,000	1.50	1.80
Pumping equipment	100,000	1.40	1.55
Piping	20,000	2.00	2.60
Storage tanks	200,000	1.20	1.40
Building costs	100,000	1.50	2.00
	$430,000		

Assume that the increase in costs is linear between 1982 and 1985 and estimate the cost of the pumping station in 1985.

13.7 K.T. Construction has been asked to give a preliminary estimate for constructing an old-age home (Engineering News-Record Index = 1,150). The building will be 250,000 ft². Last year similarly styled buildings (200,000 ft²) cost $17 per square foot (index = 925).
   a. What is the order-of-magnitude estimate if the published estimating factor is 0.85?
   b. If the actual estimating factor was 0.90, what is the cost of the error?

13.8 Consider a company that builds apartment buildings on a contract basis. Discuss the importance of bidding. What would happen if the bid were too high or too low? How accurate should such a estimate be? Is it better to bid too high or too low?

13.9 Assume that you are a small concrete contractor and you have been asked to bid on the concrete work (floor, apron, and footings) for a large airport hangar, construction to begin in 2 years. Construction costs have increased at an annual average rate over the past 3 years as follows:

Labor	12%
Materials	8%
Construction overhead	10%
Administration	10%
Marketing	8%

Costs for a similar job today are as follows:

Labor	$360,115
Overhead	125,000
Administration	70,000
Marketing	38,013

The contract involves 25,000 yards of concrete and 200 tons of steel.

Concrete $40 per yard today
Steel $400 per ton today

This contract can lead to several similar contracts and greatly improve your cash flow. However, a 10% loss on the job would almost certainly bankrupt the company. Prepare a cost estimate for the job considering all data presented and any additional factors that you consider essential.

13.10 A company has completed a time study on the production of a certain product they manufacture. Three operators have been studied and the times recorded.

Operator	Times (minutes)					
1	10.0	10.5	11.0	9.5	12.0	10.2
2	8.0	7.5	9.0	11.0	7.0	8.5
3	12.0	11.5	10.0	9.0	4.0	11.0

The operator base wage is $10.00 per hour and fringe benefits cost the company 25% of the hourly rate. The material cost per unit is $2.00. The overhead rate is 40% of the base wage.
a. Calculate the manufacturing cost for this product.
b. What comments do you have regarding the data given (e.g., validity)?

13.11 The Flexigrow Manufacturing Company has determined the following breakdown as a basis for allocating costs. The company uses the straight-line method for allocating depreciation (equation 4.2). The salvage value of all machinery is assumed to be zero.

Dept.	Floor Space	Horsepower of Machinery	Direct Labor Personnel	Machinery First Cost	Machinery Life (years)
A	5,000	200	10	$100,000	10
B	3,000	50	10	50,000	10
C	20,000	100	15	200,000	15
D	10,000	50	25	250,000	10

Factory expense items are estimated as follows:

1. Depreciation of building (allocate by floor area)	$100,000
2. Heat, light, water, insurance (allocate by floor area)	20,000
3. Miscellaneous (allocate by floor area)	5,000
4. Production supervision (allocate by workers)	50,000

5. Average annual wages per worker:

Department A	10,000
Department B	12,000
Department C	11,000
Department D	12,000

  a. Find the overhead rate (ratio of dollars of overhead to dollars of direct labor cost) for the total factory.
  b. Find the overhead rate for each department.
  c. Find the factory cost for a product that requires 2 hours to manufacture assuming 2,000 hours per year and:
     (1) Direct labor department A = 20%.
         Direct labor department B = 50%.
         Direct labor department D = 30%.
     (2) Direct material costs = $5.00 per unit.
  d. What are the advantages and disadvantages of using shop overhead for estimating product costs?

13.12 George intends to build a fence around his hobby farm. This entails post holes, waterproofing posts, setting posts, rails and fence boards, and painting the entire fence. The post holes are to be set every 8 feet along the perimeter. Prepare an estimate to build the fence. Poles cost $3, two 8-foot rails cost $1.50, 8 feet of fence boards cost $10, and 1 gallon of paint costing $20 will do 40 feet of fence one side only. Post holes will be dug and set by hand.

13.13 George established a large fencing company with a department devoted to each operation involved in building fences. He has established that a fence can be built at a rate of 4 feet every hour at a labor rate of $32 an hour. In addition, George adds a 30% profit margin on each job. If George were to have estimated the cost of his own fence using the material costs above, what would his estimate have been?

13.14 Duncan Fertilizers Ltd. believes that the sales of fertilizers are correlated to the sales expenses. Given the following information, is the assumption valid? Use linear regression.

Sales Expense (thousands of dollars) (x)	Sales (millions of tons) (y)
87	21.6
80	20.0
94	23.8
75	19.1
78	19.0
83	21.9

What is the expected sales if the estimated sales expense for the next year is $84,200?

13.15 Global Corporation is planning to build a cement factory with a capacity of 20,000 tons/day. A similar plant with a capacity of 12,500 tons/day cost $19,000,000 3 years ago (index = 900). The latest index value is 1,200. Use a cost capacity factor of 0.85. What is the estimated cost of the proposed plant?

13.16 Ace Rental Company rents air compressors by the day for $100. The company owns 100 compressors. A simplified revenue versus expense statement for this year is as follows.

Account	Amount
Machine-days rented	11,400
Revenue	$1,140,000
Expense	
Labor	$205,000
Rent	20,000
Maintenance	67,000
Depreciation	120,000
Interest	240,000
Total expenses	$652,000

Next year business demand is expected to be constant as long as rental rates remain constant. With each additional 1% rate increase, business will decline by 5% (business will increase by the same amount for every 1% decrease in price up to about 7% decrease in price). Labor has already settled for an 8% increase, rent is going up 9%, and maintenance costs will rise 8%. Between selling old compressors and buying new ones, depreciation expense will be $11,000 per month and the interest expense will be approximately the same as that of the previous year. The owners of the company do not wish to expand. However, they want to have at least the same taxable income as this year. What should the rental rate be for the next year?

CHAPTER FOURTEEN

The Budgeting Process

The budget is a tangible statement of an organization's goals and objectives and the means by which they are to be accomplished. The budget process includes considerations of the capital structure, sources of capital, capital rationing, and the timing of acquisition of assets. In this chapter we explain and describe cash, capital, and operating budgets. The important topic of leasing, which is an alternative way of obtaining capital assets, is also discussed.

14.1 THE PURPOSE OF BUDGETING

Although all organizations plan, there are considerable differences in the way in which they plan. Some people do their planning entirely in their heads, others make notes and rough estimates on scrap paper, and still others express their plans in quantitative terms and commit these plans to paper in an orderly and systematic fashion. The process engaged in by the latter group is called *budgeting,* for a budget is merely a plan expressed in quantitative terms. We are concerned primarily with budgets that are expressed in monetary terms, although some budgets are expressed in units of product, number of employees, units of time, or other nonmonetary quantities. In addition to its use in planning, the budget is also used for control and for coordination.

The *master budget* summarizes the objectives of all subsystems within an organization: marketing, production, distribution, engineering, and finance. It quantifies the expectations regarding future sales, cash flows, financial position, and supporting plans. These expectations are the culmination of a series of decisions resulting from a careful look at the organization's future. The master budget usually represents the best practical approximation to a formal model of the total organization stating its objectives, input requirements, and expected outputs.

When administered wisely, budgets (1) force management to plan, (2) provide definite expectations that result in an effective framework for judging subsequent performance, and (3) promote communication and coordination among the various segments of the organization. The key point is that budgets provide a discipline that brings planning to the forefront as a key management responsibility at all levels throughout the organization.

Many managers claim that the uncertainties associated with their business make budgets impractical and of little value. However, one can always find some companies in every industry that make effective use of budgets. Such companies are usually among the industry leaders, and they regard budgeting as an essential tool. Managers must contend with uncertainties, either with or without a budget. The budget represents an explicit effort to quantify these uncertainties. The advocates of budgeting maintain that the benefits from budgeting nearly always exceed the costs. All organizations do not and should not budget to the same degree, but some form of budget program will be helpful in every organization.

14.1.1 Types of Budgets

Most companies, except the very smallest, should and do have some sort of budget, but a great many do not have a truly comprehensive master budgeting system. Such a system consists of three types of budgets: (1) an operating budget, showing planned operations for the forthcoming period; (2) a cash budget, showing the anticipated sources and uses of cash; and (3) a capital budget, showing planned changes in fixed assets.

An *operating budget* usually consists of two parts, a program budget and a responsibility budget. These parts represent two ways of portraying the overall operating plan for the business, two different methods of slicing the pie; therefore, both arrive at the same figure for projected net income and return on investment.

The *program budget,* as one would expect, describes the major programs the company plans to undertake. Such a budget might be arranged, for example, by product lines and show the anticipated revenue and costs associated with each product. This type of budget is useful to an executive examining the overall balance among the various subsystems within the organization. It helps to answer such questions as these: Is the profit margin on each product line satisfactory? Is production capacity in balance with the size and capability of the sales organization? Can we afford to spend so much for research? Are adequate funds available? And so on. A negative answer to any of these questions indicates a possible problem area and the necessity for revising the overall plan. This systematic approach allows management to develop a workable integrated total system.

The *responsibility budget* sets forth plans in terms of the persons responsible for the successful completion of each plan. It is therefore primarily a control device, since it is a statement of expected or standard performance against which actual performance can later be measured. In the factory, for example, there may be a responsibility budget for each department, showing the costs that are controllable by the supervisor of the department. There may also be a budget showing costs for each product, including both direct costs and allocated costs. The figures on both sets of budgets add up to total factory costs, but the product-cost budget would not be useful for control purposes, since the costs shown on it could not ordinarily be related to the responsibility of specific persons.

14.1.2 Principles of Budgeting

The general principles relevant to the management control process apply to the budgeting process. The budget should be sponsored by management. The organization must regard it primarily as a tool of management, not as primarily an accounting device. Responsible supervisors should participate in the process of setting the budget figures and should agree that the budget goals are reasonable. If they do agree, their attitude during the budget period is likely to be: "I said I could meet this goal, and I will do my best to live up to this promise." If they are not consulted, their attitude toward the budget is likely to be one of indifference and/or resentment.

The budget figures should represent reasonably attainable goals, not so high as to be frustrating, yet not so low as to encourage complacency.

In a comparison of actual performance with budgeted performance, attention should be focused on *significant exceptions,* that is, figures that are significantly different from those expected.

The review of budget estimates by successively higher levels of management should be thorough. Perfunctory review is a signal that management is really not dedicated to the budget process.

Final approval of the budget should be specific, and this approval should be communicated to the organization in writing. An attempt to operate on the doctrine "silence gives consent" inevitably leads to misunderstanding.

14.2 THE BUDGET PROCESS

The preparation of a budget can be studied as both an accounting process and a management process. From an accounting standpoint, the end result of the recording and summarizing operations is a set of financial statements, a balance sheet and an income statement, identical in format with those resulting from the process of recording historical events. The only difference is that the budget figures are estimates of what will happen in the future rather than historical data on what has happened in the past.

From a management standpoint, the budgeting process is so closely associated with the operation of a business that a complete description of the factors and considerations involved cannot be covered herein. Rather than attempt such a description, we shall merely indicate some of the important general considerations in the budget process.

14.2.1 Choice of Time Periods

A useful time period for a budget is 1 year. Probably the majority of companies prepare budgets once a year, but some companies follow the practice of preparing a new budget every quarter. Each quarter, the quarter just completed is dropped, the figures for the three quarters originally covered in the previous budget are revised if necessary, and a new budget for the fourth quarter is added. This is called a *rolling budget*.

Within the year, the budget may be broken down by months; that is, the month usually constitutes the basic time period for comparison of actual results with the budget. In many companies, only the data for the next 3 months or the next 6 months are shown in detail, and totals are given by quarters for the remainder of the year.

For the purpose of showing the general direction in which the company plans to move, and the long-run implications of management policies as to expansion, new products, new facilities, and so on, a useful tool is the long-range budget, covering a period of 3, 5, or even more years ahead. This is not prepared in as much detail as the annual budget, and it is usually prepared entirely at headquarters without the participation of operating personnel. Such a budget is especially useful in preparation plans for obtaining new, permanent capital.

14.2.2 Organization for Preparation of Budgets

A budget committee, consisting of several members of the top-management group, may oversee the work of preparing the budget. This committee will set the general guidelines that the organization is to follow, coordinate the separate budgets prepared by the various organizational units and resolve differences among them, and submit the final budget to the president and to the board of directors for approval. In a small company, this work is done personally by the president or by his or her immediate line subordinate. Instructions go down through the regular chain of command, and the budget comes back for successive reviews and approvals through the same channels. *The essential point is that decisions about the budget are made by the line organization, and the final approval is given by the head of that organization, the president or the board.*

The line organization is usually assisted in its preparation of the budget by a staff unit headed by the budget director, preferably reporting to the controller. If the director reports to the president, as is sometimes the case, the person is likely to be performing the functions described above for the budget committee and therefore acting in a line rather than a staff capacity.

As a staff member, the budget director's functions are to disseminate instructions about the mechanics of budget preparation (the forms and how to fill them out), to provide data on past performance that are useful in preparation of the budget, to make computations on the basis of decisions reached by the line organization, to assemble the budget figures, and to see that everyone submits figures on time. Thus, although the budget organization may do a very large fraction of the budget work, it is not the crucial part. However, the significant decisions are always made by the line organization. Once the members of the line organization have reached an agreement on labor productivity and wage rates, for example, the budget director can calculate all the detailed figures for labor costs by products and by responsibility centers; this is a considerable job of computation, but it is entirely based on the judgment of the line supervisors.

14.3 USES OF THE BUDGET

The usefulness of the budget as the process of making and coordinating plans is apparent from the description above. The budget is also useful as a communication device and as a standard with which to compare actual performance.

14.3.1 The Budget as a Communication Device

Management's plans will not be carried out (except by accident) unless the organization understands what the plans are. Adequate understanding includes not only a knowledge of programs and objectives (e.g., how many units are to be manufactured, what methods and machines are to be used, how much material is to be purchased, what selling prices are to be) but also a knowledge about policies and restrictions to which the organization is expected to adhere. Examples of these types of information follow: the maximum amounts that may be spent for such items as advertising, maintenance, administrative costs, and the like; wage rates and hours of work; desired budget levels; and so on. A most useful device for communicating quantitative information concerning these objectives and limitations is the approved budget.

14.3.2 The Commitment Concept

When a budget is prepared in accordance with the procedures outlined above, the final document may be regarded as a sort of contract or two-way commitment between management and the operating supervisors. By agreeing to the budget estimates, the supervisor in effect says to management: "I can and will operate my department in accordance with the plan described in the budget." By approving the budget estimates, management in effect says to the supervisor: "If you operate your department in accordance with this plan, you will do what we consider to be a good job." Both of these statements contain the implicit qualification of "adjusted for changes in circumstances" since both parties recognize that actual events, such as price levels and general business conditions, may not correspond to those assumed when the budget was prepared and that these changes will inevitably affect the plans set forth in the budget. In judging whether the commitment is in fact being carried out as the year progresses, management must take these changes into account.

14.3.3 The Budget as a Standard

A carefully prepared budget is the best possible standard against which to measure actual performance, and it is increasingly being used for this purpose. Until fairly recently, the general practice was to compare current results with results for last month or with results for the same period a year ago; and this is still the basic means of comparison in many companies. Such a historical standard has the fundamental weakness that it does not take account either of changes in the underlying forces at work or of the planned program for the current year.

For example, in a favorable market situation, a certain company increased its selling prices and hence increased its net income in 1982 by 25% over the net income of 1981. If 1982's results are compared with 1981's, there is shown an apparent cause for rejoicing. However, the company had planned to increase profits by 35%, and performance when measured against the plan was not so good. The company quite properly took steps to find out, and if possible to correct, the factors accounting for the difference between actual and budgeted results.

In general, it is more significant to answer the question, "Why didn't we do what we planned to do?" than the question, "Why is this year different from last year?" Presumably, the principal factors accounting for the difference between this year and last year were taken into consideration in the preparation of the budget.

14.4 THE OPERATING BUDGET

The preparation of the operating budget for any given year includes the preparation of several basic budget schedules. For a typical manufacturing company these include:

1. Sales budget
2. Production budget
3. Direct-material purchases budget
4. Direct-labor budget
5. Factory-overhead budget
6. Ending-inventory budget
7. Cost-of-goods-sold budget
8. Selling and administrative expense budget

From these individual budgets are developed: (1) a budgeted income statement, and (2) a budgeted balance sheet.

14.4.1 Sales Forecasting

The sales forecast is the starting point for the budget because inventory levels and production (and hence costs) are generally related to the rate of sales activity. The sales budget is the result of a series of management decisions.

The chief sales officer has direct responsibility for the preparation of the sales budget. The task of preparation forces the officer to crystallize his or her plans.

The sales forecast is made after consideration of the following factors:

1. Past sales volume
2. General economic and industry conditions
3. Relationship of sales to such economic indicators as gross national product, personal income, employment, prices, and industrial production
4. Relative product profitability
5. Market research studies
6. Pricing policies
7. Advertising and other promotion
8. Quality of sales force

9. Competition
10. Seasonal variations
11. Production capacity
12. Long-term sales trends for various products

An effective aid to accurate forecasting is to approach the same goal by several methods; each forecast acts as a check on the others. The three methods described below are usually combined in some fashion that is suitable for a specific company.

14.4.2 Sales Staff Procedures

As is the case for all budgets, those responsible should have an active role in sales-budget formulation. If possible, the budget data should flow from individual sales personnel or district sales managers up to the chief sales officer. A valuable benefit from the budgeting process is the holding of discussions, which generally result in adjustments and which tend to broaden participants' thinking.

Previous sales volumes are usually the springboard for sales predictions. Sales executives examine historical sales behavior and relate it to other historical data, such as economic indicators, advertising, pricing policies, and competitive conditions. Current information is assembled, production capacity is considered, and then the outlook is derived for the ensuing months (years in long-run sales budgets).

One of the common difficulties in budgeting sales is the widespread aversion of sales executives to figures. The usefulness of budgeting must be sold to sales personnel. The best sales executives may not particularly enjoy working with figures, but they realize that intelligent decisions cannot be made without concrete information. Market research is a sales executive's tool that helps to eliminate hunches and guessing.

14.4.3 Statistical Approaches

Trend, cycle projection, and correlation analysis are useful supplementary techniques. Correlations between sales and economic indicators help make sales forecasts more reliable, especially if fluctuations in certain economic indicators precede fluctuations in company sales. However, no firm should rely entirely on this approach. Too much reliance on statistical evidence is dangerous, because statistics are based on historical patterns. Once the sales budget has been formulated, the production and other related budgets can be prepared.

Final approval for the budget comes from top management. The approved budget is then transmitted to the organization. This act communicates to the organization the approved objectives and the plans for reaching these objectives during the coming year.

14.5 THE CASH BUDGET

The operating budget is usually prepared in terms of revenues and expenses. For financial planning purposes it must be translated in terms of cash receipts and cash disbursements. This translation is the *cash budget*. The financial people use the cash budget to make plans to ensure that the company has enough, but not too much, cash on hand during the year ahead.

There are two approaches to the preparation of a cash budget:

1. Start with the budgeted balance sheet and income statements and adjust the figures thereon to reveal the sources and use of cash.
2. Project directly each of the items that results in cash receipts or cash disbursements. A cash budget prepared by this means is shown in Table 14.1 for the first 6 months of 19XX. Some useful points in connection with this technique are mentioned below.

TABLE 14.1 Cash budget for the XYZ Company, 19XX (thousands of dollars).

	Jan.	Feb.	Mar.	Apr.	May	June
Net sales	1,000	1,500	1,300	1,150	1,850	1,540
Cash balance beginning of period	350	345	110	180	110	120
Cash Receipts						
Accounts receivable collected	1,200	1,150	1,500	1,250	1,225	1,700
Miscellaneous receipts	25	30	40	20	25	25
Total receipts	1,225	1,180	1,540	1,270	1,250	1,725
Total cash available	1,575	1,525	1,650	1,450	1,360	1,845
Disbursements						
Operating expenses	400	800	400	500	650	625
Material purchases	350	425	425	320	450	200
Administrative costs	100	120	110	110	120	110
Selling costs	120	180	150	145	175	160
Fringe benefits	150	190	160	155	185	165
Taxes		200	25		160	170
Equipment purchases			200			
Dividends	110			110		
Total disbursements	1,230	1,915	1,470	1,340	1,740	1,430
Cash balance end of month before bank loans or repayment	345	(390)	180	110	(380)	415
Bank loans or (repayments)		500			500	(300)
Cash balance end of month	345	110	180	110	120	115

Collections of accounts receivable is estimated by applying a "lag" factor to estimated sales or shipments. This factor may be simply based on the assumption that the cash from this month's sales will be collected next month; or there may be a more elaborate assumption, for example, that 20% of this month's sales will be collected this month, 50% next month, 20% in the third month, and 10% in the fourth month. The approach used will vary from company to company based on actual experience.

Other operating expenses are often taken directly from the expense budget, since the timing of cash disbursements is likely to correspond closely to the time the expense occurs. Noncash expenses such as depreciation are not included.

The cash flow projections represent the basis for cash planning. In February, for example, Table 14.1 indicates that the company plans to borrow $500,000. It plans to borrow an additional $500,000 in May and to start repaying these bank loans in June when cash receipts exceed cash disbursements.

14.6 THE CAPITAL BUDGET

The *capital budget* is essentially a list of what management believes are worthwhile projects for the acquisition of new capital assets together with the estimated cost of each project. Proposals for such projects may originate anywhere in the organization. The capital budget is usually prepared separately from the operating budget, and in many companies it is prepared at different times. However, these two budgets are closely coordinated.

Each proposal, except those for minor amounts, is accompanied by a justification. For some projects, the expected return on investment can be estimated. For others, such as the remodeling of employee recreation rooms, no estimate of earnings is possible, and these are justified on the basis of improved morale, safety, appearance, convenience, or other subjective grounds. A lump sum is usually included in the capital budget for projects that are not large enough to warrant individual consideration by top management.

As proposals come up through the organization, they are screened at various levels, and only the sufficiently attractive ones flow up to the top and appear in the final capital budget. On this document, they are usually arranged in order of desirability, and the estimated expenditures are broken down by years, or by quarters, so that the funds required in each time period are shown. At the final review meeting, which is usually at the board-of-director level, not only are the individual projects discussed, but the total amount requested on the budget is compared with total funds available. Many apparently worthwhile projects may not be approved simply because the funds are not available.

Approval of the capital budget usually means approval of the projects in principle but does not constitute final authority to proceed with them. For this authority, a specific authorization request is prepared for the project, spelling out the proposal in more detail, perhaps with firm bids or price quotations on the new assets. These authorization requests are approved at various levels in the organization, depending on their size and character. For example, each shop supervisor may be authorized to buy production tools or similar items costing not more than $500 each, provided that the total for the year does not exceed $3,000; and at the other extreme, all projects costing more than $500,000 and all projects for new products whatever their cost may require approval of the board of directors. In between, there is a scale of amounts that various levels of management may authorize without the approval of their superiors.

14.7 RANKING CAPITAL INVESTMENT OPPORTUNITIES

The demand schedule for funds is constructed from the budgetary requests made by departments and divisions within the organization. The investment funds requested for the coming period for each project, together with the merit of each, are listed usually in the form of the expected rate of return. These requests may then be laddered as shown in Table 14.2.

TABLE 14.2 Laddering the demands for capital budgeting purposes

Request	Department	Amount	Rate of Return (%)	Cumulative Amount
Computer system	Accounts	$150,000	52	$ 150,000
Mat'l-handling equip.	Mat'l. mgmt.	400,000	47	550,000
Extrusion equip.	Production	106,000	42	656,000
Shop extension	Production	280,000	40	936,000
Tractor-trailer unit	Mat'l. mgmt.	90,000	37	1,026,000
Modify power plant	Maintenance	175,000	33	1,201,000
Microcomputer system	Sales	95,000	25	1,296,000
Multislide press	Production	200,000	20	1,496,000
Word processor	Admin.	30,000	17	1,526,000
Storage shed	Mat'l. mgmt.	85,000	13	1,611,000
Pave lot	Mat'l. mgmt.	120,000	10	1,731,000

To satisfy all the requests for capital expenditures would require $1,731,000. Suppose that the company only has $1,500,000 available, at a weighted-average cost of capital of 12%. Further assume that this is all the money available because either the owners and creditors are not able or willing to advance additional funds, or the management does not wish to obtain additional assets. This is the capital-rationing situation. The money is allocated by working down the table accepting projects until the funds are exhausted. This would mean that the multislide press is the last project accepted, and the three below it would be cut off.

The rate of return of the last project accepted is 20%. If some other project were accepted, there would not be sufficient funds to accept this project. In other words, an opportunity to earn 20% would be forgone. Thus, in this capital-rationing situation, 20% becomes the MARR value for the company and is called the *cutoff rate of return* or the *cutoff cost of capital.*

Alternatively, if $2,000,000 were available, the company would still reject the bottom request, pave lot, because its 10% rate of return is less than the 12% weighted-average cost of capital. For this capital-surplus situation the weighted-average cost is the MARR value.

Figure 14.1 shows the cumulative requests and the available supply. If the gap between the weighted-average rate and the cutoff rate is large, there will be pressure on the firm to increase its supply of capital. However, one should anticipate that in a well-managed firm that is constantly seeking opportunities for improvement, there will always be more opportunities than money and that the cutoff rate will be the MARR value.

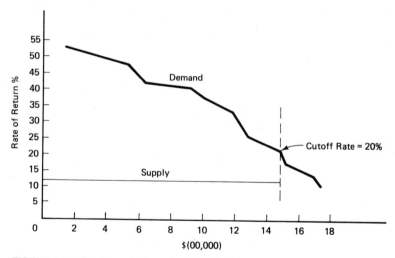

FIGURE 14.1 Capital requests and capital supply.

14.8 PRESENT VALUE VERSUS INTERNAL RATE OF RETURN

The two most common methods of ranking proposals for capital budgeting purposes are the present-value and the internal-rate-of-return methods. Important differences exist between these two methods that must be recognized if consistent results are to be maintained when applying either method.

When two or more investments are mutually exclusive, the two methods may give contradictory results if not properly applied (note Example 6.9). To illustrate the nature of the problem, consider the following example.

Example 14.1

Problem: The HM Company is considering two mutually exclusive proposals which have the following cash flows. Each proposal requires an initial investment of $25,000 and generates the following after-tax cash flows. MARR = 10%.

	ATCF for Year:			
	1	*2*	*3*	*4*
Proposal A	$10,000	$10,000	$10,000	$10,000
Proposal B	1,000	5,000	10,000	30,000

Solution: Using present-value analysis:

Proposal A

$$\text{PE} = \$10,000(P/A, 10\%, 4) - \$25,000 = \$6,170$$

Proposal B

$$\text{PE} = \$1,000(P/F, 10\%, 1) + \$5,000(P/F, 10\%, 2)$$
$$+ \$10,000(P/F, 10\%, 3) + \$30,000(P/F, 10\%, 4)$$
$$- \$25,000$$
$$= \$909 + \$4,132 + \$7513 + \$20,490 = \$8,044$$

Accept Proposal B.
   Using rate-of-return analysis:

Proposal A

Try 20%:

$$\text{PE} = \$10,000(P/A, 20\%, 4) - \$25,000 = \$890$$

Try 25%:

$$\text{PE} = \$10,000(P/A, 25\%, 4) - \$25,000 = \$-1,380$$
$$i = 20\% + 5\% (890/2,270) = 22\%$$

Proposal B

Try 20%:

$$PE = \$1,000(P/F, 20\%, 1) + \$5,000(P/F, 20\%, 2)$$
$$+ \$10,000(P/F, 20\%, 3) + \$30,000(P/F, 20\%, 4)$$
$$- \$25,000$$
$$= \$833 + \$3,472 + \$5,787 + \$14,469 = \$-439$$

Try 15%:

$$PE = \$870 + \$3781 + \$6575 + \$17,154 = \$3,380$$
$$i = 15\% + 5\%(3,380/3,819) = 19.4\%$$

Accept proposal A.

Applying present-value analysis, we select proposal B. Applying rate-of-return analysis, we select proposal A.

The conflict between these two methods is due to differences in the implicit compounding of interest rates. The rate-of-return method implies a reinvestment rate equal to the internal rate of return, whereas the present-value method implies a reinvestment rate equal to the MARR specified.

Many authors suggest that the present-value method is the preferred and more theoretically correct method. However, when correctly applied, the rate-of-return method gives results consistent with those of the present-value method. An incremental analysis is necessary when using the rate-of-return method.

	ATCF for Year:				
	0	*1*	*2*	*3*	*4*
Proposal A	$25,000	$ 10,000	$ 10,000	$10,000	$ 10,000
Proposal B	25,000	1,000	5,000	10,000	30,000
Proposal B − A	0	−9,000	−5,000	0	+20,000

Using present value:

$$PE = \$-9,000(P/F, 10\%, 1) - \$5,000(P/F, 10\%, 2)$$
$$+ \$20,000(P/F, 10\%, 4)$$
$$= \$-8,182 - \$4,132 + \$13,660 = \$1,346$$

Accept proposal B.

Using rate of return, try 12%:

$$PE = \$-9,000(P/F, 12\%, 1) - \$5,000(P/F, 10\%, 2)$$
$$+ \$20,000(P/F, 10\%, 4)$$
$$= \$-8,036 - \$3,986 + \$12,710 = \$688$$

Try 15%:

$$PE = \$-7,826 - \$3,781 + \$11,436 = \$-171$$
$$i = 12\% + 3\%(688/859) = 14.4\%$$

The rate of return associated with the incremental cash flows in alternative B over alternative A generates a rate of return of 14.4%. This rate exceeds a MARR of 10%. Therefore, proposal B should be accepted.

Thus the internal-rate-of-return method can be modified to deal with the special case of mutually exclusive investment proposals. Given such a modification, it provides a decision identical to that given by the net-present-value method. Many financial managers feel that the rate-of-return method is easier to visualize and to interpret than is the present-value method. In addition, one does not have to specify a required rate of return in the calculations. To the extent that the required rate of return is but a rough estimate, the use of the rate-of-return method may permit a more realistic comparison of projects.

Example 14.2

Problem: Global Industries is considering several capital investment opportunities. Each involves a present expenditure that will produce uniform after-tax cash flows over the next 6 years, then be retired with zero salvage value. Alternatives with the same prefix number are a mutually exclusive group. The mutually exclusive groups may be considered to be independent of each other. Global Industries has many investments available at 8%.

Alternative	First Cost	ATCF
1A	$20,000	$ 6,000
1B	30,000	8,200
1C	40,000	10,600
2A	10,000	2,500
2B	30,000	7,200
2C	40,000	9,300
3A	10,000	2,500
3B	20,000	5,150
3C	30,000	7,400

Solution

1. Use rate of return as the method of analysis to rank the investment alternatives. This method requires that the ROR be calculated for each increment of investment.

   a. Investment 1A over doing nothing:

   $$PE = 0 = \$-20,000 + \$6,000(P/A, i, 6)$$
   $$(P/A, i, 6) = 20,000/6,000 = 3,333$$
   $$i = 20\%$$

   b. Investment 1B over doing nothing:

   $$PE = 0 = \$-30,000 + \$8,200(P/A, i, 6)$$
   $$(P/A, i, 6) = 30,000/8,200 = 3.6585$$
   $$(P/A, 15\%, 6) = 3.7845$$
   $$(P/A, 20\%, 6) = 3.3255$$
   $$i = 15\% + 5\%(0.126/0.459)$$
   $$= 16.4\%$$

   c. Investment 1C over doing nothing:

   $$PE = 0 = \$-40,000 + \$10,600(P/A, i, 6)$$
   $$(P/A, i, 6) = 40,000/10,600 = 3.7736$$
   $$(P/A, 15\%, 6) = 3.7845$$
   $$(P/A, 20\%, 6) = 3.3255$$
   $$i = 15\% + 5\%(0.0109/0.459)$$
   $$= 15.1\%$$

   d. Investment 1B − 1A:

   $$PE = 0 = \$-10,000 + \$2,200(P/A, i, 6)$$
   $$(P/A, i, 6) = \$10,000/\$2,200 = 4.5455$$
   $$(P/A, 8\%, 6) = 4.6229$$
   $$(P/A, 10\%, 6) = 4.3553$$
   $$i = 8\% + 2\%(0.0774/0.2676)$$
   $$= 8.6\%$$

e. Investment 1C − 1A:

$$PE = 0 = \$-20{,}000 + \$4{,}600(P/A, i, 6)$$
$$(P/A, i, 6) = \$20{,}000/4{,}600 = 4.3478$$
$$i = 10\%$$

f. Investment 1C − 1B:

$$PE = 0 = \$-10{,}000 + \$2{,}400(P/A, i, 6)$$
$$(P/A, i, 6) = \$10{,}000/\$2{,}400 = 4.1667$$
$$(P/A, 10\%, 6) = 4.355$$
$$(P/A, 12\%, 6) = 4.111$$
$$i = 10\% + 2\%(0.1883/0.2550)$$
$$= 11.5\%$$

In a similar manner the incremental rate of return on the remaining investments is calculated.

| | | | ROR (%) on the Incremental Investment Compared to: | | |
Alternative	First Cost	ATCF	O	A	B
1A	$20,000	$ 6,000	20.0		
1B	30,000	8,200	16.4	8.6	
1C	40,000	10,600	15.1	10.0	11.5
2A	10,000	2,500	12.5		
2B	30,000	7,200	10.5	10.8	
2C	40,000	9,300	10.4	9.6	7.0
3A	10,000	2,500	12.5		
3B	20,000	5,150	14.1	15.0	
3C	30,000	7,400	12.3	12.0	9.3

2. Draw the network diagram for each mutually exclusive group and show the investments on the decision path with a dashed line.

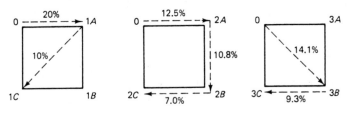

3. List the incremental investments in tabular format that lie on the decision path.

Alternative	Incremental Investment	ROR (%)
1A	$20,000	20.0
1C − 1A	20,000	10.0
2A	10,000	12.5
2B − 2A	20,000	10.8
2C − 2B	10,000	7.0
3B	20,000	14.1
3C − 3B	10,000	9.3

4. Rearrange the investments in the table above in descending order of ROR showing cumulative investment total (i.e., ladder the alternatives).

Alternative	Incremental Investment	Investments Accepted	Cumulative Investment	Cutoff ROR (%)
Nil	$ 0	None	$ 0	Over 20
1A	20,000	1A	20,000	14.1–20
3B	20,000	1A, 3B	40,000	12.5–14.1
2A	10,000	1A, 2A, 3B	50,000	10.8–12.5
2B − 2A	20,000	1A, 2B, 3B	70,000	10.0–10.8
1C − 1A	20,000	1C, 2B, 3B	90,000	9.3–10.0
3C − 3B	10,000	1C, 2B, 3C	100,000	8.0–9.3
Others	Depends on funds available	1C, 2B, 3C Others		7.0–8.0
2C − 2B	10,000			Under 7

5. The company is now in a position to select the investments that will maximize profits. For example, if $50,000 is available, investments 1A, 2A, and 3B will be selected. The cutoff rate of return is between 10.8 and 12.5%.

From the analysis above we can see that the cutoff rate of return is dependent on both the investment alternatives available and the funds available for investment. The MARR value used to compare alternatives within any given organization should be close to the cutoff rate of return for that specific organization. Therefore, the importance of a formalized budgeting system that includes an effective tracking system to provide good follow-up on all investments cannot be overemphasized.

14.9 THE PAYBACK METHOD

The *payback method* is one additional method used to evaluate investment alternatives. The payback (payoff) period associated with this method can be defined as the number of years required for the annual before-tax cash flows from a project to be equivalent to the initial investment using a 0% rate of return.

From a group of mutually exclusive projects the objective is to select the project with the minimum payback period. This project is then accepted if the investment in one of the projects in the group is mandatory or if the payback period for this project is less than or equal to some acceptable payback period set by the company.

To evaluate a group of independent projects the alternatives are ranked in ascending order of prospective payback period. Projects are then approved according to this ranking and the funds available for investment.

We recommend that the payback method, although still somewhat popular due to the ease of calculation, be avoided, for the following reasons:

1. Cash flows beyond the payback period are ignored in the analysis. This procedure may have a significant impact on the analysis.
2. The timing of cash flows within the payback period is ignored. For example, assume that the following cash flows for two mutually exclusive projects were projected to occur.

	BTCF for Year:				
	0	*1*	*2*	*3*	*4*
Project A	$ – 10,000	$5,000	$4,000	$ 500	$ 500
Project B	– 10,000	0	3,000	3,000	4,000

Using the payback method, the payback periods for these two alternatives are identical (4 years). The payback period thus ignores any time value of money.

3. The payback period discriminates against long-lived projects. This discrimination can readily occur where large projects are under consideration that require significant funds to bring the project to the operational stage. In addition, 1 or 2 years of operation are often required to reach peak production creating a relatively long payback period.

14.10 LEASING

Lease financing has grown at a rapid pace over the past 20 years and represents a substantial source of capital to many industrial organizations. A lease is a contractual agreement between the owner of the asset (lessor) and the user (lessee), which sets forth in detail the obligations and the rights of both parties. The major provisions deal with the life of the lease, renewal options (if any), and the rental conditions, such as which of the two parties is responsible for such operating costs as maintenance, insurance, and taxes. A lease may also contain a provision giving a lessee the option to purchase the assets and appoints a time during the lease at which this option may be exercised.

Lease contracts are divided into two broad categories: financial leases and operating leases. A *financial lease* involves an agreement in which the lessee agrees to make a stipulated series of periodic payments over a span of time which covers the major portion or the entire useful life of the asset (usually for a period of 3 years or more). The lessor normally expects to recover the full price of the asset, which otherwise would have been sold plus interest and other costs that the lessee agrees to bear over the interval of time in which the lease is enforced.

An *operating lease* is, for a period of time, usually substantially less than the useful life of the asset. Normally, after a short period of time, and upon reasonable notice agreeable to both the lessee and the lessor, the lease can be terminated and the assets returned to the lessor. During this interval the lessor generally expects to recover less than the whole purchase price of the asset. The operating lease usually differs from a straight rental in the duration of the commitment.

In this section several common methods of evaluating financial leases will be discussed with a view to presenting a systematic approach to determining the true cost of a financial lease. The financial lease is as much a source of capital as either debt capital or equity capital. Similarly, the cost of a financial lease varies somewhat from corporation to corporation in the same way that the cost of debt capital or the cost of equity capital may vary. However, the method of evaluating this cost to any corporation may be approached in a systematic manner.

Before proceeding with an analysis of a leasing situation to portray the true cost of financial leasing, several areas of finance require a brief discussion. This discussion will help to place leasing in its proper perspective in the total financial budgeting system.

14.10.1 Similarities between Financial Leasing and Debt Financing

Financial leasing and debt financing have many aspects in common. They both represent a contractual agreement to meet a fixed commitment. Most financial leases are similar to debt in that the cost of the lease or the cost of debt is fully deductible from operating revenues to arrive at taxable income.

Financial leasing may be subject to a higher interest rate than is debt financing, due primarily to the fact that most leasing agreements involve a third party, the lessor, who raises the necessary capital to finance the leasing agreement from basically the same source from which the company would raise this capital directly through a bond issue. If this is true, one may wonder why large corporations, which represent the majority of leasing agreements, would revert to financial leasing. There are at least two reasons that may help to explain this situation:

1. Leasing frees company funds for investments in projects critical to the growth of the company that cannot be financed through a leasing arrangement.
2. Most leasing agreements require less financial planning than is necessary to float a bond issue and therefore can be undertaken on relatively short notice.

14.10.2 Additional Factors Affecting Leasing Decisions

There are several other factors that companies do consider in evaluating a lease versus buy decision. They are:

1. Method of write-off
2. Interest rates and credit availability
3. Salvage values
4. Obsolescence

Method of Write-off

For tax purposes leasing is considered as an operating cost. Therefore, all lease payments, including interest, can be deducted directly from the before-tax cash flow to arrive at taxable income. Thus the lease contract may allow a faster write-off than is allowed through depreciation allowance and may, in fact, be an advantage over buying even when 100% debt capital is involved.

Interest Rate and Credit Availability

Financial leases do not appear on the balance sheet and may therefore allow a company to operate with a larger debt-equity structure than would normally be possible using debt capital versus leasing. However, this assumption may be a fallacy as the financial analyst does try to determine the amount of lease capital involved in the total capital structure of a company. The problem is that few valid data are available to use as a basis to evaluate the impact of leasing on the financial structure. Second, it is difficult to separate the capital investment component of the lease from the operating cost components (e.g., maintenance) of the lease.

As a result, the analyst often adopts a conservative approach and assumes that the corporation has considerably more capital funds invested in leasing than actually exists. This assumption results in the analyst judging the company to be of a higher risk nature than is warranted. This assumption may more than offset any increase in interest rates due to the involvement of a third party in a leasing arrangement.

Each individual company should monitor and evaluate their rating with financial institutions to maintain a favorable status.

Salvage Values

With leasing, the lessor owns the asset at the expiration of the lease. The market value of the asset on termination of the lease is a factor that may be significant in the analysis.

Obsolescence

Rapid obsolescence of the asset may favor leasing versus buy, but in all likelihood, the obsolescence factor is built into the lease payment. However, the lessor may be in a better position to market an asset that experiences high obsolescence (e.g., computers and airplanes) than an individual company, resulting in a lease arrangement that is more desirable to the firm than an outright purchase.

14.10.3 The Composite Cost of Capital

There are many theoretical and empirical studies that have been undertaken over the years that supply some interesting insights into the composite cost of capital. For example, Modigliani and Miller (see Appendix D, reference 35) support the view that under certain conditions the composite cost of capital tends to be independent of the debt ratio and therefore eliminates any need for a corporation to seek an optimal debt ratio. Barges (see Appendix D, reference 4) has produced an interesting study which would indicate that the composite cost of capital decreases over a range in which the debt/equity ratio is low, remains stable through a certain range of debt/equity ratio, and then tends to increase as the debt/equity ratio increases beyond a certain point. This view indicates that an optimal debt/equity ratio does exist and that the owner of equity capital is very reluctant to see the amount of debt increase beyond this point because the extra leverage is more than offset by the increased variability (increased risk) of earnings per share. This view is accepted by the authors and is portrayed in Figure 14.2. The difficult task of determining when a company is operating in range 1, range 2, or range 3 is very dependent on the industry within which a company operates and the financial stability of each company within that industry.

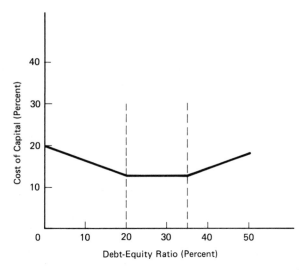

FIGURE 14.2 Effect of debt/equity ratio on the cost of capital to the firm.

14.10.4 The Cost of Leasing to the Small-to-Medium-Size Company

Debt financing, at least through a bond issue, is not available to many small to medium-size companies. Therefore, many of these companies resort to leasing as an alternative to debt financing.

Assuming for the moment that a financial lease and debt capital for all practical purposes are interchangeable, the question arises as to what extent the company should become involved in financial lease agreements. Theoretically, lease capital should be used to produce a debt-equity structure that falls somewhere in range 2 of Figure 14.2, where the cost of capital remains constant. However, one cannot overlook management's utility function where any debt capital is involved. In other words, in practice, the amount of debt capital that any company will have, particularly a small company, is somewhat dependent on management's aversion to debt.

The important factor that exists is that lease financing is available to the small to medium-size company, and in view of this fact, these companies should determine the extent to which this lease capital is advantageous.

14.10.5 Combining the Investment and Financing Decisions

Should the investment and financing decisions be combined in the analysis of problems involving lease alternatives? There are two approaches worthy of discussion.

1. We can think of the total funds available to a company as a finite pool of funds having a composite cost based on the specific mix of capital in the pool. Similarly, we can think of the total projects to be evaluated over the year as a pool of projects competing for the total funds available in the pool. Based on this line of reasoning it becomes impossible to associate a specific project with individual sources of funds. Using this approach, the analyst needs only to determine the cost of capital in the pool and use this cost of capital as the basis for analysis.

   A reasonable assumption using this approach is that the cost of capital will remain constant over a reasonable period of time. That is, the firm is operating in range 2 of Figure 14.2. The analyst will develop the composite cost of the capital using a specific cost for equity capital, debt capital, and lease capital.

2. We can associate specific financing with each individual investment. For example, the company may finance investment A through a specific bank loan or through a lease arrangement, whereas the company may finance investment B through internally generated funds versus through a lease arrangement. Using this approach, it may be desirable to finance investment A through a bank loan and investment B with composite capital. However, if the method of financing were reversed, the investment alternatives selected could also be reversed.

   There is a limit to the amount of funds that a firm will borrow (either through debt or leasing), and to try to associate these funds with individual investments will in all likelihood lead to inconsistencies in evaluating investment alternatives. Therefore, it is strongly suggested that the financing decision be separated from the investment decision and that a constant cost of capital be used as a basis for all evaluations within a specific budgeting period.

Example 14.3 illustrates some of the inconsistencies that arise when the financing and evaluation steps are combined.

Example 14.3

Problem: Assume that a firm plans to acquire an asset that will cost $50,000. Operating revenues are expected to be $35,000 per year, operating costs $10,000 per year, each year over the next 3 years. The salvage value at the end of 3 years is estimated to be $20,000. Use the straight-line method (equation 4.2). $t = 50\%$, $i_d = 8\%$, $i_e = 20\%$, $r = 50\%$, and $i_a = 12\%$. The asset can be leased for year-end payments of $27,000 per year (lease payments include operating costs and would normally be monthly but for simplicity annual payments are used), or financed entirely through a bank loan (100% debt capital). Which method should be chosen?

Solution: Using the annual revenue requirements approach, the cost to finance using debt capital is

$$D = (\$50,000 - \$20,000)/3 = \$10,000$$

$$AER = AEOC + [P(A/P, 8\%, 3) - NSV(A/F, 8\%, 3)$$

$$- t(AED)]/(1 - t)$$

$$= \$10,000 + [\$50,000(0.3880) - \$20,000(0.3080)$$

$$- 0.5(\$10,000)]/0.50$$

$$= \$10,000 + (\$19,400 - \$6160 - \$5,000)/0.50$$

$$= \$10,000 + \$16,480 = \$26,480$$

The annual revenue requirement of leasing $= \$27,000$.

Conclusion: Buy the asset using a bank loan.

Problem: The alternatives available are to lease the asset or to purchase the asset using internally generated funds at the company's cost of capital. Which method should be chosen?

Solution: Finance using the company's cost of capital:

$$i_a = 12\%$$

$$AER = \$10,000 + [\$50,000(0.4164) - \$20,000(0.2964)$$

$$- 0.5(\$10,000)]/0.5 = \$29,784$$

Conclusion: Lease the asset.

Thus we see that by associating specific funds with a specific investment decision, the alternative selected is very dependent on our choice of source of funds. This approach undermines consistent company decision making and project comparison.

14.10.6 A Recommended Method for Evaluating Leases

Most financial leases represent a contractual obligation to make the lease payments for a fixed period of time. The lease contract under this fixed obligation is very similar to financing with 100% debt capital. Therefore, this type of financing represents an additional risk to the existing debt and equity holders. It seems reasonable to assume that any additional risk to the present debt and equity holders within the company created through a lease arrangement may be offset by replacing the lease capital by composite capital and determining the additional revenue that must be generated each year. This revenue is added to the lease payment to determine the true annual cost of the lease.

Therefore, we have two key assumptions associated with our recommended method.

1. The cost of capital remains constant within any budgeting period. (This cost can be calculated knowing r, i_d, i_e, and t.)
2. Lease capital and debt capital are interchangeable. Thus we can replace the lease capital with debt capital.

Solution Procedure

1. Replace the lease with debt capital.
2. Replace the debt capital with composite capital using the company's debt/equity ratio (r).
3. Determine the additional cost of each year of financing with composite capital versus 100% debt capital.
4. Add this additional cost each year to the lease payment. This total value represents the true annual cost of the lease.
5. We can now calculate the annual revenue requirements of the lease and compare this to the annual revenue requirements of financing with composite capital.

Example 14.4

Problem: The B-J Corporation is interested in either buying a new milling machine or leasing the machine for $25,000 a year (lease payments occur at year end). Assume that the company is locked into the lease arrangement. $P = \$80,000$, SV $= \$35,000$, OC $= \$50,000$ per year, $n = 3$ years, $r = 40\%$, $i_d = 10\%$, $i_e = 21.33\%$, and $t = 45\%$. Use the straight-line method (equation 4.2).

Solution:

$$i_a = ri_d(1 - t) + (1 - r)i_e = 15\%$$

AER to Buy

$$D = (\$80,000 - \$35,000)/3 = \$15,000$$

$$AER = \$50,000 + [\$80,000(A/P, 15\%, 3) - \$35,000(A/F, 15\%, 3)$$

$$- 0.45(\$15,000)]/0.55$$

$$= \$83,109$$

AER of the Lease

1. Replace the lease capital with debt capital.

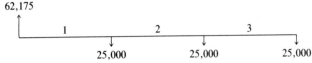

   a. Initially, the company is obligated to make three lease payments:

   $$\$25,000(P/A, 10\%, 3) = \$62,175$$

   b. At the end of 1 year the obligation reduces to two lease payments:

   $$\$25,000(P/A, 10\%, 2) = \$43,400$$

   c. At the end of 2 years the obligation reduces to one lease payment:

   $$\$25,000(P/A, 10\%, 1) = \$22,727$$

2. Replace the debt capital with composite capital using the company's debt/equity ratio ($r = 40\%$).

	Year		
	1	*2*	*3*
Debt	$24,870	$17,360	$ 9,091
Equity	37,305	26,040	13,636
	$62,175	$43,400	$22,727

3. Determine the additional cost each year of financing with composite capital versus 100% debt capital. Three methods of calculating this cost are presented.

   a. *Method 1:* For year 1:

Interest on debt capital ($24,870)10%	$ 2,487
Return on equity capital ($37,305)21.33%	7,957
Income tax: $[t/(1 - t)]$ (return on equity capital)	6,510
	$16,954
Less: Cost of 100% debt financing: ($62,175)10%	6,218
	$10,736

$10,736 is the additional cost that should be added to the lease payment in year 1. The values in the following table were developed in a similar manner.

	Year		
	1	2	3
Interest on debt capital	$ 2,487	$ 1,736	$ 909
Return on equity capital	7,957	5,554	2,909
Income tax	6,510	4,543	2,380
	$16,954	$11,833	$6,198
Less: 100% debt financing	$ 6,218	$ 4,340	$2,273
	$10,736	$ 7,493	$3,925

b. *Method 2:* The additional cost of financing with composite capital versus 100% debt capital may be determined by calculating the incremental cost of equity capital. $i_e = 21.33\%$, $i_d = 10\%$, and incremental cost $= 11.33\%$. The total incremental cost in year 1 is:

Incremental cost of equity capital $=(\$37,305)0.1133$	$ 4,227
Income tax $= t/(1 - t)$ (return on equity capital) $=$	
$(0.818)(37,305)(0.2133)$	6,509
	$10,736

In a similar manner the remaining table values were developed.

	Year		
	1	2	3
Incremental cost of equity capital	$ 4,227	$2,950	$1,545
Income tax	6,509	4,543	2,380
	$10,736	$7,493	$3,925

c. *Method 3:* This method is the easiest to apply but perhaps is not the easiest to understand. Calculate the before-tax cost of composite capital:

$$\text{BTRR} = i_a/(1 - t)$$

$$= 0.15/0.55$$

$$= 27.27\%$$

$$i_d = 10\%$$

$$\text{BTRR} - i_d = 27.27\% - 10\%$$

$$= 17.27\%$$

The additional cost of replacing debt capital with composite capital on a year-to-year basis is:

$$\text{Year 1: } \$25,000(P/A, 10\%, 3)(0.1727) = \$10,736$$

$$\text{Year 2: } \$25,000(P/A, 10\%, 2)(0.1727) = \ \ \$7,493$$

$$\text{Year 3: } \$25,000(P/A, 10\%, 1)(0.1727) = \ \ \$3,925$$

4. These costs are now added to the lease payment each year to arrive at a total annual cost of the lease.

	Year		
	1	2	3
Lease payment	$25,000	$25,000	$25,000
Additional cost	10,736	7,493	3,925
	$35,736	$32,493	$28,925

5. AER of the lease arrangement:

$$\text{AER} = \$50,000 + [\$35,736(P/F, 15\%, 1) + \$32,493(P/F, 15\%, 2)$$

$$+ \ \$28,925(P/F, 15\%, 3)](A/P, 15\%, 3) = \$82,701$$

Conclusion: Lease the milling machine.

We can see by the example that the actual cost of leasing will normally be somewhat higher than the lease payment. It is important to determine this additional cost when considering a lease arrangement.

14.11 SUMMARY

Budgets are used for planning and for control. The operating budget usually consists of two parts with identical totals, the program budget and the responsibility budget. The latter, and especially the underlying variable budgets, represent the principal basis for subsequent comparisons with actual costs.

The line organization, assisted by a budget committee, makes the decisions on what the plans are to be, and the staff organization, directed by a budget director, assists the line organization by providing data, making calculations, and assembling and transmitting the budget figures. Chronologically, the important steps in the budget process are as follows: setting guidelines, preparation of the sales budget, preparation of other operating budgets, coordination of the separate budgets, and review and approval.

The cash budget translates revenues and expenses into receipts and disbursements and thus facilitates financial planning.

The capital budget is a list of presumably worthwhile projects for the acquisition of new capital assets. Often it is prepared separately from the operating budget. Approval of the capital budget constitutes approval only in principle, and a subsequent specific authorization is usually required before work on the project can begin. All organizations require budgets to give substance to their plans.

Leasing represents a significant source of capital to many organizations. The firm should give consideration to the additional costs, over and above the lease payment, in comparing leasing to other investment alternatives. The suggested methodology outlined in this chapter is based on two realistic assumptions:

1. Lease capital and debt capital are interchangeable.
2. The cost of capital within an individual organization does not vary, for all practical purposes, within any single budgeting period.

PROBLEMS

14.1 Why should a company have a formalized capital budgeting system?

14.2 Should an effective capital budgeting system include a means of tracking investments that are made? Explain.

14.3 Rate of return and present value are two common methods used to evaluate investment proposals. Why do these two methods sometimes result in conflicting answers?

14.4 In the preceding chapters we have used a MARR (i_a) value calculated as follows (ignoring inflation):

$$i_a = ri_d(1 - t) + (1 - r)i_e$$

This value was based on an external cost of capital. In actual practice the cutoff rate of return that will normally be used as a discount rate will be higher than the i_a value calculated above. Why?

14.5 Two mutually exclusive investment proposals are under consideration.

	Proposal	
	A	B
Initial cost	$50,000	$10,000
Estimated salvage value in 10 years	$10,000	$3,000
Estimated annual operating costs	$15,000	$5,000
Estimated annual revenues	$40,000	$14,000
Estimated economic life	10 years	10 years

$t = 46\%$. Use the straight-line method (equation 4.2).

a. Select the preferable alternative using the present-worth method. MARR = 15%.
b. Select the preferable alternative using the rate-of-return method. Calculate the rate of return on each individual investment.
c. Select the preferable alternative using ROR. Do an incremented analysis.
d. Which method is correct from part (b) or part (c)?

14.6 Two mutually exclusive investment proposals are under consideration.

			BTCF for Year:				
	0	1	2	3	4	5	SV5
A	$-20,000	$7,000	$7,000	$ 7,000	$ 7,000	$17,000	$ 8,000
B	-25,000	5,000	8,000	10,000	20,000	10,000	10,000

a. Compare these two proposals using the payback method.
b. Compare these two proposals on a before-tax basis using the rate-of-return approach.

14.7 Compare the proposals in Problem 14.6 on an after-tax basis using the payback method and the ROR method. $t = 46\%$ and $r = 0$. Use the straight-line method (equation 4.2).

14.8 At the conclusion of the annual budget cycle each department submits its requests for capital acquisitions for the following year. The budget office then groups these requests according to prospective rate of return and prepares a corporate demand schedule. For example, departments X, Y, and Z can invest a total of $240,000 at prospective rates of return from 40 to 44.9%. The complete demand schedule for the corporation is as follows:

Request	Estimated Rate of Return (%)
$ 150,000	55+
220,000	50–55
130,000	45–50
240,000	40–45
370,000	35–40
690,000	30–35
2,350,000	25–30
1,400,000	20–25
1,600,000	15–20
2,100,000	10–15
1,300,000	5–10

For the coming year, the company has available $4,500,000 at a cost of 16%. What cutoff rate of return should be established for next year's budget? What suggestions might be made with regard to future budgets?

14.9 Prepare a capital expenditure demand schedule for an organization with the following proposals.

1. A robot welder costing $125,000 for the main assembly line, to replace three existing semiautomatic machines. The saving on incremental investment has a 45% rate of return. The old machines have a net realizable value of $47,000. They can be moved to the fitting shop to replace still older machines which can be disposed of for $5,000. The rate of return on this incremental investment is 20%.

2. An automatic conveyor for one of the subassembly lines for $60,000. The expected rate of return is 35%. The conveyor will replace the dollies currently in use; the dollies will be sold as scrap for $1,000.

3. A mobile yard crane for $37,000 to replace the current ancient one. The expected rate of return on incremental investment is 23%. The old crane can be sold for $8,000 or retained as standby equipment. If retained, the estimated rate of return is 15%.

The company's cost of capital is 12% and it appears that $150,000 will be available. What is the cutoff rate of return, and what recommendations would you make to the company?

14.10 The following two mutually exclusive groups of proposals are under study. Assume that each alternative has infinite life and zero salvage value.

Proposal	Total Investment	ATCF
A1	$10,000	$ 2,300
A2	20,000	7,000
A3	30,000	9,500
A4	40,000	15,000
A5	50,000	18,000
B1	10,000	3,500
B2	20,000	5,000
B3	40,000	12,500
B4	50,000	15,500
B5	60,000	19,500

a. Calculate the relevant rates of return and make a choice table for the various MARR possibilities.

b. Ladder the investments and decide which proposals you would choose if:

(1) $100,000 is available.

(2) Funds are unlimited.

14.11 Cost-reduction proposals in the ABC Company are under consideration. The following two mutually exclusive groups of proposals are available. All proposals have a 10-year life and zero salvage value. You cannot select more than one proposal from group A or group B.

Proposal	Investment	ATCF
A1	$10,000	$2,000
A2	25,000	8,500
A3	15,000	4,600
A4	20,000	5,000
A5	30,000	9,600
B1	20,000	5,000
B2	25,000	6,000
B3	22,000	5,200
B4	24,000	5,500

a. Calculate the relevant rates of return and make a choice table for the various MARR possibilities.

b. Which of these proposals would you recommend? What is the cutoff rate of return:

(1) If $45,000 is available?

(2) If funds are unlimited?

14.12 Global Inc. is considering a number of investment opportunities. The initial costs and annual after-tax cash flows over the next 10 years are listed below. The opportunities are equivalent in all respects, including risk, except for size and prospective rate of return. Alternatives 1A, 1B, and 1C represent a mutually exclusive set; 2A, 2B, and 2C are a mutually exclusive set; and 3A, 3B, and 3C are a mutually exclusive set. All alternatives have 100% salvage. The company has many investments available at 12%.

Proposal	Investment	ATCF
1A	$10,000	$ 1,500
1B	20,000	4,000
1C	40,000	10,000
2A	10,000	2,700
2B	30,000	7,000
2C	40,000	7,800
3C	30,000	5,800
3A	10,000	2,300
3B	20,000	5,200

a. Draw network diagrams.
b. Ladder alternatives for capital budgeting purposes.
c. What investments should be made if the following funds are available:
   (1) $70,000?
   (2) $200,000?

14.13 Santa Claus and Co.'s current lease on their North Pole premises is expiring, so they must locate in new premises. After exhaustive analysis they have decided to construct their own plant. The following proposals have been developed for consideration. Alternatives with the same prefix letter are a mutually exclusive group.

Site		*ROR (%) on Incremental Investment Compared to:*			
Alternative	Investment	0	A1	A2	A3
A1: Garden of Sno					
White's Castle	$100,000	10			
A2: South Pole	150,000	8	4		
A3: Dark Side					
of the Moon	200,000	11	12	20	
A4: Never-Never Land	250,000	12	13	18	16

Building		*ROR (%) on Incremental Investment Compared to:*				
Alternative	Investment	0	B1	B2	B3	B4
B1: House of straw	$ 75,000	20				
B2: House of sticks	100,000	21	24			
B3: House of bricks	125,000	22	25	26		
B4: Gingerbread house	150,000	23	26	27	38	
B5: Rock candy castle	175,000	23	25.2	25.7	23	23

The proposed move provided Santa's helpers with the opportunity to request shiny new equipment. Currently, all operations are done manually. The fairies, whose concern was receiving and raw materials storage, requested:

Warehouse		*ROR (%) Incremental Investment Compared to:*	
Alternative	Investment	0	C1
C1: 1 lift truck	$ 75,000	30	
C2: 2 lift trucks	150,000	28	27

The elves, who manufacture the toys, requested:

Manufacturing Alternative	Investment	ROR (%) Incremental Investment Compared to: O	D1	D2
D1: Manual extruder	$ 20,000	15		
D2: Semiautomatic extruder	100,000	20	21	
D3: Automatic extruder	200,000	17	17.2	14

The reindeer (shipping department) also had their requests:

Shipping Alternative	Investment	ROR (%) on Incremental Investment Compared to: O	R1
R1: Jet-assisted sleigh	$ 75,000	25	
R2: Atomic sleigh	150,000	25	25

Navigation equipment:

Navigation Alternative	Investment	ROR (%) on Incremental Investment Compared to: O
N1: Navigation computer	$100,000	22

1. Santa only has $500,000 to invest.
2. Investment in a site and a building are mandatory.
   a. Draw the network diagram for each mutually exclusive group and show the investments on the decision path with a dashed line.
   b. List the incremental investments in tabular format that are on the decision path.
   c. Rearrange the investments in the table in descending order of ROR. Note that the site and building investment are mandatory.
   d. What projects should be invested in?
   e. What is the implied MARR (cutoff)?
   f. What additional recommendations would you make to Santa?

14.14 The Black Corporation, which has a 40% tax rate, can purchase a $100,000 milling machine, which can be depreciated using the straight-line method (equation 4.2). The machine is expected to have a useful life to the company of 3 years and an expected salvage value at the end of 3 years equal to $60,000. The machine can be leased for $30,000

per year. Assume end-of-year lease payments. Operating costs are estimated to be $40,000 per year. i_e = 24%, i_d = 10%, and r = 50%.
a. Should the machine be purchased or leased if we assume that there is no added risk due to leasing?
b. Should the machine be purchased or leased if we assume that the cost of capital remains constant (that is, if there is an added risk to leasing)?

14.15 The ABC Company has the alternative of leasing a duplicating machine or buying the machine. Information pertaining to these alternatives is as follows:

	Buy	Lease
Initial installed cost	$8,000	
Annual lease payment (end of year)		$2,400
Annual operating costs (end of year)	$500	—
Estimated salvage value	$2,000	
Estimated life	5 years	5 years

r = 0.40, i_d = 10%, t = 40%, i_e = 16%, and i_a = 12%. Use the SL method (equation 4.2). Operating costs are included in the lease payment. Should the machine be leased or purchased?

14.16 The XYZ Corporation is considering the replacement of certain equipment. Data pertaining to the choices are as follows:

	Present Equip.	Proposed Equip. Buy	Proposed Equip. Lease
Annual operating costs	$9,000	$ 2,000	$1,500
Current value (book value = market value)	5,000	32,000	—
Rental costs per year guaranteed for the next 5 years	—	—	5,000
Salvage value 5 years hence	Negligible	8,000	—

r = 40%, i_d = 10%, t = 50%, and i_e = 13.3%. Use the straight-line method (equation 4.2). Compare the annual revenue requirements for status-quo versus lease versus buy.

14.17 A company is considering the alternative of buying a piece of production equipment listed at $40,000 or leasing it for a period of 7 years and then exercising the option to buy. The rental is 25% of the list price for each of the first 3 years and 10% of list price each year thereafter. The lease carries an option to buy at the end of the contract for 25% of list. The lessee is to pay for transportation and installation (estimated at $1,500) and for maintenance, property taxes, and insurance. If the equipment is purchased, it will be depreciated by the straight-line method (equation 4.2) over a 15-year service life using a zero salvage value. The assumed tax rate is 50%, the debt ratio is 50%, the interest rate on borrowed capital is 8%, and MARR is 10%. Evaluate this lease–buy alternative.

14.18 A gas transmission company is considering the alternative of buying a mobile line testing unit or leasing this equipment. A lease arrangement can be made for 2 years with the option to purchase the equipment at the end of 2 years. The economic life of the equipment is 5 years. The estimated salvage value at the end of 5 years is zero.

1. The initial cost if purchased is $100,000.
2. The cost to purchase in 2 years is $25,000.
3. Lease payments are $50,000 per year. Payments are at year's end.
4. All operating costs are paid by the lessee.
5. Use straight-line depreciation, (equation 4.2) $r = 40\%$, $i_d = 10\%$, $i_a = 15\%$, $t = 50\%$.
   Evaluate this lease–buy alternative.

CHAPTER FIFTEEN

Presentation, Implementation, and Tracking Investment Decisions

Economy studies are prepared for various purposes and the recommendations resulting from each study are usually presented by managers who did not perform the study. Therefore, it is essential that the study include all relevant information required by the manager in a format that allows the managers to evaluate the report objectively and be confident in its presentation.

15.1 THE ECONOMY STUDY

Every economy study should include:

1. A precise statement regarding the purpose of the study
2. An evaluation of all alternatives considered
3. The source and use of all input data
4. The method(s) used to convert input data into results
5. The study results clearly presented
6. The plan of action recommended

The specific format may vary slightly from company to company. However, it should be clear that the integrity of the results of any economy study are dependent on:

1. A thorough evaluation of all relevant alternatives
2. The validity of the data used
3. The validity of the study techniques employed

15.1.1 The Purpose of the Study

The purpose of every economy study should be clearly stated and agreed to by both the analyst and the manager concerned. An economy study is often initiated by a brief statement from the manager either orally or in writing. The time spent by the analyst in defining the purpose of the study and ensuring that this purpose is agreeable to the manager depends on the scope and complexity of the study. Several meetings with the manager may be necessary before proceeding beyond this point.

The measures of effectiveness with respect to such factors as company policy, social impact, and environmental effects should be stated. Limitations on the study such as cost, time, and geographic areas should be clarified early in the study.

A little extra time spent at this stage of the study will often result in major time savings as the study progresses and the elimination of much misunderstanding and frustration. The final report should be very explicit regarding the purpose, measures of effectiveness, and constraints associated with the study.

15.1.2 Alternatives Studied

All too often, during the review process, the manager raises the question: "Did you explore this possibility?"

Once the specific purpose of the economy study has been clearly defined, the analyst should list as many alternative solutions to the problem as possible. These alternatives should be outlined in sufficient detail to enable the manager to pass judgment on their acceptability as solutions to the problem. Usually, the possible solutions can be narrowed to a select few in a relatively short time, due to such factors as financial, technical, environmental, and policy constraints. Communication with the manager concerned at this point in the study is highly desirable. The alternatives retained must be presented in a format that allows the manager to make realistic and economically sound comparisons.

15.1.3 Input Data

The results of any study are only as valid as the input data on which these results are based. The data used usually come from many sources, not all having the same degree of reliability. Therefore, the degree of uncertainty and possible risks associated with the input data should be stated in the report in an explicit manner. If the study is to have meaning, the manager must be confident that all data used have been properly verified regarding source and reliability.

There are many input parameters that must be considered in our economy study, such as:

1. Capital costs
2. Operating revenues
3. Operating costs
4. Depreciation allowance
5. Income tax
6. Cost of capital
7. Inflation/recession
8. Capital gains
9. Recaptured depreciation allowance
10. Loss on disposal
11. Tax credits

There are many sources of information regarding these input parameters. Many of these sources were discussed in Chapter 13. A selected few are listed below:

1. Company records
2. Government agencies
3. Sales representatives
4. Equipment suppliers
5. Trade journals
6. Supply catalogs

The reliability and validity of the data depend not only on their source but also on how the data were originally developed. Government indices concerning such factors as inflation and GNP are developed utilizing a basic format. The same applies to such factors as building material cost indices developed for a specific area. The analyst must know what goes into these indices and their applicability to the specific study under question. It is imperative that the study clearly state the source of the input data, its base, and validity regarding the study.

15.1.4 Study Methodology

The three basic models that are usually used to compare alternatives within an economy study are rate of return, present equivalent, and annual equivalent revenue requirements. The application of these models has been discussed thoroughly in previous chapters and will not be repeated here.

One important consideration that will be discussed is the timing of cash flows. An investment alternative may meet the requirements for acceptability based on any one of the three models above and still result in serious financial problems to the company. For example, consider a very large project that requires several years and a significant investment before the project becomes operational. This type of investment results in a significant drain on company finances before any net revenues are generated. This drain on finances may result in the abandonment of the project and possible bankruptcy. The analyst must fully consider the timing of cash flows in the study.

As discussed in Chapter 12, single-valued estimates are adequate for many studies, but under some circumstances multivalued estimates are desirable. Company policy and the background of the analyst will dictate to a degree the methodology used in the economy study. However, the analyst should not hesitate to call on the expertise of others when the circumstances suggest that help is needed.

15.1.5 Presenting the Economy Study

Communication, both orally and in writing, represents the key to a successful economy study. Both the written and oral reports are directed to a specific audience. It is important to know the makeup of this audience. Different managers have different goals and therefore have a need for different degrees of detail.

Considerable effort may be devoted to the economy study and the written report only to find that the oral presentation lacked the necessary finesse to convince the audience. This deficiency in the oral presentation may be due to several reasons. However, from the analyst's standpoint the problem can be caused by poor representation of input data and results in tabular and graphical format: for example, the use of the simple cash flow diagram to portray the timing, quantity, and direction of the flow of funds for each alternative being analyzed. In many cases, such a diagram is an excellent method of comparing several alternatives in a concise and easily comprehendible form.

15.1.6 A Checklist for an Economy Study

Checklists provide a means of keeping on track. They are especially useful in fields such as economic analysis where there is the possibility of becoming too immersed in the sophistication of the analysis and overlooking the overall decision process. The following are a list of questions that should be asked continuously throughout the preparation of a study. The purpose of asking them is not to elicit answers but to provide thought and discussion and to keep the analysis focused on its goals.

Purpose

1. For whom is the study done?
2. Is the problem clearly defined?
3. Is the scope of the study clearly specified?
4. Is the report designed to provide information or conclusions?
5. What happens to the report after it is completed?

Alternatives

1. Are all reasonable alternatives listed?
2. Are the alternatives technically feasible?
3. Who provided the alternatives?
4. Is it possible to compare the alternatives with economic analysis, or are some dependent on nonquantifiable items?
5. Why is this study being done now?

Data

1. What are the sources of data?
2. What assumptions are made to derive the data?
3. Are the data verifiable from external sources?
4. What are the obvious sources of error or uncertainty?
5. What effect will external influences (weather, government policy, recession) have on the data?

Method

1. By what criteria will the study be judged?
2. Is standard methodology (AER, PE, ROR) used?
3. Is the methodology understood by the person for whom the study is done?

4. Are variable or uncertain data subjected to a sensitivity analysis?
5. Is the magnitude of the project such that it requires utility consideration?
6. Is the detail of the methodology appropriate to the accuracy and completeness of the data?

Presentation

1. Who is the audience? Will they understand the methodology and jargon?
2. Who will be presenting the study. Are they fully briefed?
3. Are the charts, diagrams, and visual aids simple and readable?
4. What would you do if required to condense the presentation to 10 minutes?
5. What is your answer to the question: What is the bottom line?

The report must have good continuity and be presented in a method that is readily understandable to the audience concerned.

15.2 A SYSTEM FOR IMPLEMENTING CAPITAL INVESTMENTS

The capital budget is part of the total budgeting process. It is essentially a list of worthwhile projects requiring the acquisition of capital assets. The screening process occurs at several levels in the organization, allowing only the more attractive projects to appear in the final capital budget.

Approval of projects for capital assets can have far-reaching implications within the organization. The planning period in many cases spans several years in order to compensate for such factors as equipment procurement lead times, building construction, cash flow problems, and personnel training.

Initial approval of the capital budget usually means approval of the projects in principle, but not necessarily final approval for implementation. A system for implementing final authorization is essential. This system will usually include specific bids or price quotations. These authorization requests originate at different levels in the organization. Care must be exercised in setting expenditure amounts at each level of supervision. Too often the major emphasis is placed on the projects with a large dollar value, leaving a large number of smaller projects to the discretion of individual supervisors without due attention to total expenditure levels.

It is essential that the capital budget be an integral part of the total budgeting system. One person should be in charge of monitoring all capital expenditures, and this person must have good communication with other budgeting personnel to ensure overall coordination of the budgeting system.

15.3 TRACKING CAPITAL EXPENDITURES

The tracking of capital expenditures is an essential ingredient of the total capital budgeting system. A comparison of actual performance to that predicted can supply very useful information for the evaluation of future projects. In addition, those projects that are not performing as expected should be reassessed for possible improvement. The end result in some cases may be the abandonment of the project.

Too often, projects that are uneconomical are carried by the more productive projects to the detriment of the company as a total entity.

15.4 SUMMARY

A successful capital budgeting system requires presentation of the economy study in a clear and concise manner. Information used in the report must be well documented with respect to source and reliability.

Once the capital budget has been approved, in principle, the implementation should be approached within a well-structured framework in a systematic manner. The system should be monitored by a responsible and knowledgeable person who coordinates capital expenditures within the total budgeting system.

Tracking of capital expenditures is essential to determine how close the actual performance is to that predicted. The necessary steps can then be taken to improve the reliability of the total capital budgeting system.

PROBLEMS

15.1 Outline the essential information that should be included in any economy study.

15.2 Name the major sources of input data used in economy studies within an organization with which you are familiar. Comment on the varying reliability associated with these data.

15.3 Why is it important to know the audience to whom a report is being presented?

15.4 Review the procedure for implementing capital expenditures within an organization with which you are familiar. What recommendations do you have for changes in this procedure?

APPENDICES

Notation

A	The end-of-period sum in a uniform series flowing at the end of each of n periods
$\bar{A}$	The uniform flow rate of money per year
AC	Administration costs
ACRS	Accelerated Cost Recovery System from the American Economic Recovery Act of 1981
ADR	Asset depreciation range
AE	Annual equivalent value
AEC	Annual equivalent cost
AED	Annual equivalent depreciation allowance
AEOC	Annual equivalent operating costs
AEOR	Annual equivalent operating revenues
AER	Annual equivalent revenue requirements
ATCF	After-tax cash flow
B/C	Benefit-cost ratios
BTCF	Before-tax cash flow
BTRR	Minimum acceptable before-tax rate of return
BV	Book value; original cost minus accumulated depreciation
BV_z	Book value in year z

C	A sum discretely flowing at the end of the first period of a series of sums of C, $C(1 + k)$, $C(1 + k)^2$, . . . , $C(1 + k)^n$
C_i	Unit production cost (Chapter 8)
C_z	Revenue requirement (annual cost for year z)
CCA	Capital cost allowance (Canadian Income Tax Act)
CM	Contribution margin
CPI	Consumer Price Index
CTF	Capital tax factor
CV	Coefficient of variation
d	Declining-balance depreciation rate (percentage)
D	Amount of depreciation allowance
D_i	Annual demand for product i (Chapter 8)
D_t	Dividend expected at time t (in Chapter 5)
D_z	Amount of depreciation in year z
DL	Direct labor cost
DM	Direct material cost
f_i	Frequency of sample x_i
F	A future sum flowing discretely at the end of the nth period
F	Forecasted average (Chapter 13)
FC	Fixed cost
G	A gradient-sum, uniform period-by-period increase or decrease in cash flows
H	Holding fraction of cost for storing products for 1 year
i	Interest rate per period
i_a	Tax-sheltered cost of composite capital
i_{af}	Inflation-adjusted after-tax cost of composite capital
i_d	Before-tax cost of debt capital
i_{dt}	Tax-sheltered cost of debt capital
i_e	Cost of equity capital
i_{ef}	Inflation-adjusted cost of equity capital
i_f	Expected inflation rate
i_y	Effective rate of interest per year
ILDC	Interest lost during construction
IT	Income tax amount
k	Rate of change in a geometrically increasing or decreasing series
m	Number of compounding periods per year
MARR	Minimum acceptable after-tax rate of return
MC	Marginal cost (Chapter 8)
	Manufacturing cost (Chapter 13)
MR	Marginal revenue
MV	Market value
n	Number of periods
N	Number of setups (product runs) per year (Chapter 8)
NSV	Net salvage value (salvage value adjusted for taxes)
OC	Operating costs

OH	Overhead
OR	Operating revenue
P	A present sum
P	Profits (Chapters 8 and 13)
P_i	Production run for product i (Chapter 8)
P_0	Estimate of the current market price of a share of stock (Chapter 5)
P_t	Amount paid for a share in year t (Chapter 5)
P_t	Basis for depreciation allowance calculation
P_v	First cost used for capital recovery purposes
PCL	Production capability limit in units
PE	Present equivalent value or present value
PEC	Present equivalent cost
PED	Present equivalent depreciation
PEOC	Present equivalent operating costs
PEOR	Present equivalent operating revenue
PER	Present equivalent of revenue requirements
Q_i	Lot size (order quantity) of product i
r	Debt ratio
R	Sample range
RDC	Research and development costs
S	Number of units sold (Chapter 8)
S	Sample standard deviation (Chapter 12)
S^2	Sample variance (Chapter 12)
S_i	Setup costs for product i (Chapter 8)
S_n	Sum of a geometric series $a + ar + ar^2 + \cdots ar^n$
SL	Straight-line depreciation method
SP	Selling price
SV	Salvage value
SYD	Sum-of-years-digits' depreciation method
t	Marginal income tax rate
TC	Total cost
TI	Taxable income
TR	Total revenue
VC	Variable cost per unit
X_i	Sample observation
x	Cost capacity factor (Chapter 13)
$\overline{X}$	Sample mean
z	Usually a subscript indicating a year or a number of years

APPENDIX B

Summary of Formulas

(3.1) $F = P(1 + i)^n$

(3.2) $F = P(F/P, i\%, n)$

(3.3) $P = F[1/(1 + i)^n]$

(3.4) $P = F(P/F, i\%, n)$

(3.5) $F = A[(1 + i)^n - 1]/i$

(3.6) $F = A(F/A, i\%, n)$

(3.7) $A = F\{i/[(1 + i)^n - 1]\}$

(3.8) $A = F(A/F, i\%, n)$

(3.9) $P = A\{[(1 + i)^n - 1]/[i(1 + i)^n]\}$

(3.10) $P = A(P/A, i\%, n)$

(3.11) $A = P\{[i(1 + i)^n]/[(1 + i)^n - 1]\}$

(3.12) $A = P(A/P, i\%, n)$

(3.13) $F = G(1/i)\{[1 + i)^n - 1]/i - n\}$

(3.14) $F = G(F/G, i\%, n)$

(3.15) $P = G(1/i)\{(1 + i)^n - 1]/[i(1 + i)^n] - [n/(1 + i)^n]\}$

(3.16) $P = G(P/G, i\%, n)$

(3.17) $A = G\{(1/i) - n/[(1 + i)^n - 1]\}$

(3.18) $A = G(A/G, i\%, n)$

(3.19) $P = C\{[1 - (1 + k)^n/(1 + i)^n]/(i - k)\}$

(3.20) $P = nC/(1 + k)$

(3.21) $P = C(P/C, i, k, n)$

(3.22) $i_y = (1 + i)^m - 1$

(3.23) $im = m[(1 + i_y)^{1/m} - 1]$

(3.24) $i = (1 + i_y)^{1/m} - 1$

(3.25) $i_y = e^{im} - 1$

(3.26) $F = Pe^{imn}$

(3.27) $F = P(F/P, im, n)$

(3.28) $P = Fe^{-imn}$

(3.29) $P = F(P/F, im, n)$

(3.30) $P = \overline{A}(e^{imn} - 1)/ime^{imn}$

(3.31) $P = \overline{A}(P/\overline{A}, im, n)$

(4.1) $ATCF = OR(1 - t) - OC(1 - t) + tD + tI$

(4.2) $D = (P - SV)/n$

(4.3) $BV = P - \dfrac{z}{n}(P - SV)$

(4.4) $D = dP(1 - d)^{n-1}$

(4.5) $BV = P(1 - d)^n$

(4.6) $D_z = (P - SV)\left(\dfrac{\text{remaining life in years}}{\text{sum of the years' digits for the total useful life}}\right)$

(4.7) $BV_z = (P - SV)\left[\dfrac{(n - z)(n - z + 1)}{n(n + 1)}\right] + SV$

(4.8) $CTF = 1 - \dfrac{t}{n}(P/A, i, n)$

(4.9) $CTF = 1 - \dfrac{td}{i + d}$

(5.1) $i_{dt} = i_d(1 - t)$

(5.2) $P_0 = \displaystyle\sum_{n=1}^{\infty} \dfrac{D_n}{(1 + i)^n}$

(5.3) $P_0 = \displaystyle\sum_{n=0}^{\infty} \dfrac{D_n}{(1 + i)^n}$

(5.4) $P_0 = \displaystyle\sum_{n=1}^{t} \dfrac{D_n}{(1 + i)^n} + \dfrac{P_t}{(1 + i)^t}$

(5.5) $P_0 = \displaystyle\sum_{n=0}^{t} \dfrac{D_n}{(1 + i)^n} + \dfrac{P_t}{(1 + i)^t}$

(5.6) $P_0 = D_1\{[1 - (1 + k)^n/(1 + i)^n]/(i - k)\}$

(5.7) $P_0 = D_1/(i - k)$

(5.8) $i = D_1/P_0 + k$

(5.9) $i_a = ri_d(1 - t) + (1 - r)i_e$

(6.1) $AE = (AEOR - AEOC)(1 - t) - P(A/P, i_a, n)$
         $\quad\quad + NSV(A/F, i_a, n) + t(AED)$

(6.2) $PE = (PEOR - PEOC)(1 - t) - P + NSV(P/F, i_a, n) + t(PED)$

(6.3) $AER = AEOC$
$$+ [P(A/P, i_a, n) - NSV(A/F, i_a, n) - t(AED)]/(1 - t)$$

(6.4) $PER = PEOC + [P - NSV(P/F, i_a, n) - t(PED)]/(1 - t)$

(6.5) $AE = (AEOR - AEOC)(1 - t) - P(A/P, i_a, n)CTF$
$$+ SV(A/F, i_a, n)CTF$$

(6.6) $PE = (PEOR - PEOC)(1 - t) - P(CTF) + SV(P/F, i_a, n)CTF$

(6.7) $AER = AEOC + [P(A/P, i_a, n)CTF - SV(A/F, i_a, n)CTF]/(1 - t)$

(6.8) $PER = PEOC + [P(CTF) - SV(P/F, i_a, n)CTF]/(1 - t)$

(7.1) $i_{af} = i_{ef} = (1 + i_e)(1 + i_f) - 1$

(7.2) $i_{af} = (ri_d)(1 - t) + (1 - r)[(1 + i_e)(1 + i_f) - 1]$

(8.1) Total revenue (TR) = Total cost (TC)

(8.2) $S \times SP = FC + VC \times S$

(8.3) $S = FC/(SP - VC)$

(8.4) $S = (FC + P)/(SP - VC)$

(8.5) $IT = [t/(1 - t)](\text{profit to the shareholder})$

(8.6) Total cost $= Ax + \dfrac{B}{x} + C$

(9.1) $P_v = (\text{cost to make operational}) + (\text{market value forgone})$
$$+ (\text{income tax consideration})$$

(10.1) $C_z = OC_z + [P_z(1 + i_a) - SV_z - tD_z]/(1 - t)$

(11.1) B/C (conventional)
$$= \frac{\text{PE of net benefits to the user}}{\text{PE of capital} + \text{operating costs to the supplier}}$$

(11.2) B/C (modified) $= \dfrac{\text{PE of the annual BTCF values}}{\text{PE of all capital investment costs}}$

(12.1) $\overline{X} = \dfrac{\Sigma X_i}{n}$

(12.2) $\overline{X} = \dfrac{\Sigma X_i f_i}{\Sigma f_i} = \dfrac{\Sigma X_i f_i}{n}$

(12.3) $S^2 = \dfrac{\sum\limits_{i=1}^{n} X_i^2 - \left(\sum\limits_{i=1}^{n} X_i\right)^2 \Big/ n}{n - 1}$

(12.4) $S = \sqrt{S^2}$

(12.5) $R = X_{max} - X_{min}$

(12.6) $CV = \dfrac{S}{\overline{X}}$

(13.1) $b = \dfrac{n \Sigma xy - (\Sigma x)(\Sigma y)}{n(\Sigma x^2) - (\Sigma x)^2}$

(13.2) $a = \dfrac{\Sigma y - b \Sigma x}{n}$

(13.3) $\quad r = \dfrac{n \Sigma xy - (\Sigma x)(\Sigma y)}{\{[n \Sigma x^2 - (\Sigma x)^2][n \Sigma y^2 - (\Sigma y)^2]\}^{1/2}}$

(13.4) $\quad C_1/C_2 = (S_1/S_2)^x$

(13.5) $\quad C_T = \left(C_E + \displaystyle\sum_{i=1}^{n} F_i C_E \right)(F_x + 1)$

(13.6) $\quad SP = RDC + MC + AC + SC + IT + P$

(13.7) $\quad ST = NT[1.0/(1.0 - \text{allowance } \%)]$

(13.8) $\quad Rate = \dfrac{\text{actual factory overhead}}{\text{actual direct labor hours or direct labor costs}}$

APPENDIX C

Glossary of Terms

Accounts Payable represent the claims of vendors and others for goods and services.

Accounts Receivable the sale of merchandise to customers for credit, that is, on account.

Acid Test Ratio (quick ratio) current assets less inventories divided by current liabilities.

Annual Equivalent uniform annual amount which is equivalent to a present sum or a series of sums at a given rate of interest.

Assets resources owned and used to generate income.

Balance Sheet (a statement of financial position) an accounting report that indicates the financial position of an entity at a particular date.

Before-Tax Rate of Return (BTRR) represents the rate of return on the before-tax cash flows.

Benefit-Cost Analysis method of evaluation of public-sector investment projects which expressly considers all the relevant benefits and costs.

Bond long-term debt instrument that promises to repay the principal at some future date (maturity) and pay interest in the meantime.

Book Value value of an asset on the accounting records. Equivalent to the original cost less accumulated depreciation to date.

Break-even Analysis   method of determining the level of operations at which ROR is usually zero, although not necessarily so.

Capital Assets   assets whose useful lives exceed 1 year. Also called fixed assets.

Capital Budgeting   process of deciding which long-term investment projects will be undertaken in a given budgeting period.

Capital Cost Allowance   depreciation allowance specified in the Canadian Income Tax Act.

Capital Gains   on the sale of capital assets the amount that the price is in excess of original cost.

Capital Rationing   constraint on the total amount which may be invested in a given budgeting period.

Capital Recovery Factor   coefficient used to calculate the annual sum required to recover the first cost of a project plus compound interest on the unrecovered balance.

Capital Structure   percentage of each type of permanent capital used by the firm: debt, preferred stock, and common stock.

Capital Tax Factor (CTF)   coefficient used to calculate the present worth of tax savings (shields) due to the depreciation allowance.

Carrybacks (forwards)   amounts that can be carried back (forward) to other fiscal periods for tax purposes.

Cash Flow   the actual revenues and costs associated with an investment alternative.

Challenger   the new machine in replacement studies.

Compound Interest   interest paid on the remaining balance (principal + interest).

Continuous Compounding   situations in which the interest is added continuously rather than discretely.

Cost of Capital   weighted average cost of the total pool of capital that is available for investment.

Current Assets   assets that will normally be turned into cash within 1 year.

Current Liabilities   liabilities that will normally be paid within 1 year.

Current Ratio   current assets divided by current liabilities.

Debt Ratio (r)   ratio of debt capital to equity capital plus debt capital.

Decisions under Certainty   decision situations in which complete information is available to the decision maker.

Decisions under Risk   decision situations in which there is a probability that the venture may have an unfavorable outcome.

Decisions under Uncertainty   decision situations in which the outcome of the venture is not, or cannot, be known.

Declining-Balance Depreciation   method of depreciation in which a constant percentage is applied to the undepreciated book value of an asset each year to determine the annual deduction from the before-tax cash flow to arrive at taxable income.

Defender   existing machine in replacement studies.

Depreciation   reduction in value of fixed assets due to obsolescence, use, or accounting convention.

Discounted Cash Flow (DCF) applying a discount factor (interest rate) to determine the present worth of a series of cash inflows and outflows.

Double Declining Balance Double the straight-line rate ignoring salvage value.

Effective Interest Rate true rate of interest computed by dividing the interest payment by the amount of money available to the borrower.

Equity equity capital (capital contributed by the owners).

Expected Value sum of the products of all possible outcomes multiplied by their respective probabilities.

Financial Analysis economic comparisons which consider both the expected returns from investment proposals and their financial costs.

Financial Lease lease which usually does not provide maintenance, is usually not cancellable, and is fully amortized over its life.

Financial Risk that portion of total risk which results from using debt.

Financial Structure the entire right-hand side of the balance sheet. It indicates how a firm has been financed.

First Cost (initial outlay) the original cash outlay required for an investment project (usually installed cost).

Fiscal Policy government policy concerning its expenditures and revenues.

Fixed Cost cost that does not vary with output.

Future Equivalent an equivalent worth at a future date, based on the time value of money, of one or more amounts at given earlier dates.

Gradient amount of change per unit of time.

Income (profit and loss) Statement a financial report that indicates the revenues and expenses of a particular period.

Incremental change in total results due to adding one additional unit.

Inflation increase in the general price level.

Installed Cost represents the total first cost used in an economy study (P).

Internal Rate of Return (IRR) rate of interest at which the present worth of expected cash inflows from an investment project equals the present worth of the project's cash outflows (can also use the annual equivalent approach).

Inventory Turnover ratio of sales to inventory, used to evaluate liquidity of inventories and managerial efficiency.

Lessee user of a leased asset.

Lessor owner of a leased asset.

Leverage changes in profits (losses) resulting from the use of debt or other fixed costs.

Liabilities debts owed.

Life-Cycle Costing (LCC) method of analysis using cash flows over the entire life of a project with the objective of minimizing the total cost, not just the first cost.

Life, Economic represents point at which AEC is minimized.

Life, Service period during which similar assets are required. For most companies this is an infinite period.

Loss on Disposal   loss incurred due to selling an asset at less than the book value.

Marginal   change in total results due to the addition (or subtraction) of one unit.

Minimum Acceptable Rate of Return (MARR)   after-tax lower limit for investment acceptability.

Monetary Policy   government program of control of the money supply and credit to achieve desired economic goals.

Multiple Rates of Return   situation which occurs when an investment project's expected future cash flows change sign more than once during the study period.

Net Worth (equity)   capital provided by the owners; includes common and preferred share capital, retained earnings, and any surplus accounts. Amount needed to make a balance sheet balance. Assets − liabilities = equity.

Nominal Interest Rate   the contractual or stated interest rate.

Operating Lease   lease which is cancellable by the lessee on due notice to the lessor (no fixed obligation).

Operating Ratio   ratio of total expenses to net sales.

Opportunity Cost   return on the best available alternative, which is forgone because another project was selected.

Payback (payoff) Period   time required to recoup the first cost of an investment from the net cash flows generated at zero interest.

Present Equivalent (worth)   value found by discounting future cash flows by an appropriate discount rate, such as the cost of capital.

Price Index   measure of the change in prices over some period of time.

Profitability Index   present worth of expected returns divided by the first cost.

Public Goods   items available to all, such as public parks, air, water, and defense.

Recapture   occurs when an asset is sold for more than book value but not more than original cost.

Retained Earnings   portion of net income not paid out in dividends.

Salvage Value   expected value of a capital asset at the end of a specified period.

Sensitivity Analysis   analyses in which key variables are changed to see what is the effect on the total.

Simple Interest   interest which is not compounded, paid only on the principal amount.

Sinking Fund   fund to which annual payments are made to amortize a bond or preferred share issue, or to accumulate funds with which to purchase a replacement asset or meet an obligation.

Social Discount Rate   discount rate which reflects the opportunity cost to society of resources used in public sector investment projects.

Straight-Line Depreciation   method of calculating depreciation in which the annual depreciation charge is the same each year; it is calculated by dividing the first cost less expected salvage value by the expected economic life.

Sunk Cost   a past investment cost; its relevance in an economy study is due to its impact on tax effects at the time of disposal of the asset.

Tax Incentives   reductions in taxes otherwise payable for certain kinds of government-approved expenditures.

Time Value of Money   effect of time on the money value of an event, taking into consideration the opportunity cost of money.

Undepreciated Capital Cost (UCC)   the book value of an asset, used as the basis for determining the annual depreciation deduction for tax purposes.

Variable Cost   cost that varies with output.

Working Capital   current assets less current liabilities.

APPENDIX D

Selected References

1. American Telephone and Telegraph Company, *Engineering Economy—A Manager's Guide to Economic Decision Making,* 3rd ed. New York: McGraw-Hill Book Company, 1971.

2. Anthony, R. N., *Management Accounting,* 3rd ed. Homewood, Ill.: Richard D. Irwin Inc., 1964.

3. Archer, S. H., and C. A. D'Ambrosio, *The Theory of Business Finance: A Book of Readings.* New York: Macmillan Publishing Company, 1967.

4. Barges, A., *The Effect of Capital Structure on the Cost of Capital.* Englewood Cliffs, N.J.: Prentice-Hall, Inc., 1963.

5. Barish, N. N., and S. Kaplan, *Economic Analysis for Engineering and Managerial Decision Making,* 2nd ed. New York: McGraw-Hill Book Company, 1978.

6. Bauer, W. F., "Monte Carlo Method," *Society for Industrial and Applied Mathematics Journal,* Vol. 6, pp. 438–451, 1958.

7. *Benefit-Cost Analysis Guide,* Treasury Board, Government of Canada, 1976; available from Printing and Publishing, Supply and Services Canada, Ottawa, Canada K1A 0S9.

8. Bernhard, R. H., "A Simpler Internal Rate of Return Uniqueness Condition

Which Dominates That of de Faro and Soares," *The Engineering Economist,* Vol. 24, No. 2, Winter 1979.

9. Bierman, H., Jr., and S. Smidt, *The Capital Budgeting Decision,* 5th ed. New York: Macmillan Publishing Company, 1980.

10. Blank, L. T., and A. J. Tarquin, *Engineering Economy,* 2nd ed. New York: McGraw-Hill Book Company, 1983.

11. Braums, M. R., "A Case Study of the Application of Cost-Benefit Analysis to Water System Consolidation by Local Government," *Engineering Economist,* Vol. 17, No. 2, 1972.

12. Brigham, E. F., A. L. Kahl, and W. F. Rentz, *Canadian Financial Management: Theory and Practice.* Holt, Rinehart and Winston of Canada Ltd., 1983.

13. Canada, J. L., and J. A. White, *Capital Investment Decision Analysis for Management and Engineering.* Englewood Cliffs, N.J.: Prentice-Hall, Inc., 1980.

14. DeGarmo, E. P., W. Sullivan, and J. R. Canada, *Engineering Economy,* 7th ed. New York: Macmillian Publishing Company, 1984.

15. de Neufville, R., and J. H. Stafford, *Systems Analysis for Engineers and Managers.* New York: McGraw-Hill Book Company, 1971.

16. Eckstern, O., *Water Resource Development: The Economics of Project Evaluation.* Cambridge, Mass.: Harvard University Press, 1958.

17. Edge, C. G., and V. B. Irvine, *A Practical Approach to the Appraisal of Capital Expenditures,* 2nd ed. Hamilton, Ontario: The Society of Management Accountants of Canada, 1981.

18. *The Engineering Economist,* a quarterly journal jointly published by the Engineering Economy Division of the American Society for Engineering Education and the Institute of Industrial Engineers, published by IIE, Norcross, Ga.

19. *Engineering News-Record,* published by McGraw-Hill, Inc., New York.

20. Fabrycky, W. J., and G. J. Thuesen, *Economic Decision Analysis,* 2nd ed. Englewood Cliffs, N.J.: Prentice-Hall, Inc., 1980.

21. Gordon, M. J., and E. Shapiro, "Capital Equipment Analysis: The Required Rate of Profit," *Management Science,* October 1956, copyright 1956, The Institute of Management Science.

22. Grant, E. L., W. G. Ireson, and R. S. Leavenworth, *Principles of Engineering Economy,* 7th ed. New York: The Ronald Press Company, 1982.

23. Hajdasinski, M. M., "A Complete Method for Separation of Internal Rates of Return," *The Engineering Economist,* Vol. 24, No. 3, Spring 1983.

24. *Harvard Business Review,* published bimonthly by Harvard University Press, Boston.

25. Hertz, D. B., "Risk Analysis in Capital Investment," *Harvard Business Review,* January–February 1964.

26. Hillier, F. S., "Derivation of Probabilistic Information for the Evaluation of Risky Investments," *Management Scientist,* Vol. 9, No. 4, April 1983.

27. *Industrial Engineering*, a monthly magazine published by the Institute of Industrial Engineers, Norcross, Ga.

28. Johnson, R. W., *Financial Management*, 2nd ed. Boston: Allyn and Bacon, Inc., 1962.

29. Jones, B. W., Inflation in Engineering Economic Analysis, New York: John Wiley and Sons Inc., 1982.

30. Lusztig, P. A., and B. Schwab, *Managerial Finance in a Canadian Setting*. Toronto: Holt, Rinehart and Winston of Canada Ltd., 1973.

31. Machinery and Allied Products Institute, *MAPI Replacement Manual*. Washington, D.C.: Machinery and Allied Products Institute, 1950.

32. Magyar, W. B., "Economic Evaluation of Engineering Projects," *The Engineering Economist*, Vol. 13, No. 2, Winter 1968, pp. 67–85.

33. Merrett, A. J., and A. Sykes, *The Finance and Analysis of Capital Projects*, Longmans, Green and Co. Ltd., London, 1963.

34. Mishan, E. J., *Cost Benefit Analysis*. London: George Allen & Unwin Ltd., 1971.

35. Modigliani, F., and M. H. Miller, "The Cost of Capital, Corporation Finance and the Theory of Investment," *American Economic Review*, Vol. 48, June 1958, pp. 261–297.

36. Newman, D. G., *Engineering Economic Analysis*, rev. ed. San Jose, Calif.: Engineering Press, 1980.

37. Ostwald, P. F., *Cost Estimating for Engineering and Management*. Englewood Cliffs, N.J.: Prentice-Hall, Inc., 1974.

38. Ostwald, P. F., "Variable Cost Estimating," *Proceedings*, American Institute of Industrial Engineers, Inc., Spring Conference, 1974, pp. 427–434.

39. Park, W. R., *Cost Engineering Analysis*. New York: John Wiley & Sons, Inc., 1973.

40. Reisman, A., *Managerial and Engineering Economics*, Boston: Allyn and Bacon, Inc., 1971.

41. Riggs, J. L., W. F. Rentz, and A. L. Kahl, *Essentials of Engineering Economics*, first Canadian edition, Scarborough, Ontario: McGraw-Hill Ryerson Ltd., 1982.

42. Schlaifer, R., *Probability and Statistics for Business Decisions*. Maidenhead, Berkshire, England: McGraw-Hill Book Company Ltd., 1959.

43. Simon, H. A., "A Behavioral Model of Rational Choice," *Quarterly Journal of Economics*, Vol. 69, 1955, pp. 99–118.

44. Simon, H. A., *Model of Man*. New York: John Wiley & Sons, Inc., 1957.

45. Smith, G. W., *Engineering Economy: Analysis of Capital Expenditures*, 3rd ed. Ames, Iowa: Iowa State University Press, 1979.

46. Sturges, H. A., "The Choice of a Class Interval," *Journal of American Statistics Association*, March 1926.

47. Swalm, R. O., "Utility Theory—Insights into Risk Taking," *Harvard Business Review*, November–December 1966, pp. 123–136.

48. Szonyi, A. J., R. G. Fenton, J. A. White, M. H. Agee, and K. E. Case, *Principles of Engineering Economic Analysis,* Canadian edition. Rexdale, Ontario: John Wiley & Sons Canada Ltd., 1982.

49. Taylor, G. A., *Managerial and Engineering Economy,* 3rd ed. New York: Van Nostrand Reinhold Company, Inc. 1980.

50. Terborgh, G., *Business Investment Management.* Washington, D.C.: Machinery and Allied Products Institute, 1967.

51. Thuesen, H. G., W. J. Fabrycky, and G. J. Thuesen, *Engineering Economy,* 6th ed. Englewood Cliffs, N.J.: Prentice-Hall, Inc., 1984.

52. Van Horne, J. C., C. R. Dipchand, and J. R. Hanrahan, *Financial Management and Policy,* Canadian 4th edition. Toronto: Prentice-Hall Canada Ltd., 1977.

53. Vernon, I. R., *Realistic Cost Estimating for Manufacturing,* Deerborn, Mich.: Society of Manufacturing Engineers, 1968.

54. Weingartner, H. M., *Mathematical Programming and the Analysis of Capital Budgeting Problems.* Englewood Cliffs, N.J.: Prentice-Hall, Inc. 1963.

55. Weston, F. J., E. F. Brigham, and P. Halpern, *Essentials of Canadian Managerial Finance.* Toronto: Holt, Rinehart and Winston of Canada Ltd., 1979.

56. White, J. A., M. H. Agee, and K. E. Case, *Principles of Engineering Economic Analysis.* New York: John Wiley & Sons, Inc., 1977.

APPENDIX E
Answers to Selected Problems

Many of these problems were worked using Supercalc and the basic formulas. Therefore, slight differences in answers may occur when problems are worked using the interest tables.

CHAPTER 2

2.4 a. $12.18/share; b. $743,600, $682,700

2.8 Year 1: $10,000;
Year 2: $12,500

CHAPTER 3

3.2 Plan (d)

3.4 a. F = $16,386; b. F = $27,385;
c. P = $41,064

3.6 a. $11,620, $11,605;
b. $28,500, $28,394;
c. $68,160, $67,275;
d. $50,760, $49,960;
e. $190, $133; f. $992, $978

3.10 a. $2,011; b. $4,046; c. $8,137;
d. $16,367

3.12 a. $12,760; b. $16,110; c. $24,880

3.14 a. $2,374; b. $2,774;
c. $3,344

3.16 a. $6,210; b. $5,335; c. $4,487

3.18 $2,181

3.20 $7,907

3.22 $1,086,222

3.24 a. $982; b. $1,327; c. $1,436

3.26 a. $395; b. $470 c. $485

3.30 a. 24%; b. 25.44%; c. 26.82%

3.32 24%; 26.82%

3.34 Alt. B

3.36 1.87% versus 1%; no

3.38 a. $8,154; b. $8,154; c. $8,089;
d. $7,781

3.40 $10,553

3.42 7.2%

3.44 12.4%

CHAPTER 4

4.4 $141; $535

4.6 $1,631

4.8 6.4%

4.10 18.4%

4.12 a. $7,300; b. $977

4.14 a. $3,791; b. $3,276

4.16 a. $842.58/month; b. $76,522.62

4.18 $323.06

4.20 $1686

4.24 Yes

4.26 $352,772

4.28 a. $33,450; b. $44,775; c. $107,300

4.30 a. $35,000; b. $65,000

4.32 a. $75,000; b. $75,000

4.38 a. $50,000; b. $25,000; c. $11,872;
d. $0

4.40 1. $D = \$30,000$, BV = $270,000:
2. $D = \$54,000$, BV = $216,000
3. $D = \$43,200$, BV = $172,800

4.44 a. $2,417; b. $37,866; c. $20,295

4.46 a. $50,000; b. $250,000

4.48 $65,971

4.50 a. $-9,152; b. $-5,787

CHAPTER 5

5.2 12.8%

5.8 a. $100,000, 100%;
b. $100,000, -100%;
c. $100,000, 20%
$100,000; -20%

5.10 6.3%

5.12 10%

5.14 No; $23.61 < $30.00

5.16 11.2%

5.18 a. $60,000; b. $189,600

CHAPTER 6

6.2 $2,285

6.4 $PEC_A = \$699,168$; $PEC_B = \$691,400$

6.6 $PEC_A = \$200,619$; $PEC_B = \$208,042$;
$PE_{B-A} = \$-7,423$

6.8 a. 15%; b. 12.9%

6.10 $PE_A = \$-19,623$; $PE_B = \$-39,774$

6.12 $PE_{B-A} = \$3,840$

6.14 a. $52,500; b. $9,975; c. $72,718
d. $6,409

6.16 MARR = 8%

6.18 a. $322,178; b. $1,079,989; c. no

6.20 a. $27,115; b. 8%

6.22 $AE_{2-1} = \$657$

6.24 a. 12.3%; b. 10%; c 13.4%

6.26 20%

6.28 AER = $120,284

6.30 a. $124,541; b = 10.0%

6.32 a. 32.8%; b. 22.8%

6.34 $i_e = 21\%$; $i_e = 41\%$

6.38 AER = $212,290

6.40 PER = $927,613

6.42 AER = $107,857

CHAPTER 7

7.2 a. $3,604; b. $3,604

7.4 a. 10.2%; b. 15%; c. 17.4%

7.6 $i_{ef} = 16.0\%$

7.8 AER $58,017

7.10 a. AER = $163,945; b. yes

7.12 PER = $1,456,990

7.14 PE = $-\$13,325$; no

7.16 $i_{ef} = 20\%$

7.18 a. PER = $132,046
b. PER = $139,938

7.20 PER = $2,228,370

7.22 AER = $45,749

7.24 $i_{af} = 15\%$; $i_e = 12.4\%$

7.26 PER = $1,962,953

CHAPTER 8

8.2 SP = $6.27/unit

8.4 Recover lost sales in 42 months
Recovery invest. in 210 months
252 months

8.6 a. P = $-75,000$; b. 4 units

8.8 a. S = 1,026 units;
b. foreign

8.10 a. A = $2.75; B = $5.75; C = $5.25
b. Item B: saving $1.25/unit;
c. Saving: B = $2.50; C = $0.20

8.12 Contrib. = $11,300; yes

8.14 P_{max} = 2.73445;
5% decrease = 2.7176

8.16 58.3 meters

CHAPTER 9

9.4 a. $50,000; b = $79,679;
c. $109,679
d. AER_D = $80,167, AER_C = $91,079

9.6 a. P_V = $721,500, P_t = $550,000;
b. P_V = $529,000; c. P_V = $546,250

9.8 a. P_V(def.) = 0, P_V(chal) = $75,000;
b. AER_D = $10,000, AER_C = $8,265;
c. n = 14.5 years

9.10 Trade-in old machine

9.12 Retain old machine

9.14 a. P_V = $139,835, P_t = $113,375;
b. P_V = $131,544; c. P_V = $106,939

9.16 a. P_V(alt.2) = $88,157,000
P_t = $152,633,000;
b. PER_1 = $462,820,000
PER_2 = $442,277,000

CHAPTER 10

10.2 No; AER_C(min) = $61,826

10.4 IF MARR ≤ 7.0%

10.6 If MARR ≤ 15%

10.8 $AE_{(2-1)}$ = +$22,778

10.10 n = 7 years; AER = $6,050

10.12 No; AER_2(min) = $55,462

10.14 3 years, AER = $154,056

10.16 AER_1 = $51,114; AER_2 = $56,242

10.18 AER_2 = $4,769; AER_3 = $5,015

10.20 2 years: AER_2 = $91,818

10.22 7 years; AER = $32,132

10.24 a. 7 years; b. 6 years

CHAPTER 11

11.6 a. B/C(conv.) = 1.07;
b. B/C(mod.) = 1.10; c. i = 9.4%

11.8 $0.97 ≈ $1.00

11.10 a. B/C(A) = 0.92; B/C(B) = 1.13
b. i(B) = 12.8% (accept)

CHAPTER 12

12.2 a. $\overline{X}$ = $342.25; b. Median = $150;
d. $E(X)$ = $243.75, e. R = $1,925;
f. 0.55

12.4 a. S_x = $5,952; b = 31%; c. 25%

12.6 AER_x = $156,619; AER_y = $176,059

12.8 a. 18%; b. 26.9%

CHAPTER 13

13.3 a. $6.56 Million; b. r = 0.9697

13.4 $1.8073 billion

13.5 $22.79/ft^2

13.10 a. $4.6228

CHAPTER 14

14.6 A = 2.9 years; B = 3.1 years

14.8 20% to 25%

14.10 b. (1) Choose A4 and B5;
(2) Choose A5 and B5

14.15 AER(lease) = $3,020;
AER(buy) = $2,874

14.16 AER(status quo) = $10,640
AER(buy)   = $11,462
AER(lease)   = $7,772

APPENDIX F

Tables

TABLE F.1 The standardized normal distribution function,* F(S).

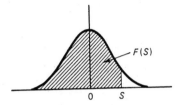

s	0.00	0.01	0.02	0.03	0.04	0.05	0.06	0.07	0.08	0.09
0.0	0.5000	0.5040	0.5080	0.5120	0.5160	0.5199	0.5239	0.5279	0.5319	0.5359
0.1	0.5398	0.5438	0.5478	0.5517	0.5557	0.5596	0.5636	0.5675	0.5714	0.5753
0.2	0.5793	0.5832	0.5871	0.5910	0.5948	0.5987	0.6026	0.6064	0.6103	0.6141
0.3	0.6179	0.6217	0.6255	0.6293	0.6331	0.6368	0.6406	0.6443	0.6480	0.6517
0.4	0.6554	0.6591	0.6628	0.6664	0.6700	0.6736	0.6772	0.6808	0.6844	0.6879
0.5	0.6915	0.6950	0.6985	0.7019	0.7054	0.7088	0.7123	0.7157	0.7190	0.7224
0.6	0.7257	0.7291	0.7324	0.7357	0.7389	0.7422	0.7454	0.7486	0.7517	0.7549
0.7	0.7580	0.7611	0.7642	0.7673	0.7703	0.7734	0.7764	0.7794	0.7823	0.7852
0.8	0.7881	0.7910	0.7939	0.7967	0.7995	0.8023	0.8051	0.8078	0.8106	0.8133
0.9	0.8159	0.8186	0.8212	0.8238	0.8264	0.8289	0.8315	0.8340	0.8365	0.8389
1.0	0.8413	0.8438	0.8461	0.8485	0.8508	0.8531	0.8554	0.8577	0.8599	0.8621
1.1	0.8643	0.8665	0.8686	0.8708	0.8729	0.8749	0.8770	0.8790	0.8810	0.8830
1.2	0.8849	0.8869	0.8888	0.8907	0.8925	0.8944	0.8962	0.8980	0.8997	0.90147
1.3	0.90320	0.90490	0.90658	0.90824	0.90988	0.91149	0.91309	0.91466	0.91621	0.91774
1.4	0.91924	0.92073	0.92220	0.92364	0.92507	0.92647	0.92785	0.92922	0.93056	0.93189
1.5	0.93319	0.93448	0.93574	0.93699	0.93822	0.93943	0.94062	0.94179	0.94295	0.94408
1.6	0.94520	0.94630	0.94738	0.94845	0.94950	0.95053	0.95154	0.95254	0.95352	0.95449
1.7	0.95543	0.95637	0.95728	0.95818	0.95907	0.95994	0.96080	0.96164	0.96246	0.96327
1.8	0.96407	0.96485	0.96562	0.96638	0.96712	0.96784	0.96856	0.96926	0.96995	0.97062
1.9	0.97128	0.97193	0.97257	0.97320	0.97381	0.97441	0.97500	0.97558	0.97615	0.97670
2.0	0.97725	0.97778	0.97831	0.97882	0.97932	0.97982	0.98030	0.98077	0.98124	0.98169
2.1	0.98214	0.98257	0.98300	0.98341	0.98382	0.98422	0.98461	0.98500	0.98537	0.98574
2.2	0.98610	0.98645	0.98679	0.98713	0.98745	0.98778	0.98809	0.98840	0.98870	0.98899
2.3	0.98928	0.98956	0.98983	$0.9^2 0097$	$0.9^2 0358$	$0.9^2 0613$	$0.9^2 0863$	$0.9^2 1106$	$0.9^2 1344$	$0.9^2 1576$
2.4	$0.9^2 1802$	$0.9^2 2024$	$0.9^2 2240$	$0.9^2 2451$	$0.9^2 2656$	$0.9^2 2857$	$0.9^2 3053$	$0.9^2 3244$	$0.9^2 3431$	$0.9^2 3613$
2.5	$0.9^2 3790$	$0.9^2 3963$	$0.9^2 4132$	$0.9^2 4297$	$0.9^2 4457$	$0.9^2 4614$	$0.9^2 4766$	$0.9^2 4915$	$0.9^2 5060$	$0.9^2 5201$
3.0	$0.9^2 8650$	$0.9^2 8649$	$0.9^2 8736$	$0.9^2 8777$	$0.9^2 8817$	$0.9^2 8856$	$0.9^2 8893$	$0.9^2 8930$	$0.9^2 8965$	$0.9^2 8999$
3.5	$0.9^3 7674$	$0.9^3 7759$	$0.9^3 7842$	$0.9^3 7922$	$0.9^3 7999$	$0.9^3 8074$	$0.9^3 8146$	$0.9^3 8215$	$0.9^3 8282$	$0.9^3 8347$
4.0	$0.9^4 6833$	$0.9^4 6964$	$0.9^4 7090$	$0.9^4 7211$	$0.9^4 7327$	$0.9^4 7439$	$0.9^4 7546$	$0.9^4 7649$	$0.9^4 7748$	$0.9^4 7843$

For example: $F(2.41) = 0.9^2 2024 = 0.992024$

* Reprinted from A. Wald, *Statistical Tables and Formulas* (New York: John Wiley & Sons, Inc., 1952), by permission of the publisher.

TABLE F.2 Table of random numbers.*

48867	33971	29678	13151	56644	49193	93469	43252	14006	47173
32267	69746	00113	51336	36551	56310	85793	53453	09744	64346
27345	03196	33877	35032	98054	48358	21788	98862	67491	4222
55753	05256	51557	90419	40716	64589	90398	37070	78318	02918
93124	50675	04507	44001	06365	77897	84566	99600	67985	49133
98658	86583	97433	10733	80495	62709	61357	66903	76730	79355
68216	94830	41248	50712	46878	87317	80545	31484	03195	14755
17901	30815	78360	78260	67866	42304	07293	61290	61301	04815
88124	21868	14942	25893	72695	56231	18918	72534	86737	77792
83464	36749	22336	50443	83576	19238	91730	39507	22717	94719
91310	99003	25704	55581	00729	22024	61319	66162	20933	67713
32739	38252	91256	77744	75080	01492	90984	63090	53087	41301
07751	66724	03290	56386	06070	67105	64219	48192	70478	84722
55228	64156	90480	97774	08055	04435	26999	42039	16589	06757
89013	51781	81116	24383	95569	97247	44437	36293	29967	16088
51828	81819	81038	89146	39192	89470	76331	56420	14527	34828
59783	85454	93327	06078	64924	07271	77563	92710	42183	12380
80267	47103	90556	16128	41490	07996	78454	47929	81586	67024
82919	44210	61607	93001	26314	26865	26714	43793	94937	28439
77019	77417	19466	14967	75521	49967	74065	09746	27881	01070
66225	61832	06242	40093	40800	76849	29929	18988	10888	40344
98534	12777	84601	56336	00034	85939	32438	09549	01855	40550
63175	70789	51345	43723	06995	11186	38615	56646	54320	39632
92362	73011	09115	78303	38901	58107	95366	17226	74626	78208
61831	44794	65079	97130	94289	73502	04857	68855	47045	06309
42502	01646	88493	48207	01283	16474	08864	68322	92454	19287
89733	86230	04903	55015	11811	98185	32014	84761	80926	14509
01336	66633	26015	66768	24846	00321	73118	15802	13549	41335
72623	56083	65799	88934	87274	19417	84897	90877	76472	52145
74004	68388	04090	35239	49379	04456	07642	68642	01026	43810
09388	54633	27684	47117	67583	42496	20703	68579	65883	10729
51771	92019	39791	60400	08585	60680	28841	09921	00520	73135
69796	30304	79836	20631	10743	00246	24979	35707	75283	39211
98417	33403	63448	90462	91645	24919	73609	26663	09380	30515
56150	18324	43011	02660	86574	86097	49399	21249	90380	94375
76199	75692	09063	72999	94672	69128	39046	15379	98450	09159
74978	98693	21433	34676	97603	48534	59205	66265	03561	83075
85769	92530	04407	53725	96963	19395	16193	51018	70333	12094
63819	65669	38960	74631	39650	39419	93707	61365	46302	26134
18892	43143	19619	43200	49613	50904	73502	19519	11667	53294
32855	17190	61587	80411	22827	38852	51952	47785	34952	93574
29435	96277	53583	92804	05027	19736	54918	66396	96547	00351
36211	67263	82064	41624	49826	17566	02476	79368	28831	02805
73514	00176	41638	01420	31850	41380	11643	06787	09011	88924
90895	93099	27850	29423	98693	71762	39928	35268	59359	20674
69719	90656	62186	50435	77015	29661	94698	56057	04388	33381
94982	81453	87162	28248	37921	21143	62673	81224	38972	92988
84136	04221	72790	04719	34914	95609	88695	60180	58790	12802
58515	80581	88442	65727	72121	40481	06001	13159	55324	93591
20681	59164	75797	08928	68381	12616	97487	84803	92457	88847

* Reproduced with permission from the Rand Corporation, *A Million Random Numbers*. (New York: The Free Press, 1955).

TABLE F.3 Effective interest rates corresponding to nominal rate im.

	Compounding Frequency					
r	Semi-annually	Quarterly	Monthly	Weekly	Daily	Continu-ously
	$\left(1+\dfrac{im}{2}\right)^2 -1$	$\left(1+\dfrac{im}{4}\right)^4 -1$	$\left(1+\dfrac{im}{12}\right)^{12} -1$	$\left(1+\dfrac{im}{52}\right)^{52} -1$	$\left(1+\dfrac{im}{365}\right)^{365} -1$	$\left(1+\dfrac{im}{\infty}\right)^{\infty} -1$
.01	.010025	.010038	.010046	.010049	.010050	.010050
.02	.020100	.020151	.020184	.020197	.020200	.020201
.03	.030225	.030339	.030416	.030444	.030451	.030455
.04	.040400	.040604	.040741	.040793	.040805	.040811
.05	.050625	.050945	.051161	.051244	.051261	.051271
.06	.060900	.061364	.061678	.061797	.061799	.061837
.07	.071225	.071859	.072290	.072455	.072469	.072508
.08	.081600	.082432	.082999	.083217	.083246	.083287
.09	.092025	.093083	.093807	.094085	.094132	.094174
.10	.102500	.103813	.104713	.105060	.105126	.105171
.11	.113025	.114621	.115718	.116144	.116231	.116278
.12	.123600	.125509	.126825	.127336	.127447	.127497
.13	.134225	.136476	.138032	.138644	.138775	.138828
.14	.144900	.147523	.149341	.150057	.150217	.150274
.15	.155625	.158650	.160755	.161582	.161773	.161834
.16	.166400	.169859	.172270	.173221	.173446	.173511
.17	.177225	.181148	.183891	.184974	.185235	.185305
.18	.188100	.192517	.195618	.196843	.197142	.197217
.19	.199025	.203971	.207451	.208828	.209169	.209250
.20	.210000	.215506	.219390	.220931	.221316	.221403
.21	.221025	.227124	.231439	.233153	.233584	.233678
.22	.232100	.238825	.243596	.245494	.245976	.246077
.23	.243225	.250609	.255863	.257957	.258492	.258600
.24	.254400	.262477	.268242	.270542	.271133	.271249
.25	.265625	.274429	.280731	.283250	.283901	.284025
.26	.276900	.286466	.293333	.296090	.296796	.296930
.27	.288225	.298588	.306050	.309049	.309821	.309964
.28	.299600	.310796	.318880	.322135	.322976	.323130
.29	.311025	.323089	.331826	.335350	.336264	.336428
.30	.322500	.335469	.344889	.348693	.349684	.349859
.31	.334025	.347936	.358068	.362168	.363238	.363425
.32	.345600	.360489	.371366	.375775	.376928	.377128
.33	.357225	.373130	.384784	.389515	.390756	.390968
.34	.368900	.385859	.398321	.403389	.404722	.404948
.35	.380625	.398676	.411979	.417399	.418827	.419068

Reprinted from G. J. Thuesen and W. J. Fabrycky, *Engineering Economy*, 6th ed. (Englewood Cliffs, NJ: Prentice-Hall, Inc., 1984).

TABLE F.4 Continuous compounding interest factors for various common values of *im*.

$$im = 2\%$$

	Discrete Flows				Continuous Flows		
	SINGLE PAYMENT		UNIFORM SERIES		UNIFORM SERIES		
	Compound Amount Factor	Present Worth Factor	Compound Amount Factor	Present Worth Factor	Compound Amount Factor	Present Worth Factor	
N	To find F Given P F/P	To find P Given F P/F	To find F Given A F/A	To find P Given A P/A	To find F Given $\overline{A}$ F/$\overline{A}$	To find P Given $\overline{A}$ P/$\overline{A}$	N
1	1.0202	0.9802	1.0000	0.9802	1.0101	0.9901	1
2	1.0408	0.9608	2.0202	1.9410	2.0405	1.9605	2
3	1.0618	0.9418	3.0610	2.8828	3.0918	2.9118	3
4	1.0833	0.9231	4.1228	3.8059	4.1644	3.8442	4
5	1.1052	0.9048	5.2061	4.7107	5.2585	4.7581	5
6	1.1275	0.8869	6.3113	5.5976	6.3748	5.6540	6
7	1.1503	0.8694	7.4388	6.4670	7.5137	6.5321	7
8	1.1735	0.8521	8.5891	7.3191	8.6755	7.3928	8
9	1.1972	0.8353	9.7626	8.1544	9.8609	8.2365	9
10	1.2214	0.8187	10.9598	8.9731	11.0701	9.0635	10
11	1.2461	0.8025	12.1812	9.7756	12.3038	9.8741	11
12	1.2712	0.7866	13.4273	10.5623	13.5625	10.6686	12
13	1.2969	0.7711	14.6985	11.3333	14.8465	11.4474	13
14	1.3231	0.7558	15.9955	12.0891	16.1565	12.2108	14
15	1.3499	0.7408	17.3186	12.8299	17.4929	12.9591	15
16	1.3771	0.7261	18.6685	13.5561	18.8564	13.6925	16
17	1.4049	0.7118	20.0456	14.2678	20.2474	14.4115	17
18	1.4333	0.6977	21.4505	14.9655	21.6665	15.1162	18
19	1.4623	0.6839	22.8839	15.6494	23.1142	15.8069	19
20	1.4918	0.6703	24.3461	16.3197	24.5912	16.4840	20
21	1.5220	0.6570	25.8380	16.9768	26.0981	17.1477	21
22	1.5527	0.6440	27.3599	17.6208	27.6354	17.7982	22
23	1.5841	0.6313	28.9126	18.2521	29.2037	18.4358	23
24	1.6161	0.6188	30.4967	18.8709	30.8037	19.0608	24
25	1.6487	0.6065	32.1128	19.4774	32.4361	19.6735	25
26	1.6820	0.5945	33.7615	20.0719	34.1014	20.2740	26
27	1.7160	0.5827	35.4435	20.6547	35.8003	20.8626	27
28	1.7507	0.5712	37.1595	21.2259	37.5336	21.4395	28
29	1.7860	0.5599	38.9102	21.7858	39.3019	22.0051	29
30	1.8221	0.5488	40.6962	22.3346	41.1059	22.5594	30
35	2.0138	0.4966	50.1824	24.9199	50.6876	25.1707	35
40	2.2255	0.4493	60.6663	27.2591	61.2770	27.5336	40
45	2.4596	0.4066	72.2528	29.3758	72.9802	29.6715	45
50	2.7183	0.3679	85.0578	31.2910	85.9141	31.6060	50
55	3.0042	0.3329	99.2096	33.0240	100.208	33.3564	55
60	3.3201	0.3012	114.850	34.5921	116.006	34.9403	60
65	3.6693	0.2725	132.135	36.0109	133.465	36.3734	65
70	4.0552	0.2466	151.238	37.2947	152.760	37.6702	70
75	4.4817	0.2231	172.349	38.4564	174.084	38.8435	75
80	4.9530	0.2019	195.682	39.5075	197.652	39.9052	80
85	5.4739	0.1827	221.468	40.4585	223.697	40.8658	85
90	6.0496	0.1653	249.966	41.3191	252.482	41.7351	90
95	6.6859	0.1496	281.461	42.0978	284.295	42.5216	95
100	7.3891	0.1353	316.269	42.8023	319.453	43.2332	100

Reprinted from Canada & White, *Capital Investment Decision Analysis for Management and Engineering* (Englewood Cliffs, NJ: Prentice-Hall, Inc., 1980).

TABLE F.4 Continuous compounding interest factors for various common values of *im*.

im = 5%

		Discrete Flows			Continuous Flows		
	SINGLE PAYMENT		UNIFORM SERIES		UNIFORM SERIES		
	Compound Amount Factor	Present Worth Factor	Compound Amount Factor	Present Worth Factor	Compound Amount Factor	Present Worth Factor	
	To find F Given P	To find P Given F	To find F Given A	To find P Given A	To find F Given $\overline{A}$	To find P Given $\overline{A}$	
N	F/P	P/F	F/A	P/A	$F/\overline{A}$	$P/\overline{A}$	N
1	1.0513	0.9512	1.0000	0.9512	1.0254	0.9754	1
2	1.1052	0.9048	2.0513	1.8561	2.1034	1.9033	2
3	1.1618	0.8607	3.1564	2.7168	3.2367	2.7858	3
4	1.2214	0.8187	4.3183	3.5355	4.4281	3.6254	4
5	1.2840	0.7788	5.5397	4.3143	5.6805	4.4240	5
6	1.3499	0.7408	6.8237	5.0551	6.9972	5.1836	6
7	1.4191	0.7047	8.1736	5.7598	8.3814	5.9062	7
8	1.4918	0.6703	9.5926	6.4301	9.8365	6.5936	8
9	1.5683	0.6376	11.0845	7.0678	11.3662	7.2474	9
10	1.6487	0.6065	12.6528	7.6743	12.9744	7.8694	10
11	1.7333	0.5769	14.3015	8.2512	14.6651	8.4610	11
12	1.8221	0.5488	16.0347	8.8001	16.4424	9.0238	12
13	1.9155	0.5220	17.8569	9.3221	18.3108	9.5591	13
14	2.0138	0.4966	19.7724	9.8187	20.2751	10.0683	14
15	2.1170	0.4724	21.7862	10.2911	22.3400	10.5527	15
16	2.2255	0.4493	23.9032	10.7404	24.5108	11.0134	16
17	2.3396	0.4274	26.1287	11.1678	26.7929	11.4517	17
18	2.4596	0.4066	28.4683	11.5744	29.1921	11.8686	18
19	2.5857	0.3867	30.9279	11.9611	31.7142	12.2652	19
20	2.7183	0.3679	33.5137	12.3290	34.3656	12.6424	20
21	2.8577	0.3499	36.2319	12.6789	37.1530	13.0012	21
22	3.0042	0.3329	39.0896	13.0118	40.0833	13.3426	22
23	3.1582	0.3166	42.0938	13.3284	43.1639	13.6673	23
24	3.3201	0.3012	45.2519	13.6296	46.4023	13.9761	24
25	3.4903	0.2865	48.5721	13.9161	49.8069	14.2699	25
26	3.6693	0.2725	52.0624	14.1887	53.3859	14.5494	26
27	3.8574	0.2592	55.7317	14.4479	57.1485	14.8152	27
28	4.0552	0.2466	59.5891	14.6945	61.1040	15.0681	28
29	4.2631	0.2346	63.6443	14.9291	65.2623	15.3086	29
30	4.4817	0.2231	67.9074	15.1522	69.6338	15.5374	30
35	5.7546	0.1738	92.7346	16.1149	95.0921	16.5245	35
40	7.3891	0.1353	124.613	16.8646	127.781	17.2933	40
45	9.4877	0.1054	165.546	17.4484	169.755	17.8920	45
50	12.1825	0.0821	218.105	17.9032	223.650	18.3583	50
55	15.6426	0.0639	285.592	18.2573	292.853	18.7214	55
60	20.0855	0.0498	372.247	18.5331	381.711	19.0043	60
65	25.7903	0.0388	483.515	18.7479	495.807	19.2245	65
70	33.1155	0.0302	626.385	18.9152	642.309	19.3961	70
75	42.5211	0.0235	809.834	19.0455	830.422	19.5296	75
80	54.5981	0.0183	1045.39	19.1469	1071.963	19.6337	80
85	70.1054	0.0143	1347.84	19.2260	1382.108	19.7147	85
90	90.0171	0.0111	1736.20	19.2875	1780.342	19.7778	90
95	115.584	0.0087	2234.87	19.3354	2291.686	19.8270	95
100	148.413	0.0067	2875.17	19.3727	2948.263	19.8652	100

Reprinted from Canada & White, *Capital Investment Decision Analysis for Management and Engineering* (Englewood Cliffs, NJ: Prentice-Hall, Inc., 1980).

TABLE F.4 Continuous compounding interest factors for various common values of *im*.

$im = 10\%$

	Discrete Flows				Continuous Flows		
	SINGLE PAYMENT		UNIFORM SERIES		UNIFORM SERIES		
	Compound Amount Factor	Present Worth Factor	Compound Amount Factor	Present Worth Factor	Compound Amount Factor	Present Worth Factor	
N	To find F Given P F/P	To find P Given F P/F	To find F Given A F/A	To find P Given A P/A	To find F Given $\overline{A}$ $F/\overline{A}$	To find P Given $\overline{A}$ $P/\overline{A}$	N
1	1.1052	0.9048	1.0000	0.9048	1.0517	0.9516	1
2	1.2214	0.8187	2.1052	1.7236	2.2140	1.8127	2
3	1.3499	0.7408	3.3266	2.4644	3.4986	2.5918	3
4	1.4918	0.6703	4.6764	3.1347	4.9182	3.2968	4
5	1.6487	0.6065	6.1683	3.7412	6.4872	3.9347	5
6	1.8221	0.5488	7.8170	4.2900	8.2212	4.5119	6
7	2.0138	0.4966	9.6391	4.7866	10.1375	5.0341	7
8	2.2255	0.4493	11.6528	5.2360	12.2554	5.5067	8
9	2.4596	0.4066	13.8784	5.6425	14.5960	5.9343	9
10	2.7183	0.3679	16.3380	6.0104	17.1828	6.3212	10
11	3.0042	0.3329	19.0563	6.3433	20.0417	6.6713	11
12	3.3201	0.3012	22.0604	6.6445	23.2012	6.9881	12
13	3.6693	0.2725	25.3806	6.9170	26.6930	7.2747	13
14	4.0552	0.2466	29.0499	7.1636	30.5520	7.5340	14
15	4.4817	0.2231	33.1051	7.3867	34.8169	7.7687	15
16	4.9530	0.2019	37.5867	7.5886	39.5303	7.9810	16
17	5.4739	0.1827	42.5398	7.7713	44.7395	8.1732	17
18	6.0496	0.1653	48.0137	7.9366	50.4965	8.3470	18
19	6.6859	0.1496	54.0634	8.0862	56.8589	8.5043	19
20	7.3891	0.1353	60.7493	8.2215	63.8906	8.6466	20
21	8.1662	0.1225	68.1383	8.3440	71.6617	8.7754	21
22	9.0250	0.1108	76.3045	8.4548	80.2501	8.8920	22
23	9.9742	0.1003	85.3295	8.5550	89.7418	8.9974	23
24	11.0232	0.0907	95.3037	8.6458	100.232	9.0928	24
25	12.1825	0.0821	106.327	8.7278	111.825	9.1791	25
26	13.4637	0.0743	118.509	8.8021	124.637	9.2573	26
27	14.8797	0.0672	131.973	8.8693	138.797	9.3279	27
28	16.4446	0.0608	146.853	8.9301	154.446	9.3919	28
29	18.1741	0.0550	163.298	8.9852	171.741	9.4498	29
30	20.0855	0.0498	181.472	9.0349	190.855	9.5021	30
35	33.1155	0.0302	305.364	9.2212	321.154	9.6980	35
40	54.5981	0.0183	509.629	9.3342	535.982	9.8168	40
45	90.0171	0.0111	846.404	9.4027	890.171	9.8889	45
50	148.413	0.0067	1401.65	9.4443	1474.13	9.9326	50
55	244.692	0.0041	2317.10	9.4695	2436.92	9.9591	55
60	403.429	0.0025	3826.43	9.4848	4024.29	9.9752	60
65	665.142	0.0015	6314.88	9.4940	6641.42	9.9850	65
70	1096.63	0.0009	10417.6	9.4997	10956.3	9.9909	70
75	1808.04	0.0006	17182.0	9.5031	18070.7	9.9945	75
80	2980.96	0.0003	28334.4	9.5051	29799.6	9.9966	80
85	4914.77	0.0002	46721.7	9.5064	49137.7	9.9980	85
90	8103.08	0.0001	77037.3	9.5072	81020.8	9.9988	90
95	13359.7	^{a}	127019	9.5076	133587	9.9993	95
100	22026.5	^{a}	209425	9.5079	220255	9.9995	100

^{a} Less than 0.001.

Reprinted from Canada & White, *Capital Investment Decision Analysis for Management and Engineering* (Englewood Cliffs, NJ: Prentice-Hall, Inc., 1980).

TABLE F.4 Continuous compounding interest factors for various common values of *im*.

$im = 15\%$

	Discrete Flows				Continuous Flows		
	SINGLE PAYMENT		UNIFORM SERIES		UNIFORM SERIES		
	Compound Amount Factor	Present Worth Factor	Compound Amount Factor	Present Worth Factor	Compound Amount Factor	Present Worth Factor	
	To find F Given P	To find P Given F	To find F Given A	To find P Given A	To find F Given $\overline{A}$	To find P Given $\overline{A}$	
N	F/P	P/F	F/A	P/A	$F/\overline{A}$	$P/\overline{A}$	N
1	1.1618	0.8607	1.0000	0.8607	1.0789	0.9286	1
2	1.3499	0.7408	2.1618	1.6015	2.3324	1.7279	2
3	1.5683	0.6376	3.5117	2.2392	3.7887	2.4158	3
4	1.8221	0.5488	5.0800	2.7880	5.4808	3.0079	4
5	2.1170	0.4724	6.9021	3.2603	7.4467	3.5176	5
6	2.4596	0.4066	9.0191	3.6669	9.7307	3.9562	6
7	2.8577	0.3499	11.4787	4.0168	12.3843	4.3337	7
8	3.3201	0.3012	14.3364	4.3180	15.4674	4.6587	8
9	3.8574	0.2592	17.6565	4.5773	19.0495	4.9384	9
10	4.4817	0.2231	21.5139	4.8004	23.2113	5.1791	10
11	5.2070	0.1920	25.9956	4.9925	28.0465	5.3863	11
12	6.0496	0.1653	31.2026	5.1578	33.6643	5.5647	12
13	7.0287	0.1423	37.2522	5.3000	40.1913	5.7182	13
14	8.1662	0.1225	44.2809	5.4225	47.7745	5.8503	14
15	9.4877	0.1054	52.4471	5.5279	56.5849	5.9640	15
16	11.0232	0.0907	61.9348	5.6186	66.8212	6.0619	16
17	12.8071	0.0781	72.9580	5.6967	78.7140	6.1461	17
18	14.8797	0.0672	85.7651	5.7639	92.5315	6.2186	18
19	17.2878	0.0578	100.645	5.8217	108.585	6.2810	19
20	20.0855	0.0498	117.933	5.8715	127.237	6.3348	20
21	23.3361	0.0429	138.018	5.9144	148.907	6.3810	21
22	27.1126	0.0369	161.354	5.9513	174.084	6.4208	22
23	31.5004	0.0317	188.467	5.9830	203.336	6.4550	23
24	36.5982	0.0273	219.967	6.0103	237.322	6.4845	24
25	42.5211	0.0235	256.565	6.0338	276.807	6.5099	25
26	49.4024	0.0202	299.087	6.0541	322.683	6.5317	26
27	57.3975	0.0174	348.489	6.0715	375.983	6.5505	27
28	66.6863	0.0150	405.886	6.0865	437.909	6.5667	28
29	77.4785	0.0129	472.573	6.0994	509.856	6.5806	29
30	90.0171	0.0111	550.051	6.1105	593.448	6.5926	30
35	190.566	0.0052	1171.36	6.1467	1263.78	6.6317	35
40	403.429	0.0025	2486.67	6.1638	2682.86	6.6501	40
45	854.059	0.0012	5271.19	6.1719	5687.06	6.6589	45
50	1808.04	0.0006	11166.0	6.1757	12046.9	6.6630	50
55	3827.63	0.0003	23645.3	6.1775	25510.8	6.6649	55
60	8103.08	0.0001	50064.1	6.1784	54013.9	6.6658	60
65	17154.2	^{a}	105993	6.1788	114355	6.6663	65
70	36315.5	^{a}	224393	6.1790	242097	6.6665	70
75	76879.9	^{a}	475047	6.1791	512526	6.6666	75
80	162755	^{a}	1005680	6.1791	1085030	6.6666	80

^{a} Less than 0.0001.

Reprinted from Canada & White, *Capital Investment Decision Analysis for Management and Engineering* (Englewood Cliffs, NJ: Prentice-Hall, Inc., 1980).

TABLE F.4 Continuous compounding interest factors for various common values of *im*.

$im = 20\%$

	Discrete Flows				Continuous Flows		
	SINGLE PAYMENT		UNIFORM SERIES		UNIFORM SERIES		
	Compound Amount Factor	Present Worth Factor	Compound Amount Factor	Present Worth Factor	Compound Amount Factor	Present Worth Factor	
N	To find F Given P F/P	To find P Given F P/F	To find F Given A F/A	To find P Given A P/A	To find F Given $\overline{A}$ $F/\overline{A}$	To find P Given $\overline{A}$ $P/\overline{A}$	N
1	1.2214	0.8187	1.0000	0.8187	1.1070	0.9063	1
2	1.4918	0.6703	2.2214	1.4891	2.4591	1.6484	2
3	1.8221	0.5488	3.7132	2.0379	4.1106	2.2559	3
4	2.2255	0.4493	5.5353	2.4872	6.1277	2.7534	4
5	2.7183	0.3679	7.7609	2.8551	8.5914	3.1606	5
6	3.3201	0.3012	10.4792	3.1563	11.6006	3.4940	6
7	4.0552	0.2466	13.7993	3.4029	15.2760	3.7670	7
8	4.9530	0.2019	17.8545	3.6048	19.7652	3.9905	8
9	6.0496	0.1653	22.8075	3.7701	25.2482	4.1735	9
10	7.3891	0.1353	28.8572	3.9054	31.9453	4.3233	10
11	9.0250	0.1108	36.2462	4.0162	40.1251	4.4460	11
12	11.0232	0.0907	45.2712	4.1069	50.1159	4.5464	12
13	13.4637	0.0743	56.2944	4.1812	62.3187	4.6286	13
14	16.4446	0.0608	69.7581	4.2420	77.2232	4.6959	14
15	20.0855	0.0498	86.2028	4.2918	95.4277	4.7511	15
16	24.5325	0.0408	106.288	4.3325	117.663	4.7962	16
17	29.9641	0.0334	130.821	4.3659	144.820	4.8331	17
18	36.5982	0.0273	160.785	4.3932	177.991	4.8634	18
19	44.7012	0.0224	197.383	4.4156	218.506	4.8881	19
20	54.5981	0.0183	242.084	4.4339	267.991	4.9084	20
21	66.6863	0.0150	296.682	4.4489	328.432	4.9250	21
22	81.4509	0.0123	363.369	4.4612	402.254	4.9386	22
23	99.4843	0.0101	444.820	4.4713	492.422	4.9497	23
24	121.510	0.0082	544.304	4.4795	602.552	4.9589	24
25	148.413	0.0067	665.814	4.4862	737.066	4.9663	25
26	181.272	0.0055	814.227	4.4917	901.361	4.9724	26
27	221.406	0.0045	995.500	4.4963	1102.03	4.9774	27
28	270.426	0.0037	1216.91	4.5000	1347.13	4.9815	28
29	330.299	0.0030	1487.33	4.5030	1646.50	4.9849	29
30	403.429	0.0025	1817.63	4.5055	2012.14	4.9876	30
35	1096.63	0.0009	4948.60	4.5125	5478.17	4.9954	35
40	2980.96	0.0003	13459.4	4.5151	14899.8	4.9983	40
45	8103.08	0.0001	36594.3	4.5161	40510.4	4.9994	45
50	22026.5	[a]	99481.4	4.5165	110127	4.9998	50
55	59874.1	[a]	270426	4.5166	299366	4.9999	55
60	162755	[a]	735103	4.5166	813769	5.0000	60

[a] Less than 0.0001.

Reprinted from Canada & White, *Capital Investment Decision Analysis for Management and Engineering* (Englewood Cliffs, NJ: Prentice-Hall, Inc., 1980).

TABLE F.4 Continuous compounding interest factors for various common values of *im*.

$$im = 25\%$$

	Discrete Flows				Continuous Flows		
	SINGLE PAYMENT		UNIFORM SERIES		UNIFORM SERIES		
	Compound Amount Factor	Present Worth Factor	Compound Amount Factor	Present Worth Factor	Compound Amount Factor	Present Worth Factor	
	To find F Given P	To find P Given F	To find F Given A	To find P Given A	To find F Given $\overline{A}$	To find P Given $\overline{A}$	
N	F/P	P/F	F/A	P/A	$F/\overline{A}$	$P/\overline{A}$	N
1	1.2840	0.7788	1.0000	0.7788	1.1361	0.8848	1
2	1.6487	0.6065	2.2840	1.3853	2.5949	1.5739	2
3	2.1170	0.4724	3.9327	1.8577	4.4680	2.1105	3
4	2.7183	0.3679	6.0497	2.2256	6.8731	2.5285	4
5	3.4903	0.2865	8.7680	2.5121	9.9614	2.8540	5
6	4.4817	0.2231	12.2584	2.7352	13.9268	3.1075	6
7	5.7546	0.1738	16.7401	2.9090	19.0184	3.3049	7
8	7.3891	0.1353	22.4947	3.0443	25.5562	3.4587	8
9	9.4877	0.1054	29.8837	3.1497	33.9509	3.5784	9
10	12.1825	0.0821	39.3715	3.2318	44.7300	3.6717	10
11	15.6426	0.0639	51.5539	3.2957	58.5705	3.7443	11
12	20.0855	0.0498	67.1966	3.3455	76.3421	3.8009	12
13	25.7903	0.0388	87.2821	3.3843	99.1614	3.8449	13
14	33.1155	0.0302	113.073	3.4145	128.462	3.8792	14
15	42.5211	0.0235	146.188	3.4380	166.084	3.9059	15
16	54.5982	0.0183	188.709	3.4563	214.393	3.9267	16
17	70.1054	0.0143	243.307	3.4706	276.422	3.9429	17
18	90.0171	0.0111	313.413	3.4817	356.068	3.9556	18
19	115.584	0.0087	403.430	3.4904	458.337	3.9654	19
20	148.413	0.0067	519.014	3.4971	589.653	3.9730	20
21	190.566	0.0052	667.427	3.5023	758.265	3.9790	21
22	244.692	0.0041	857.993	3.5064	974.768	3.9837	22
23	314.191	0.0032	1102.69	3.5096	1252.76	3.9873	23
24	403.429	0.0025	1416.88	3.5121	1609.72	3.9901	24
25	518.013	0.0019	1820.30	3.5140	2068.05	3.9923	25
26	665.142	0.0015	2338.31	3.5155	2656.57	3.9940	26
27	854.059	0.0012	3003.46	3.5167	3412.23	3.9953	27
28	1096.63	0.0009	3857.52	3.5176	4382.53	3.9964	28
29	1408.10	0.0007	4954.15	3.5183	5628.42	3.9972	29
30	1808.04	0.0006	6362.26	3.5189	7228.17	3.9978	30
35	6310.69	0.0002	22215.2	3.5203	25238.8	3.9994	35
40	22026.5	[a]	77547.5	3.5207	88101.9	3.9998	40
45	76879.9	[a]	270676	3.5208	307516	3.9999	45
50	268337	[a]	944762	3.5208	1073350	4.0000	50

[a] Less than 0.0001.

Reprinted from Canada & White, *Capital Investment Decision Analysis for Management and Engineering* (Englewood Cliffs, NJ: Prentice-Hall, Inc., 1980).

TABLE F.5 Capital tax factors for declining balance.
INCOME TAX RATE = 0.25.

The Half-Year Rule

	CCA Rate					
I	4%	6%	8%	10%	20%	30%
1%	.8010	.7868	.7789	.7739	.7631	.7593
2%	.8350	.8143	.8020	.7937	.7750	.7679
3%	.8592	.8358	.8208	.8105	.7858	.7760
4%	.8774	.8529	.8365	.8249	.7957	.7837
5%	.8915	.8669	.8498	.8373	.8048	.7908
6%	.9028	.8785	.8612	.8482	.8131	.7976
7%	.9121	.8884	.8710	.8578	.8209	.8039
8%	.9198	.8968	.8796	.8663	.8280	.8099
9%	.9263	.9041	.8872	.8739	.8347	.8156
10%	.9318	.9105	.8939	.8807	.8409	.8210
11%	.9366	.9161	.9000	.8869	.8467	.8261
12%	.9408	.9211	.9054	.8925	.8521	.8310
13%	.9446	.9256	.9102	.8976	.8572	.8356
14%	.9479	.9296	.9147	.9022	.8620	.8400
15%	.9508	.9332	.9187	.9065	.8665	.8442
16%	.9534	.9365	.9224	.9105	.8707	.8482
17%	.9558	.9395	.9258	.9141	.8747	.8520
18%	.9580	.9423	.9289	.9175	.8785	.8557
19%	.9600	.9448	.9318	.9207	.8820	.8592
20%	.9618	.9471	.9345	.9236	.8854	.8625
21%	.9635	.9493	.9370	.9264	.8886	.8657
22%	.9650	.9513	.9393	.9289	.8917	.8688
23%	.9664	.9531	.9415	.9313	.8946	.8717
24%	.9677	.9548	.9435	.9336	.8974	.8746
25%	.9690	.9565	.9455	.9357	.9000	.8773
26%	.9701	.9580	.9472	.9377	.9025	.8799
27%	.9712	.9594	.9489	.9396	.9049	.8824
28%	.9722	.9607	.9505	.9414	.9072	.8848
29%	.9731	.9620	.9520	.9431	.9094	.8872
30%	.9740	.9631	.9534	.9447	.9115	.8894

TABLE F.5 Capital tax factors for declining balance.
INCOME TAX RATE = 0.40.

The Half-Year Rule

	CCA Rate					
I	4%	6%	8%	10%	20%	30%
1%	.6816	.6588	.6462	.6382	.6209	.6148
2%	.7359	.7029	.6831	.6699	.6399	.6287
3%	.7748	.7372	.7133	.6968	.6572	.6417
4%	.8038	.7646	.7385	.7198	.6731	.6538
5%	.8265	.7870	.7597	.7397	.6876	.6653
6%	.8445	.8057	.7779	.7571	.7010	.6761
7%	.8593	.8214	.7936	.7724	.7134	.6863
8%	.8716	.8349	.8074	.7860	.7249	.6959
9%	.8820	.8466	.8195	.7982	.7355	.7050
10%	.8909	.8568	.8303	.8091	.7455	.7136
11%	.8986	.8658	.8399	.8190	.7547	.7218
12%	.9054	.8738	.8486	.8279	.7634	.7296
13%	.9113	.8809	.8564	.8361	.7715	.7370
14%	.9166	.8874	.8635	.8436	.7792	.7440
15%	.9213	.8932	.8699	.8504	.7863	.7507
16%	.9255	.8984	.8759	.8568	.7931	.7571
17%	.9293	.9032	.8813	.8626	.7995	.7632
18%	.9328	.9076	.8863	.8680	.8055	.7691
19%	.9360	.9117	.8909	.8731	.8112	.7747
20%	.9389	.9154	.8952	.8778	.8167	.7800
21%	.9416	.9188	.8992	.8822	.8218	.7851
22%	.9440	.9220	.9030	.8863	.8267	.7900
23%	.9463	.9250	.9064	.8901	.8313	.7948
24%	.9484	.9277	.9097	.8937	.8358	.7993
25%	.9503	.9303	.9127	.8971	.8400	.8036
26%	.9522	.9327	.9156	.9004	.8440	.8078
27%	.9539	.9350	.9183	.9034	.8479	.8119
28%	.9555	.9371	.9208	.9063	.8516	.8157
29%	.9570	.9391	.9232	.9090	.8551	.8195
30%	.9584	.9410	.9255	.9115	.8585	.8231

TABLE F.5 Capital tax factors for declining balance.

INCOME TAX RATE = 0.46.

The Half-Year Rule

			CCA Rate			
I	4%	6%	8%	10%	20%	30%
1%	.6338	.6077	.5931	.5839	.5641	.5570
2%	.6963	.6584	.6356	.6204	.5859	.5730
3%	.7410	.6978	.6703	.6513	.6058	.5879
4%	.7744	.7293	.6992	.6777	.6240	.6019
5%	.8004	.7551	.7237	.7006	.6408	.6151
6%	.8212	.7765	.7446	.7206	.6562	.6275
7%	.8382	.7946	.7627	.7383	.6704	.6392
8%	.8523	.8102	.7785	.7539	.6836	.6503
9%	.8643	.8236	.7925	.7679	.6959	.6608
10%	.8745	.8353	.8048	.7805	.7073	.6707
11%	.8834	.8457	.8159	.7918	.7179	.6801
12%	.8912	.8549	.8259	.8021	.7279	.6890
13%	.8980	.8631	.8348	.8115	.7372	.6975
14%	.9041	.8705	.8430	.8201	.7460	.7056
15%	.9095	.8771	.8504	.8280	.7543	.7133
16%	.9143	.8832	.8572	.8353	.7621	.7207
17%	.9187	.8887	.8635	.8420	.7694	.7277
18%	.9227	.8938	.8693	.8482	.7764	.7344
19%	.9264	.8984	.8746	.8540	.7829	.7409
20%	.9297	.9027	.8795	.8594	.7892	.7470
21%	.9328	.9066	.8841	.8645	.7951	.7529
22%	.9356	.9103	.8884	.8692	.8007	.7585
23%	.9382	.9137	.8924	.8736	.8061	.7640
24%	.9406	.9169	.8961	.8778	.8111	.7692
25%	.9429	.9199	.8996	.8817	.8160	.7742
26%	.9450	.9226	.9029	.8854	.8206	.7790
27%	.9470	.9253	.9060	.8889	.8251	.7836
28%	.9488	.9277	.9090	.8922	.8293	.7881
29%	.9505	.9300	.9117	.8953	.8333	.7924
30%	.9521	.9322	.9143	.8983	.8372	.7965

TABLE F.5 Capital tax factors for declining balance.
INCOME TAX RATE = 0.50.

The Half-Year Rule

			CCA Rate			
I	*4%*	*6%*	*8%*	*10%*	*20%*	*30%*
1%	.6020	.5735	.5578	.5477	.5262	.5185
2%	.6699	.6287	.6039	.5874	.5499	.5358
3%	.7184	.6715	.6417	.6210	.5715	.5521
4%	.7548	.7058	.6731	.6497	.5913	.5673
5%	.7831	.7338	.6996	.6746	.6095	.5816
6%	.8057	.7571	.7224	.6963	.6263	.5951
7%	.8241	.7768	.7421	.7155	.6417	.6079
8%	.8395	.7937	.7593	.7325	.6561	.6199
9%	.8525	.8083	.7744	.7477	.6694	.6313
10%	.8636	.8210	.7879	.7614	.6818	.6420
11%	.8733	.8323	.7999	.7737	.6934	.6523
12%	.8817	.8423	.8107	.7849	.7042	.6620
13%	.8891	.8512	.8205	.7951	.7144	.6712
14%	.8957	.8592	.8293	.8045	.7239	.6800
15%	.9016	.8665	.8374	.8130	.7329	.6884
16%	.9069	.8730	.8448	.8210	.7414	.6964
17%	.9117	.8790	.8516	.8283	.7494	.7040
18%	.9160	.8845	.8579	.8350	.7569	.7113
19%	.9200	.8896	.8637	.8414	.7641	.7183
20%	.9236	.8942	.8690	.8472	.7708	.7250
21%	.9269	.8985	.8740	.8527	.7773	.7314
22%	.9300	.9025	.8787	.8578	.7834	.7375
23%	.9329	.9062	.8830	.8627	.7892	.7434
24%	.9355	.9097	.8871	.8672	.7947	.7491
25%	.9379	.9129	.8909	.8714	.8000	.7545
26%	.9402	.9159	.8945	.8754	.8050	.7598
27%	.9423	.9188	.8979	.8792	.8099	.7648
28%	.9443	.9214	.9010	.8828	.8145	.7697
29%	.9462	.9239	.9040	.8862	.8189	.7743
30%	.9480	.9263	.9069	.8894	.8231	.7788

TABLE F.5 Capital tax factors for declining balance.
INCOME TAX RATE = 0.25.

			Rate			
I	*4%*	*6%*	*8%*	*10%*	*20%*	*30%*
1%	.8000	.7857	.7778	.7727	.7619	.7581
2%	.8333	.8125	.8000	.7917	.7727	.7656
3%	.8571	.8333	.8182	.8077	.7826	.7727
4%	.8750	.8500	.8333	.8214	.7917	.7794
5%	.8889	.8636	.8462	.8333	.8000	.7857
6%	.9000	.8750	.8571	.8437	.8077	.7917
7%	.9091	.8846	.8667	.8529	.8148	.7973
8%	.9167	.8929	.8750	.8611	.8214	.8026
9%	.9231	.9000	.8824	.8684	.8276	.8077
10%	.9286	.9062	.8889	.8750	.8333	.8125
11%	.9333	.9118	.8947	.8810	.8387	.8171
12%	.9375	.9167	.9000	.8864	.8438	.8214
13%	.9412	.9211	.9048	.8913	.8485	.8256
14%	.9444	.9250	.9091	.8958	.8529	.8295
15%	.9474	.9286	.9130	.9000	.8571	.8333
16%	.9500	.9318	.9167	.9038	.8611	.8370
17%	.9524	.9348	.9200	.9074	.8649	.8404
18%	.9545	.9375	.9231	.9107	.8684	.8437
19%	.9565	.9400	.9259	.9138	.8718	.8469
20%	.9583	.9423	.9286	.9167	.8750	.8500
21%	.9600	.9444	.9310	.9194	.8780	.8529
22%	.9615	.9464	.9333	.9219	.8810	.8558
23%	.9630	.9483	.9355	.9242	.8837	.8585
24%	.9643	.9500	.9375	.9265	.8864	.8611
25%	.9655	.9516	.9394	.9286	.8889	.8636
26%	.9667	.9531	.9412	.9306	.8913	.8661
27%	.9677	.9545	.9429	.9324	.8936	.8684
28%	.9688	.9559	.9444	.9342	.8958	.8707
29%	.9697	.9571	.9459	.9359	.8980	.8729
30%	.9706	.9583	.9474	.9375	.9000	.8750

TABLE F.5 Capital tax factors for declining balance.
INCOME TAX RATE = 0.40.

			Rate			
I	*4%*	*6%*	*8%*	*10%*	*20%*	*30%*
1%	.6800	.6571	.6444	.6364	.6190	.6129
2%	.7333	.7000	.6800	.6667	.6364	.6250
3%	.7714	.7333	.7091	.6923	.6522	.6364
4%	.8000	.7600	.7333	.7143	.6667	.6471
5%	.8222	.7818	.7538	.7333	.6800	.6571
6%	.8400	.8000	.7714	.7500	.6923	.6667
7%	.8545	.8154	.7867	.7647	.7037	.6757
8%	.8667	.8286	.8000	.7778	.7143	.6842
9%	.8769	.8400	.8118	.7895	.7241	.6923
10%	.8857	.8500	.8222	.8000	.7333	.7000
11%	.8933	.8588	.8316	.8095	.7419	.7073
12%	.9000	.8667	.8400	.8182	.7500	.7143
13%	.9059	.8737	.8476	.8261	.7576	.7209
14%	.9111	.8800	.8545	.8333	.7647	.7273
15%	.9158	.8857	.8609	.8400	.7714	.7333
16%	.9200	.8909	.8667	.8462	.7778	.7391
17%	.9238	.8957	.8720	.8519	.7838	.7447
18%	.9273	.9000	.8769	.8571	.7895	.7500
19%	.9304	.9040	.8815	.8621	.7949	.7551
20%	.9333	.9077	.8857	.8667	.8000	.7600
21%	.9360	.9111	.8897	.8710	.8049	.7647
22%	.9385	.9143	.8933	.8750	.8095	.7692
23%	.9407	.9172	.8968	.8788	.8140	.7736
24%	.9429	.9200	.9000	.8824	.8182	.7778
25%	.9448	.9226	.9030	.8857	.8222	.7818
26%	.9467	.9250	.9059	.8889	.8261	.7857
27%	.9484	.9273	.9086	.8919	.8298	.7895
28%	.9500	.9294	.9111	.8947	.8333	.7931
29%	.9515	.9314	.9135	.8974	.8367	.7966
30%	.9529	.9333	.9158	.9000	.8400	.8000

TABLE F.5 Capital tax factors for declining balance.
INCOME TAX RATE = 0.46.

			Rate			
I	*4%*	*6%*	*8%*	*10%*	*20%*	*30%*
1%	.6320	.6057	.5911	.5818	.5619	.5548
2%	.6933	.6550	.6320	.6167	.5818	.5687
3%	.7371	.6933	.6655	.6462	.6000	.5818
4%	.7700	.7240	.6933	.6714	.6167	.5941
5%	.7956	.7491	.7169	.6933	.6320	.6057
6%	.8160	.7700	.7371	.7125	.6462	.6167
7%	.8327	.7877	.7547	.7294	.6593	.6270
8%	.8467	.8029	.7700	.7444	.6714	.6368
9%	.8585	.8160	.7835	.7579	.6828	.6462
10%	.8686	.8275	.7956	.7700	.6933	.6550
11%	.8773	.8376	.8063	.7810	.7032	.6634
12%	.8850	.8467	.8160	.7090	.7125	.6714
13%	.8918	.8547	.8248	.8000	.7212	.6791
14%	.8978	.8620	.8327	.8083	.7294	.6864
15%	.9032	.8686	.8400	.8160	.7371	.6933
16%	.9080	.8745	.8467	.8231	.7444	.7000
17%	.9124	.8800	.8528	.8296	.7514	.7064
18%	.9164	.8850	.8585	.8357	.7579	.7125
19%	.9200	.8896	.8637	.8414	.7641	.7184
20%	.9233	.8938	.8686	.8467	.7700	.7240
21%	.9264	.8978	.8731	.8516	.7756	.7294
22%	.9292	.9014	.8773	.8562	.7810	.7346
23%	.9319	.9048	.8813	.8606	.7860	.7396
24%	.9343	.9080	.8850	.8647	.7909	.7444
25%	.9366	.9110	.8885	.8686	.7956	.7491
26%	.9387	.9137	.8918	.8722	.8000	.7536
27%	.9406	.9164	.8949	.8757	.8043	.7579
28%	.9425	.9188	.8978	.8789	.8083	.7621
29%	.9442	.9211	.9005	.8821	.8122	.7661
30%	.9459	.9233	.9032	.8850	.8160	.7700

TABLE F.5 Capital tax factors for declining balance.
INCOME TAX RATE = 0.50.

			Rate			
I	*4%*	*6%*	*8%*	*10%*	*20%*	*30%*
1%	.6000	.5714	.5556	.5455	.5238	.5161
2%	.6667	.6250	.6000	.5833	.5455	.5312
3%	.7143	.6667	.6364	.6154	.5652	.5455
4%	.7500	.7000	.6667	.6429	.5833	.5588
5%	.7778	.7273	.6923	.6667	.6000	.5714
6%	.8000	.7500	.7143	.6875	.6154	.5833
7%	.8182	.7692	.7333	.7059	.6296	.5946
8%	.8333	.7857	.7500	.7222	.6429	.6053
9%	.8462	.8000	.7647	.7368	.6552	.6154
10%	.8571	.8125	.7778	.7500	.6667	.6250
11%	.8667	.8235	.7895	.7619	.6774	.6341
12%	.8750	.8333	.8000	.7727	.6875	.6429
13%	.8824	.8421	.8095	.7826	.6970	.6512
14%	.8889	.8500	.8182	.7917	.7059	.6591
15%	.8947	.8571	.8261	.8000	.7143	.6667
16%	.9000	.8636	.8333	.8077	.7222	.6739
17%	.9048	.8696	.8400	.8148	.7297	.6809
18%	.9091	.8750	.8462	.8214	.7368	.6875
19%	.9130	.8800	.8519	.8276	.7436	.6939
20%	.9167	.8846	.8571	.8333	.7500	.7000
21%	.9200	.8889	.8621	.8387	.7561	.7059
22%	.9231	.8929	.8667	.8437	.7619	.7115
23%	.9259	.8966	.8710	.8485	.7674	.7170
24%	.9286	.9000	.8750	.8529	.7727	.7222
25%	.9310	.9032	.8788	.8571	.7778	.7273
26%	.9333	.9062	.8824	.8611	.7826	.7321
27%	.9355	.9091	.8857	.8649	.7872	.7368
28%	.9375	.9118	.8889	.8684	.7917	.7414
29%	.9394	.9143	.8919	.8718	.7959	.7458
30%	.9412	.9167	.8947	.8750	.8000	.7500

TABLE F.6 1% Interest factors for discrete compounding.

	SINGLE PAYMENT		EQUAL PAYMENT SERIES				Uniform gradient-series factor
	Compound-amount factor	Present-worth factor	Compound-amount factor	Sinking-fund factor	Present-worth factor	Capital-recovery factor	
n	To find F Given P $F/P, i, n$	To find P Given F $P/F, i, n$	To find F Given A $F/A, i, n$	To find A Given F $A/F, i, n$	To find P Given A $P/A, i, n$	To find A Given P $A/P, i, n$	To find A Given G $A/G, i, n$
1	1.010	0.9901	1.000	1.0000	0.9901	1.0100	0.0000
2	1.020	0.9803	2.010	0.4975	1.9704	0.5075	0.4975
3	1.030	0.9706	3.030	0.3300	2.9410	0.3400	0.9934
4	1.041	0.9610	4.060	0.2463	3.9020	0.2563	1.4876
5	1.051	0.9515	5.101	0.1960	4.8534	0.2060	1.9801
6	1.062	0.9421	6.152	0.1626	5.7955	0.1726	2.4710
7	1.072	0.9327	7.214	0.1386	6.7282	0.1486	2.9602
8	1.083	0.9235	8.286	0.1207	7.6517	0.1307	3.4478
9	1.094	0.9143	9.369	0.1068	8.5660	0.1168	3.9337
10	1.105	0.9053	10.462	0.0956	9.4713	0.1056	4.4179
11	1.116	0.8963	11.567	0.0865	10.3676	0.0965	4.9005
12	1.127	0.8875	12.683	0.0789	11.2551	0.0889	5.3815
13	1.138	0.8787	13.809	0.0724	12.1338	0.0824	5.8607
14	1.149	0.8700	14.947	0.0669	13.0037	0.0769	6.3384
15	1.161	0.8614	16.097	0.0621	13.8651	0.0721	6.8143
16	1.173	0.8528	17.258	0.0580	14.7179	0.0680	7.2887
17	1.184	0.8444	18.430	0.0543	15.5623	0.0643	7.7613
18	1.196	0.8360	19.615	0.0510	16.3983	0.0610	8.2323
19	1.208	0.8277	20.811	0.0481	17.2260	0.0581	8.7017
20	1.220	0.8196	22.019	0.0454	18.0456	0.0554	9.1694
21	1.232	0.8114	23.239	0.0430	18.8570	0.0530	9.6354
22	1.245	0.8034	24.472	0.0409	19.6604	0.0509	10.0998
23	1.257	0.7955	25.716	0.0389	20.4558	0.0489	10.5626
24	1.270	0.7876	26.973	0.0371	21.2434	0.0471	11.0237
25	1.282	0.7798	28.243	0.0354	22.0232	0.0454	11.4831
26	1.295	0.7721	29.526	0.0339	22.7952	0.0439	11.9409
27	1.308	0.7644	30.821	0.0325	23.5596	0.0425	12.3971
28	1.321	0.7568	32.129	0.0311	24.3165	0.0411	12.8516
29	1.335	0.7494	33.450	0.0299	25.0658	0.0399	13.3045
30	1.348	0.7419	34.785	0.0288	25.8077	0.0388	13.7557
31	1.361	0.7346	36.133	0.0277	26.5423	0.0377	14.2052
32	1.375	0.7273	37.494	0.0267	27.2696	0.0367	14.6532
33	1.389	0.7201	38.869	0.0257	27.9897	0.0357	15.0995
34	1.403	0.7130	40.258	0.0248	28.7027	0.0348	15.5441
35	1.417	0.7059	41.660	0.0240	29.4086	0.0340	15.9871
40	1.489	0.6717	48.886	0.0205	32.8347	0.0305	18.1776
45	1.565	0.6391	56.481	0.0177	36.0945	0.0277	20.3273
50	1.645	0.6080	64.463	0.0155	39.1961	0.0255	22.4363
55	1.729	0.5785	72.852	0.0137	42.1472	0.0237	24.5049
60	1.817	0.5505	81.670	0.0123	44.9550	0.0223	26.5333
n	$(1+i)^n$	$\dfrac{1}{(1+i)^n}$	$\dfrac{(1+i)^n-1}{i}$	$\dfrac{i}{(1+i)^n-1}$	$\dfrac{(1+i)^n-1}{i(1+i)^n}$	$\dfrac{i(1+i)^n}{(1+i)^n-1}$	$\dfrac{1}{i}-\dfrac{n}{(1+i)^n-1}$

TABLE F.6 2% Interest factors for discrete compounding.

	SINGLE PAYMENT		EQUAL PAYMENT SERIES				Uniform gradient-series factor
	Compound-amount factor	Present-worth factor	Compound-amount factor	Sinking-fund factor	Present-worth factor	Capital-recovery factor	
n	To find F Given P $F/P, i, n$	To find P Given F $P/F, i, n$	To find F Given A $F/A, i, n$	To find A Given F $A/F, i, n$	To find P Given A $P/A, i, n$	To find A Given P $A/P, i, n$	To find A Given G $A/G, i, n$
1	1.020	0.9804	1.000	1.0000	0.9804	1.0200	0.0000
2	1.040	0.9612	2.020	0.4951	1.9416	0.5151	0.4951
3	1.061	0.9423	3.060	0.3268	2.8839	0.3468	0.9868
4	1.082	0.9239	4.122	0.2426	3.8077	0.2626	1.4753
5	1.104	0.9057	5.204	0.1922	4.7135	0.2122	1.9604
6	1.126	0.8880	6.308	0.1585	5.6014	0.1785	2.4423
7	1.149	0.8706	7.434	0.1345	6.4720	0.1545	2.9208
8	1.172	0.8535	8.583	0.1165	7.3255	0.1365	3.3961
9	1.195	0.8368	9.755	0.1025	8.1622	0.1225	3.8681
10	1.219	0.8204	10.950	0.0913	8.9826	0.1113	4.3367
11	1.243	0.8043	12.169	0.0822	9.7869	0.1022	4.8021
12	1.268	0.7885	13.412	0.0746	10.5754	0.0946	5.2643
13	1.294	0.7730	14.680	0.0681	11.3484	0.0881	5.7231
14	1.319	0.7579	15.974	0.0626	12.1063	0.0826	6.1786
15	1.346	0.7430	17.293	0.0578	12.8493	0.0778	6.6309
16	1.373	0.7285	18.639	0.0537	13.5777	0.0737	7.0799
17	1.400	0.7142	20.012	0.0500	14.2919	0.0700	7.5256
18	1.428	0.7002	21.412	0.0467	14.9920	0.0667	7.9681
19	1.457	0.6864	22.841	0.0438	15.6785	0.0638	8.4073
20	1.486	0.6730	24.297	0.0412	16.3514	0.0612	8.8433
21	1.516	0.6598	25.783	0.0388	17.0112	0.0588	9.2760
22	1.546	0.6468	27.299	0.0366	17.6581	0.0566	9.7055
23	1.577	0.6342	28.845	0.0347	18.2922	0.0547	10.1317
24	1.608	0.6217	30.422	0.0329	18.9139	0.0529	10.5547
25	1.641	0.6095	32.030	0.0312	19.5235	0.0512	10.9745
26	1.673	0.5976	33.671	0.0297	20.1210	0.0497	11.3910
27	1.707	0.5859	35.344	0.0283	20.7069	0.0483	11.8043
28	1.741	0.5744	37.051	0.0270	21.2813	0.0470	12.2145
29	1.776	0.5631	38.792	0.0258	21.8444	0.0458	12.6214
30	1.811	0.5521	40.568	0.0247	22.3965	0.0447	13.0251
31	1.848	0.5413	42.379	0.0236	22.9377	0.0436	13.4257
32	1.885	0.5306	44.227	0.0226	23.4683	0.0426	13.8230
33	1.922	0.5202	46.112	0.0217	23.9886	0.0417	14.2172
34	1.961	0.5100	48.034	0.0208	24.4986	0.0408	14.6083
35	2.000	0.5000	49.994	0.0200	24.9986	0.0400	14.9961
40	2.208	0.4529	60.402	0.0166	27.3555	0.0366	16.8885
45	2.438	0.4102	71.893	0.0139	29.4902	0.0339	18.7034
50	2.692	0.3715	84.579	0.0118	31.4236	0.0318	20.4420
55	2.972	0.3365	98.587	0.0102	33.1748	0.0302	22.1057
60	3.281	0.3048	114.052	0.0088	34.7609	0.0288	23.6961
n	$(1+i)^n$	$\dfrac{1}{(1+i)^n}$	$\dfrac{(1+i)^n-1}{i}$	$\dfrac{i}{(1+i)^n-1}$	$\dfrac{(1+i)^n-1}{i(1+i)^n}$	$\dfrac{i(1+i)^n}{(1+i)^n-1}$	$\dfrac{1}{i}-\dfrac{n}{(1+i)^n-1}$

TABLE F.6 3% Interest factors for discrete compounding.

	SINGLE PAYMENT		EQUAL PAYMENT SERIES				Uniform gradient-series factor
	Compound-amount factor	Present-worth factor	Compound-amount factor	Sinking-fund factor	Present-worth factor	Capital-recovery factor	
	To find F Given P	To find P Given F	To find F Given A	To find A Given F	To find P Given A	To find A Given P	To find A Given G
n	$F/P, i, n$	$P/F, i, n$	$F/A, i, n$	$A/F, i, n$	$P/A, i, n$	$A/P, i, n$	$A/G, i, n$
1	1.030	0.9709	1.000	1.0000	0.9709	1.0300	0.0000
2	1.061	0.9426	2.030	0.4926	1.9135	0.5226	0.4926
3	1.093	0.9152	3.091	0.3235	2.8286	0.3535	0.9803
4	1.126	0.8885	4.184	0.2390	3.7171	0.2690	1.4631
5	1.159	0.8626	5.309	0.1884	4.5797	0.2184	1.9409
6	1.194	0.8375	6.468	0.1546	5.4172	0.1846	2.4138
7	1.230	0.8131	7.662	0.1305	6.2303	0.1605	2.8819
8	1.267	0.7894	8.892	0.1125	7.0197	0.1425	3.3450
9	1.305	0.7664	10.159	0.0984	7.7861	0.1284	3.8032
10	1.344	0.7441	11.464	0.0872	8.5302	0.1172	4.2565
11	1.384	0.7224	12.808	0.0781	9.2526	0.1081	4.7049
12	1.426	0.7014	14.192	0.0705	9.9540	0.1005	5.1485
13	1.469	0.6810	15.618	0.0640	10.6350	0.0940	5.5872
14	1.513	0.6611	17.086	0.0585	11.2961	0.0885	6.0211
15	1.558	0.6419	18.599	0.0538	11.9379	0.0838	6.4501
16	1.605	0.6232	20.157	0.0496	12.5611	0.0796	6.8742
17	1.653	0.6050	21.762	0.0460	13.1661	0.0760	7.2936
18	1.702	0.5874	23.414	0.0427	13.7535	0.0727	7.7081
19	1.754	0.5703	25.117	0.0398	14.3238	0.0698	8.1179
20	1.806	0.5537	26.870	0.0372	14.8775	0.0672	8.5229
21	1.860	0.5376	28.676	0.0349	15.4150	0.0649	8.9231
22	1.916	0.5219	30.537	0.0328	15.9369	0.0628	9.3186
23	1.974	0.5067	32.453	0.0308	16.4436	0.0608	9.7094
24	2.033	0.4919	34.426	0.0291	16.9356	0.0591	10.0954
25	2.094	0.4776	36.459	0.0274	17.4132	0.0574	10.4768
26	2.157	0.4637	38.553	0.0259	17.8769	0.0559	10.8535
27	2.221	0.4502	40.710	0.0246	18.3270	0.0546	11.2256
28	2.288	0.4371	42.931	0.0233	18.7641	0.0533	11.5930
29	2.357	0.4244	45.219	0.0221	19.1885	0.0521	11.9558
30	2.427	0.4120	47.575	0.0210	19.6005	0.0510	12.3141
31	2.500	0.4000	50.003	0.0200	20.0004	0.0500	12.6678
32	2.575	0.3883	52.503	0.0191	20.3888	0.0491	13.0169
33	2.652	0.3770	55.078	0.0182	20.7658	0.0482	13.3616
34	2.732	0.3661	57.730	0.0173	21.1318	0.0473	13.7018
35	2.814	0.3554	60.462	0.0165	21.4872	0.0465	14.0375
40	3.262	0.3066	75.401	0.0133	23.1148	0.0433	15.6502
45	3.782	0.2644	92.720	0.0108	24.5187	0.0408	17.1556
50	4.384	0.2281	112.797	0.0089	25.7289	0.0389	18.5575
55	5.082	0.1968	136.072	0.0074	26.7744	0.0374	19.8600
60	5.892	0.1697	163.053	0.0061	27.6756	0.0361	21.0674
n	$(1+i)^n$	$\dfrac{1}{(1+i)^n}$	$\dfrac{(1+i)^n-1}{i}$	$\dfrac{i}{(1+i)^n-1}$	$\dfrac{(1+i)^n-1}{i(1+i)^n}$	$\dfrac{i(1+i)^n}{(1+i)^n-1}$	$\dfrac{1}{i}-\dfrac{n}{(1+i)^n-1}$

TABLE F.6 4% Interest factors for discrete compounding.

	SINGLE PAYMENT		EQUAL PAYMENT SERIES				Uniform gradient-series factor
	Compound-amount factor	Present-worth factor	Compound-amount factor	Sinking-fund factor	Present-worth factor	Capital-recovery factor	
n	To find F Given P $F/P, i, n$	To find P Given F $P/F, i, n$	To find F Given A $F/A, i, n$	To find A Given F $A/F, i, n$	To find P Given A $P/A, i, n$	To find A Given P $A/P, i, n$	To find A Given G $A/G, i, n$
1	1.040	0.9615	1.000	1.0000	0.9615	1.0400	0.0000
2	1.082	0.9246	2.040	0.4902	1.8861	0.5302	0.4902
3	1.125	0.8890	3.122	0.3204	2.7751	0.3604	0.9739
4	1.170	0.8548	4.246	0.2355	3.6299	0.2755	1.4510
5	1.217	0.8219	5.416	0.1846	4.4518	0.2246	1.9216
6	1.265	0.7903	6.633	0.1508	5.2421	0.1908	2.3857
7	1.316	0.7599	7.898	0.1266	6.0021	0.1666	2.8433
8	1.369	0.7307	9.214	0.1085	6.7328	0.1485	3.2944
9	1.423	0.7026	10.583	0.0945	7.4353	0.1345	3.7391
10	1.480	0.6756	12.006	0.0833	8.1109	0.1233	4.1773
11	1.539	0.6496	13.486	0.0742	8.7605	0.1142	4.6090
12	1.601	0.6246	15.026	0.0666	9.3851	0.1066	5.0344
13	1.665	0.6006	16.627	0.0602	9.9857	0.1002	5.4533
14	1.732	0.5775	18.292	0.0547	10.5631	0.0947	5.8659
15	1.801	0.5553	20.024	0.0500	11.1184	0.0900	6.2721
16	1.873	0.5339	21.825	0.0458	11.6523	0.0858	6.6720
17	1.948	0.5134	23.698	0.0422	12.1657	0.0822	7.0656
18	2.026	0.4936	25.645	0.0390	12.6593	0.0790	7.4530
19	2.107	0.4747	27.671	0.0361	13.1339	0.0761	7.8342
20	2.191	0.4564	29.778	0.0336	13.5903	0.0736	8.2091
21	2.279	0.4388	31.969	0.0313	14.0292	0.0713	8.5780
22	2.370	0.4220	34.248	0.0292	14.4511	0.0692	8.9407
23	2.465	0.4057	36.618	0.0273	14.8569	0.0673	9.2973
24	2.563	0.3901	39.083	0.0256	15.2470	0.0656	9.6479
25	2.666	0.3751	41.646	0.0240	15.6221	0.0640	9.9925
26	2.772	0.3607	44.312	0.0226	15.9828	0.0626	10.3312
27	2.883	0.3468	47.084	0.0212	16.3296	0.0612	10.6640
28	2.999	0.3335	49.968	0.0200	16.6631	0.0600	10.9909
29	3.119	0.3207	52.966	0.0189	16.9837	0.0589	11.3121
30	3.243	0.3083	56.085	0.0178	17.2920	0.0578	11.6274
31	3.373	0.2965	59.328	0.0169	17.5885	0.0569	11.9371
32	3.508	0.2851	62.701	0.0160	17.8736	0.0560	12.2411
33	3.648	0.2741	66.210	0.0151	18.1477	0.0551	12.5396
34	3.794	0.2636	69.858	0.0143	18.4112	0.0543	12.8325
35	3.946	0.2534	73.652	0.0136	18.6646	0.0536	13.1199
40	4.801	0.2083	95.026	0.0105	19.7928	0.0505	14.4765
45	5.841	0.1712	121.029	0.0083	20.7200	0.0483	15.7047
50	7.107	0.1407	152.667	0.0066	21.4822	0.0466	16.8123
55	8.646	0.1157	191.159	0.0052	22.1086	0.0452	17.8070
60	10.520	0.0951	237.991	0.0042	22.6235	0.0442	18.6972
n	$(1+i)^n$	$\dfrac{1}{(1+i)^n}$	$\dfrac{(1+i)^n-1}{i}$	$\dfrac{i}{(1+i)^n-1}$	$\dfrac{(1+i)^n-1}{i(1+i)^n}$	$\dfrac{i(1+i)^n}{(1+i)^n-1}$	$\dfrac{1}{i}-\dfrac{n}{(1+i)^n-1}$

TABLE F.6 5% Interest factors for discrete compounding.

	SINGLE PAYMENT		EQUAL PAYMENT SERIES				Uniform gradient-series factor
	Compound-amount factor	Present-worth factor	Compound-amount factor	Sinking-fund factor	Present-worth factor	Capital-recovery factor	
n	To find F Given P $F/P, i, n$	To find P Given F $P/F, i, n$	To find F Given A $F/A, i, n$	To find A Given F $A/F, i, n$	To find P Given A $P/A, i, n$	To find A Given P $A/P, i, n$	To find A Given G $A/G, i, n$
1	1.050	0.9524	1.000	1.0000	0.9524	1.0500	0.0000
2	1.103	0.9070	2.050	0.4878	1.8594	0.5378	0.4878
3	1.158	0.8638	3.153	0.3172	2.7233	0.3672	0.9675
4	1.216	0.8277	4.310	0.2320	3.5460	0.2820	1.4391
5	1.276	0.7835	5.526	0.1810	4.3295	0.2310	1.9025
6	1.340	0.7462	6.802	0.1470	5.0757	0.1970	2.3579
7	1.407	0.7107	8.142	0.1228	5.7864	0.1728	2.8052
8	1.477	0.6768	9.549	0.1047	6.4632	0.1547	3.2445
9	1.551	0.6446	11.027	0.0907	7.1078	0.1407	3.6758
10	1.629	0.6139	12.587	0.0795	7.7217	0.1295	4.0991
11	1.710	0.5847	14.207	0.0704	8.3064	0.1204	4.5145
12	1.796	0.5568	15.917	0.0628	8.8633	0.1128	4.9219
13	1.886	0.5303	17.713	0.0565	9.3936	0.1065	5.3215
14	1.980	0.5051	19.599	0.0510	9.8987	0.1010	5.7133
15	2.079	0.4810	21.579	0.0464	10.3797	0.0964	6.0973
16	2.183	0.4581	23.658	0.0423	10.8378	0.0923	6.4736
17	2.292	0.4363	25.840	0.0387	11.2741	0.0887	6.8423
18	2.407	0.4155	28.132	0.0356	11.6896	0.0856	7.2034
19	2.527	0.3957	30.539	0.0328	12.0853	0.0828	7.5569
20	2.653	0.3769	33.066	0.0303	12.4622	0.0803	7.9030
21	2.786	0.3590	35.719	0.0280	12.8212	0.0780	8.2416
22	2.925	0.3419	38.505	0.0260	13.1630	0.0760	8.5730
23	3.072	0.3256	41.430	0.0241	13.4886	0.0741	8.8971
24	3.225	0.3101	44.502	0.0225	13.7987	0.0725	9.2140
25	3.386	0.2953	47.727	0.0210	14.0940	0.0710	9.5238
26	3.556	0.2813	51.113	0.0196	14.3752	0.0696	9.8266
27	3.733	0.2679	54.669	0.0183	14.6430	0.0683	10.1224
28	3.920	0.2551	58.403	0.0171	14.8981	0.0671	10.4114
29	4.116	0.2430	62.323	0.0161	15.1411	0.0661	10.6939
30	4.322	0.2314	66.439	0.0151	15.3725	0.0651	10.9691
31	4.538	0.2204	70.761	0.0141	15.5928	0.0641	11.2381
32	4.765	0.2099	75.299	0.0133	15.8027	0.0633	11.5005
33	5.003	0.1999	80.064	0.0125	16.0026	0.0625	11.7566
34	5.253	0.1904	85.067	0.0118	16.1929	0.0618	12.0063
35	5.516	0.1813	90.320	0.0111	16.3742	0.0611	12.2498
40	7.040	0.1421	120.800	0.0083	17.1591	0.0583	13.3775
45	8.985	0.1113	159.700	0.0063	17.7741	0.0563	14.3644
50	11.467	0.0872	209.348	0.0048	18.2559	0.0548	15.2233
55	14.636	0.0683	272.713	0.0037	18.6335	0.0537	15.9665
60	18.679	0.0535	353.584	0.0028	18.9293	0.0528	16.6062
n	$(1+i)^n$	$\dfrac{1}{(1+i)^n}$	$\dfrac{(1+i)^n-1}{i}$	$\dfrac{i}{(1+i)^n-1}$	$\dfrac{(1+i)^n-1}{i(1+i)^n}$	$\dfrac{i(1+i)^n}{(1+i)^n-1}$	$\dfrac{1}{i}-\dfrac{n}{(1+i)^n-1}$

TABLE F.6 6% Interest factors for discrete compounding.

	SINGLE PAYMENT		EQUAL PAYMENT SERIES				Uniform gradient-series factor
	Compound-amount factor	Present-worth factor	Compound-amount factor	Sinking-fund factor	Present-worth factor	Capital-recovery factor	
n	To find F Given P $F/P, i, n$	To find P Given F $P/F, i, n$	To find F Given A $F/A, i, n$	To find A Given F $A/F, i, n$	To find P Given A $P/A, i, n$	To find A Given P $A/P, i, n$	To find A Given G $A/G, i, n$
1	1.060	0.9434	1.000	1.0000	0.9434	1.0600	0.0000
2	1.124	0.8900	2.060	0.4854	1.8334	0.5454	0.4854
3	1.191	0.8396	3.184	0.3141	2.6730	0.3741	0.9612
4	1.262	0.7921	4.375	0.2286	3.4651	0.2886	1.4272
5	1.338	0.7473	5.637	0.1774	4.2124	0.2374	1.8836
6	1.419	0.7050	6.975	0.1434	4.9173	0.2034	2.3304
7	1.504	0.6651	8.394	0.1191	5.5824	0.1791	2.7676
8	1.594	0.6274	9.897	0.1010	6.2098	0.1610	3.1952
9	1.689	0.5919	11.491	0.0870	6.8017	0.1470	3.6133
10	1.791	0.5584	13.181	0.0759	7.3601	0.1359	4.0220
11	1.898	0.5268	14.972	0.0668	7.8869	0.1268	4.4213
12	2.012	0.4970	16.870	0.0593	8.3839	0.1193	4.8113
13	2.133	0.4688	18.882	0.0530	8.8527	0.1130	5.1920
14	2.261	0.4423	21.015	0.0476	9.2950	0.1076	5.5635
15	2.397	0.4173	23.276	0.0430	9.7123	0.1030	5.9260
16	2.540	0.3937	25.673	0.0390	10.1059	0.0990	6.2794
17	2.693	0.3714	28.213	0.0355	10.4773	0.0955	6.6240
18	2.854	0.3504	30.906	0.0324	10.8276	0.0924	6.9597
19	3.026	0.3305	33.760	0.0296	11.1581	0.0896	7.2867
20	3.207	0.3118	36.786	0.0272	11.4699	0.0872	7.6052
21	3.400	0.2942	39.993	0.0250	11.7641	0.0850	7.9151
22	3.604	0.2775	43.392	0.0231	12.0416	0.0831	8.2166
23	3.820	0.2618	46.996	0.0213	12.3034	0.0813	8.5099
24	4.049	0.2470	50.816	0.0197	12.5504	0.0797	8.7951
25	4.292	0.2330	54.865	0.0182	12.7834	0.0782	9.0722
26	4.549	0.2198	59.156	0.0169	13.0032	0.0769	9.3415
27	4.822	0.2074	63.706	0.0157	13.2105	0.0757	9.6030
28	5.112	0.1956	68.528	0.0146	13.4062	0.0746	9.8568
29	5.418	0.1846	73.640	0.0136	13.5907	0.0736	10.1032
30	5.744	0.1741	79.058	0.0127	13.7648	0.0727	10.3422
31	6.088	0.1643	84.802	0.0118	13.9291	0.0718	10.5740
32	6.453	0.1550	90.890	0.0110	14.0841	0.0710	10.7988
33	6.841	0.1462	97.343	0.0103	14.2302	0.0703	11.0166
34	7.251	0.1379	104.184	0.0096	14.3682	0.0696	11.2276
35	7.686	0.1301	111.435	0.0090	14.4983	0.0690	11.4319
40	10.286	0.0972	154.762	0.0065	15.0463	0.0665	12.3590
45	13.765	0.0727	212.744	0.0047	15.4558	0.0647	13.1413
50	18.420	0.0543	290.336	0.0035	15.7619	0.0635	13.7964
55	24.650	0.0406	394.172	0.0025	15.9906	0.0625	14.3411
60	32.988	0.0303	533.128	0.0019	16.1614	0.0619	14.7910
n	$(1+i)^n$	$\dfrac{1}{(1+i)^n}$	$\dfrac{(1+i)^n-1}{i}$	$\dfrac{i}{(1+i)^n-1}$	$\dfrac{(1+i)^n-1}{i(1+i)^n}$	$\dfrac{i(1+i)^n}{(1+i)^n-1}$	$\dfrac{1}{i}-\dfrac{n}{(1+i)^n-1}$

TABLE F.6 7% Interest factors for discrete compounding.

	SINGLE PAYMENT		EQUAL PAYMENT SERIES				Uniform gradient-series factor
	Compound-amount factor	Present-worth factor	Compound-amount factor	Sinking-fund factor	Present-worth factor	Capital-recovery factor	
	To find F Given P	To find P Given F	To find F Given A	To find A Given F	To find P Given A	To find A Given P	To find A Given G
n	$F/P, i, n$	$P/F, i, n$	$F/A, i, n$	$A/F, i, n$	$P/A, i, n$	$A/P, i, n$	$A/G, i, n$
1	1.070	0.9346	1.000	1.0000	0.9346	1.0700	0.0000
2	1.145	0.8734	2.070	0.4831	1.8080	0.5531	0.4831
3	1.225	0.8163	3.215	0.3111	2.6243	0.3811	0.9549
4	1.311	0.7629	4.440	0.2252	3.3872	0.2952	1.4155
5	1.403	0.7130	5.751	0.1739	4.1002	0.2439	1.8650
6	1.501	0.6664	7.163	0.1398	4.7665	0.2098	2.3032
7	1.606	0.6228	8.654	0.1156	5.3893	0.1856	2.7304
8	1.718	0.5820	10.260	0.0975	5.9713	0.1675	3.1466
9	1.838	0.5439	11.978	0.0835	6.5152	0.1535	3.5517
10	1.967	0.5084	13.816	0.0724	7.0236	0.1424	3.9461
11	2.105	0.4751	15.784	0.0634	7.4987	0.1334	4.3296
12	2.252	0.4440	17.888	0.0559	7.9427	0.1259	4.7025
13	2.410	0.4150	20.141	0.0497	8.3577	0.1197	5.0649
14	2.579	0.3878	22.550	0.0444	8.7455	0.1144	5.4167
15	2.759	0.3625	25.129	0.0398	9.1079	0.1098	5.7583
16	2.952	0.3387	27.888	0.0359	9.4467	0.1059	6.0897
17	3.159	0.3166	30.840	0.0324	9.7632	0.1024	6.4110
18	3.380	0.2959	33.999	0.0294	10.0591	0.0994	6.7225
19	3.617	0.2765	37.379	0.0268	10.3356	0.0968	7.0242
20	3.870	0.2584	40.996	0.0244	10.5940	0.0944	7.3163
21	4.141	0.2415	44.865	0.0223	10.8355	0.0923	7.5990
22	4.430	0.2257	49.006	0.0204	11.0613	0.0904	7.8725
23	4.741	0.2110	53.436	0.0187	11.2722	0.0887	8.1369
24	5.072	0.1972	58.177	0.0172	11.4693	0.0872	8.3923
25	5.427	0.1843	63.249	0.0158	11.6536	0.0858	8.6391
26	5.807	0.1722	68.676	0.0146	11.8258	0.0846	8.8773
27	6.214	0.1609	74.484	0.0134	11.9867	0.0834	9.1072
28	6.649	0.1504	80.698	0.0124	12.1371	0.0824	9.3290
29	7.114	0.1406	87.347	0.0115	12.2777	0.0815	9.5427
30	7.612	0.1314	94.461	0.0106	12.4091	0.0806	9.7487
31	8.145	0.1228	102.073	0.0098	12.5318	0.0798	9.9471
32	8.715	0.1148	110.218	0.0091	12.6466	0.0791	10.1381
33	9.325	0.1072	118.933	0.0084	12.7538	0.0784	10.3219
34	9.978	0.1002	128.259	0.0078	12.8540	0.0778	10.4987
35	10.677	0.0937	138.237	0.0072	12.9477	0.0772	10.6687
40	14.974	0.0668	199.635	0.0050	13.3317	0.0750	11.4234
45	21.002	0.0476	285.749	0.0035	13.6055	0.0735	12.0360
50	29.457	0.0340	406.529	0.0025	13.8008	0.0725	12.5287
55	41.315	0.0242	575.929	0.0017	13.9399	0.0717	12.9215
60	57.946	0.0173	813.520	0.0012	14.0392	0.0712	13.2321
n	$(1+i)^n$	$\dfrac{1}{(1+i)^n}$	$\dfrac{(1+i)^n-1}{i}$	$\dfrac{i}{(1+i)^n-1}$	$\dfrac{(1+i)^n-1}{i(1+i)^n}$	$\dfrac{i(1+i)^n}{(1+i)^n-1}$	$\dfrac{1}{i}-\dfrac{n}{(1+i)^n-1}$

TABLE F.6 8% Interest factors for discrete compounding.

	SINGLE PAYMENT		EQUAL PAYMENT SERIES				Uniform gradient-series factor
	Compound-amount factor	Present-worth factor	Compound-amount factor	Sinking-fund factor	Present-worth factor	Capital-recovery factor	
	To find F Given P	To find P Given F	To find F Given A	To find A Given F	To find P Given A	To find A Given P	To find A Given G
n	$F/P, i, n$	$P/F, i, n$	$F/A, i, n$	$A/F, i, n$	$P/A, i, n$	$A/P, i, n$	$A/G, i, n$
1	1.080	0.9259	1.000	1.0000	0.9259	1.0800	0.0000
2	1.166	0.8573	2.080	0.4808	1.7833	0.5608	0.4808
3	1.260	0.7938	3.246	0.3080	2.5771	0.3880	0.9488
4	1.360	0.7350	4.506	0.2219	3.3121	0.3019	1.4040
5	1.469	0.6806	5.867	0.1705	3.9927	0.2505	1.8465
6	1.587	0.6302	7.336	0.1363	4.6229	0.2163	2.2764
7	1.714	0.5835	8.923	0.1121	5.2064	0.1921	2.6937
8	1.851	0.5403	10.637	0.0940	5.7466	0.1740	3.0985
9	1.999	0.5003	12.488	0.0801	6.2469	0.1601	3.4910
10	2.159	0.4632	14.487	0.0690	6.7101	0.1490	3.8713
11	2.332	0.4289	16.645	0.0601	7.1390	0.1401	4.2395
12	2.518	0.3971	18.977	0.0527	7.5361	0.1327	4.5958
13	2.720	0.3677	21.495	0.0465	7.9038	0.1265	4.9402
14	2.937	0.3405	24.215	0.0413	8.2442	0.1213	5.2731
15	3.172	0.3153	27.152	0.0368	8.5595	0.1168	5.5945
16	3.426	0.2919	30.324	0.0330	8.8514	0.1130	5.9046
17	3.700	0.2703	33.750	0.0296	9.1216	0.1096	6.2038
18	3.996	0.2503	37.450	0.0267	9.3719	0.1067	6.4920
19	4.316	0.2317	41.446	0.0241	9.6036	0.1041	6.7697
20	4.661	0.2146	45.762	0.0219	9.8182	0.1019	7.0370
21	5.034	0.1987	50.423	0.0198	10.0168	0.0998	7.2940
22	5.437	0.1840	55.457	0.0180	10.2008	0.0980	7.5412
23	5.871	0.1703	60.893	0.0164	10.3711	0.0964	7.7786
24	6.341	0.1577	66.765	0.0150	10.5288	0.0950	8.0066
25	6.848	0.1460	73.106	0.0137	10.6748	0.0937	8.2254
26	7.396	0.1352	79.954	0.0125	10.8100	0.0925	8.4352
27	7.988	0.1252	87.351	0.0115	10.9352	0.0915	8.6363
28	8.627	0.1159	95.339	0.0105	11.0511	0.0905	8.8289
29	9.317	0.1073	103.966	0.0096	11.1584	0.0896	9.0133
30	10.063	0.0994	113.283	0.0088	11.2578	0.0888	9.1897
31	10.868	0.0920	123.346	0.0081	11.3498	0.0881	9.3584
32	11.737	0.0852	134.214	0.0075	11.4350	0.0875	9.5197
33	12.676	0.0789	145.951	0.0069	11.5139	0.0869	9.6737
34	13.690	0.0731	158.627	0.0063	11.5869	0.0863	9.8208
35	14.785	0.0676	172.317	0.0058	11.6546	0.0858	9.9611
40	21.725	0.0460	259.057	0.0039	11.9246	0.0839	10.5699
45	31.920	0.0313	386.506	0.0026	12.1084	0.0826	11.0447
50	46.902	0.0213	573.770	0.0018	12.2335	0.0818	11.4107
55	68.914	0.0145	848.923	0.0012	12.3186	0.0812	11.6902
60	101.257	0.0099	1253.213	0.0008	12.3766	0.0808	11.9015
n	$(1+i)^n$	$\dfrac{1}{(1+i)^n}$	$\dfrac{(1+i)^n-1}{i}$	$\dfrac{i}{(1+i)^n-1}$	$\dfrac{(1+i)^n-1}{i(1+i)^n}$	$\dfrac{i(1+i)^n}{(1+i)^n-1}$	$\dfrac{1}{i}-\dfrac{n}{(1+i)^n-1}$

TABLE F.6 9% Interest factors for discrete compounding.

	SINGLE PAYMENT		EQUAL PAYMENT SERIES				Uniform gradient-series factor
	Compound-amount factor	Present-worth factor	Compound-amount factor	Sinking-fund factor	Present-worth factor	Capital-recovery factor	
n	To find F Given P $F/P, i, n$	To find P Given F $P/F, i, n$	To find F Given A $F/A, i, n$	To find A Given F $A/F, i, n$	To find P Given A $P/A, i, n$	To find A Given P $A/P, i, n$	To find A Given G $A/G, i, n$
1	1.090	0.9174	1.000	1.0000	0.9174	1.0900	0.0000
2	1.188	0.8417	2.090	0.4785	1.7591	0.5685	0.4785
3	1.295	0.7722	3.278	0.3051	2.5313	0.3951	0.9426
4	1.412	0.7084	4.573	0.2187	3.2397	0.3087	1.3925
5	1.539	0.6499	5.985	0.1671	3.8897	0.2571	1.8282
6	1.677	0.5963	7.523	0.1329	4.4859	0.2229	2.2498
7	1.828	0.5470	9.200	0.1087	5.0330	0.1987	2.6574
8	1.993	0.5019	11.208	0.0907	5.5348	0.1807	3.0512
9	2.172	0.4604	13.021	0.0768	5.9953	0.1668	3.4312
10	2.367	0.4224	15.193	0.0658	6.4177	0.1558	3.7978
11	2.580	0.3875	17.560	0.0570	6.8052	0.1470	4.1510
12	2.813	0.3555	20.141	0.0497	7.1607	0.1397	4.4910
13	3.066	0.3262	22.953	0.0436	7.4869	0.1336	4.8182
14	3.342	0.2993	26.019	0.0384	7.7862	0.1284	5.1326
15	3.642	0.2745	29.361	0.0341	8.0607	0.1241	5.4346
16	3.970	0.2519	33.003	0.0303	8.3126	0.1203	5.7245
17	4.328	0.2311	36.974	0.0271	8.5436	0.1171	6.0024
18	4.717	0.2120	41.301	0.0242	8.7556	0.1142	6.2687
19	5.142	0.1945	46.018	0.0217	8.9501	0.1117	6.5236
20	5.604	0.1784	51.160	0.0196	9.1286	0.1096	6.7675
21	6.109	0.1637	56.765	0.0176	9.2923	0.1076	7.0006
22	6.659	0.1502	62.873	0.0159	9.4424	0.1059	7.2232
23	7.258	0.1378	69.532	0.0144	9.5802	0.1044	7.4358
24	7.911	0.1264	76.790	0.0130	9.7066	0.1030	7.6384
25	8.623	0.1160	84.701	0.0118	9.8226	0.1018	7.8316
26	9.399	0.1064	93.324	0.0107	9.9290	0.1007	8.0156
27	10.245	0.0976	102.723	0.0097	10.0266	0.0997	8.1906
28	11.167	0.0896	112.968	0.0089	10.1161	0.0989	8.3572
29	12.172	0.0822	124.135	0.0081	10.1983	0.0981	8.5154
30	13.268	0.0754	136.308	0.0073	10.2737	0.0973	8.6657
31	14.462	0.0692	149.575	0.0067	10.3428	0.0967	8.8083
32	15.763	0.0634	164.037	0.0061	10.4063	0.0961	8.9436
33	17.182	0.0582	179.800	0.0056	10.4645	0.0956	9.0718
34	18.728	0.0534	196.982	0.0051	10.5178	0.0951	9.1933
35	20.414	0.0490	215.711	0.0046	10.5668	0.0946	9.3083
40	31.409	0.0318	337.882	0.0030	10.7574	0.0930	9.7957
45	48.327	0.0207	525.859	0.0019	10.8812	0.0919	10.1603
50	74.358	0.0135	815.084	0.0012	10.9617	0.0912	10.4295
55	114.408	0.0088	1260.092	0.0008	11.0140	0.0908	10.6261
60	176.031	0.0057	1944.792	0.0005	11.0480	0.0905	10.7683
n	$(1+i)^n$	$\dfrac{1}{(1+i)^n}$	$\dfrac{(1+i)^n-1}{i}$	$\dfrac{i}{(1+i)^n-1}$	$\dfrac{(1+i)^n-1}{i(1+i)^n}$	$\dfrac{i(1+i)^n}{(1+i)^n-1}$	$\dfrac{1}{i}-\dfrac{n}{(1+i)^n-1}$

TABLE F.6 10% Interest factors for discrete compounding.

	SINGLE PAYMENT		EQUAL PAYMENT SERIES				Uniform gradient-series factor
	Compound-amount factor	Present-worth factor	Compound-amount factor	Sinking-fund factor	Present-worth factor	Capital-recovery factor	
n	To find F Given P $F/P, i, n$	To find P Given F $P/F, i, n$	To find F Given A $F/A, i, n$	To find A Given F $A/F, i, n$	To find P Given A $P/A, i, n$	To find A Given P $A/P, i, n$	To find A Given G $A/G, i, n$
1	1.100	0.9091	1.000	1.0000	0.9091	1.1000	0.0000
2	1.210	0.8265	2.100	0.4762	1.7355	0.5762	0.4762
3	1.331	0.7513	3.310	0.3021	2.4869	0.4021	0.9366
4	1.464	0.6830	4.641	0.2155	3.1699	0.3155	1.3812
5	1.611	0.6209	6.105	0.1638	3.7908	0.2638	1.8101
6	1.772	0.5645	7.716	0.1296	4.3553	0.2296	2.2236
7	1.949	0.5132	9.487	0.1054	4.8684	0.2054	2.6216
8	2.144	0.4665	11.436	0.0875	5.3349	0.1875	3.0045
9	2.358	0.4241	13.579	0.0737	5.7590	0.1737	3.3724
10	2.594	0.3856	15.937	0.0628	6.1446	0.1628	3.7255
11	2.853	0.3505	18.531	0.0540	6.4951	0.1540	4.0641
12	3.138	0.3186	21.384	0.0468	6.8137	0.1468	4.3884
13	3.452	0.2897	24.523	0.0408	7.1034	0.1408	4.6988
14	3.798	0.2633	27.975	0.0358	7.3667	0.1358	4.9955
15	4.177	0.2394	31.772	0.0315	7.6061	0.1315	5.2789
16	4.595	0.2176	35.950	0.0278	7.8237	0.1278	5.5493
17	5.054	0.1979	40.545	0.0247	8.0216	0.1247	5.8071
18	5.560	0.1799	45.599	0.0219	8.2014	0.1219	6.0526
19	6.116	0.1635	51.159	0.0196	8.3649	0.1196	6.2861
20	6.728	0.1487	57.275	0.0175	8.5136	0.1175	6.5081
21	7.400	0.1351	64.003	0.0156	8.6487	0.1156	6.7189
22	8.140	0.1229	71.403	0.0140	8.7716	0.1140	6.9189
23	8.954	0.1117	79.543	0.0126	8.8832	0.1126	7.1085
24	9.850	0.1015	88.497	0.0113	8.9848	0.1113	7.2881
25	10.835	0.0923	98.347	0.0102	9.0771	0.1102	7.4580
26	11.918	0.0839	109.182	0.0092	9.1610	0.1092	7.6187
27	13.110	0.0763	121.100	0.0083	9.2372	0.1083	7.7704
28	14.421	0.0694	134.210	0.0075	9.3066	0.1075	7.9137
29	15.863	0.0630	148.631	0.0067	9.3696	0.1067	8.0489
30	17.449	0.0573	164.494	0.0061	9.4269	0.1061	8.1762
31	19.194	0.0521	181.943	0.0055	9.4790	0.1055	8.2962
32	21.114	0.0474	201.138	0.0050	9.5264	0.1050	8.4091
33	23.225	0.0431	222.252	0.0045	9.5694	0.1045	8.5152
34	25.548	0.0392	245.477	0.0041	9.6086	0.1041	8.6149
35	28.102	0.0356	271.024	0.0037	9.6442	0.1037	8.7086
40	45.259	0.0221	442.593	0.0023	9.7791	0.1023	9.0962
45	72.890	0.0137	718.905	0.0014	9.8628	0.1014	9.3741
50	117.391	0.0085	1163.909	0.0009	9.9148	0.1009	9.5704
55	189.059	0.0053	1880.591	0.0005	9.9471	0.1005	9.7075
60	304.482	0.0033	3034.816	0.0003	9.9672	0.1003	9.8023
n	$(1+i)^n$	$\dfrac{1}{(1+i)^n}$	$\dfrac{(1+i)^n-1}{i}$	$\dfrac{i}{(1+i)^n-1}$	$\dfrac{(1+i)^n-1}{i(1+i)^n}$	$\dfrac{i(1+i)^n}{(1+i)^n-1}$	$\dfrac{1}{i}-\dfrac{n}{(1+i)^n-1}$

TABLE F.6 12% Interest factors for discrete compounding.

	SINGLE PAYMENT		EQUAL PAYMENT SERIES				Uniform gradient-series factor
	Compound-amount factor	Present-worth factor	Compound-amount factor	Sinking-fund factor	Present-worth factor	Capital-recovery factor	
n	To find F Given P $F/P, i, n$	To find P Given F $P/F, i, n$	To find F Given A $F/A, i, n$	To find A Given F $A/F, i, n$	To find P Given A $P/A, i, n$	To find A Given P $A/P, i, n$	To find A Given G $A/G, i, n$
1	1.120	0.8929	1.000	1.0000	0.8929	1.1200	0.0000
2	1.254	0.7972	2.120	0.4717	1.6901	0.5917	0.4717
3	1.405	0.7118	3.374	0.2964	2.4018	0.4164	0.9246
4	1.574	0.6355	4.779	0.2092	3.0374	0.3292	1.3589
5	1.762	0.5674	6.353	0.1574	3.6048	0.2774	1.7746
6	1.974	0.5066	8.115	0.1232	4.1114	0.2432	2.1721
7	2.211	0.4524	10.089	0.0991	4.5638	0.2191	2.5515
8	2.476	0.4039	12.300	0.0813	4.9676	0.2013	2.9132
9	2.773	0.3606	14.776	0.0677	5.3283	0.1877	3.2574
10	3.106	0.3220	17.549	0.0570	5.6502	0.1770	3.5847
11	3.479	0.2875	20.655	0.0484	5.9377	0.1684	3.8953
12	3.896	0.2567	24.133	0.0414	6.1944	0.1614	4.1897
13	4.364	0.2292	28.029	0.0357	6.4236	0.1557	4.4683
14	4.887	0.2046	32.393	0.0309	6.6282	0.1509	4.7317
15	5.474	0.1827	37.280	0.0268	6.8109	0.1468	4.9803
16	6.130	0.1631	42.753	0.0234	6.9740	0.1434	5.2147
17	6.866	0.1457	48.884	0.0205	7.1196	0.1405	5.4353
18	7.690	0.1300	55.750	0.0179	7.2497	0.1379	5.6427
19	8.613	0.1161	63.440	0.0158	7.3658	0.1358	5.8375
20	9.646	0.1037	72.052	0.0139	7.4695	0.1339	6.0202
21	10.804	0.0926	81.699	0.0123	7.5620	0.1323	6.1913
22	12.100	0.0827	92.503	0.0108	7.6447	0.1308	6.3514
23	13.552	0.0738	104.603	0.0096	7.7184	0.1296	6.5010
24	15.179	0.0659	118.155	0.0085	7.7843	0.1285	6.6407
25	17.000	0.0588	133.334	0.0075	7.8431	0.1275	6.7708
26	19.040	0.0525	150.334	0.0067	7.8957	0.1267	6.8921
27	21.325	0.0469	169.374	0.0059	7.9426	0.1259	7.0049
28	23.884	0.0419	190.699	0.0053	7.9844	0.1253	7.1098
29	26.750	0.0374	214.583	0.0047	8.0218	0.1247	7.2071
30	29.960	0.0334	241.333	0.0042	8.0552	0.1242	7.2974
31	33.555	0.0298	271.293	0.0037	8.0850	0.1237	7.3811
32	37.582	0.0266	304.848	0.0033	8.1116	0.1233	7.4586
33	42.092	0.0238	342.429	0.0029	8.1354	0.1229	7.5303
34	47.143	0.0212	384.521	0.0026	8.1566	0.1226	7.5965
35	52.800	0.0189	431.664	0.0023	8.1755	0.1223	7.6577
40	93.051	0.0108	767.091	0.0013	8.2438	0.1213	7.8988
45	163.988	0.0061	1358.230	0.0007	8.2825	0.1207	8.0572
50	289.002	0.0035	2400.018	0.0004	8.3045	0.1204	8.1597
n	$(1+i)^n$	$\dfrac{1}{(1+i)^n}$	$\dfrac{(1+i)^n-1}{i}$	$\dfrac{i}{(1+i)^n-1}$	$\dfrac{(1+i)^n-1}{i(1+i)^n}$	$\dfrac{i(1+i)^n}{(1+i)^n-1}$	$\dfrac{1}{i}-\dfrac{n}{(1+i)^n-1}$

TABLE F.6 15% Interest factors for discrete compounding.

	SINGLE PAYMENT		EQUAL PAYMENT SERIES				Uniform gradient-series factor
	Compound-amount factor	Present-worth factor	Compound-amount factor	Sinking-fund factor	Present-worth factor	Capital-recovery factor	
n	To find F Given P $F/P, i, n$	To find P Given F $P/F, i, n$	To find F Given A $F/A, i, n$	To find A Given F $A/F, i, n$	To find P Given A $P/A, i, n$	To find A Given P $A/P, i, n$	To find A Given G $A/G, i, n$
1	1.150	0.8696	1.000	1.0000	0.8696	1.1500	0.0000
2	1.323	0.7562	2.150	0.4651	1.6257	0.6151	0.4651
3	1.521	0.6575	3.473	0.2880	2.2832	0.4380	0.9071
4	1.749	0.5718	4.993	0.2003	2.8550	0.3503	1.3263
5	2.011	0.4972	6.742	0.1483	3.3522	0.2983	1.7228
6	2.313	0.4323	8.754	0.1142	3.7845	0.2642	2.0972
7	2.660	0.3759	11.067	0.0904	4.1604	0.2404	2.4499
8	3.059	0.3269	13.727	0.0729	4.4873	0.2229	2.7813
9	3.518	0.2843	16.786	0.0596	4.7716	0.2096	3.0922
10	4.046	0.2472	20.304	0.0493	5.0188	0.1993	3.3832
11	4.652	0.2150	24.349	0.0411	5.2337	0.1911	3.6550
12	5.350	0.1869	29.002	0.0345	5.4206	0.1845	3.9082
13	6.153	0.1625	34.352	0.0291	5.5832	0.1791	4.1438
14	7.076	0.1413	40.505	0.0247	5.7245	0.1747	4.3624
15	8.137	0.1229	47.580	0.0210	5.8474	0.1710	4.5650
16	9.358	0.1069	55.717	0.0180	5.9542	0.1680	4.7523
17	10.761	0.0929	65.075	0.0154	6.0472	0.1654	4.9251
18	12.375	0.0808	75.836	0.0132	6.1280	0.1632	5.0843
19	14.232	0.0703	88.212	0.0113	6.1982	0.1613	5.2307
20	16.367	0.0611	102.444	0.0098	6.2593	0.1598	5.3651
21	18.822	0.0531	118.810	0.0084	6.3125	0.1584	5.4883
22	21.645	0.0462	137.632	0.0073	6.3587	0.1573	5.6010
23	24.891	0.0402	159.276	0.0063	6.3988	0.1563	5.7040
24	28.625	0.0349	184.168	0.0054	6.4338	0.1554	5.7979
25	32.919	0.0304	212.793	0.0047	6.4642	0.1547	5.8834
26	37.857	0.0264	245.712	0.0041	6.4906	0.1541	5.9612
27	43.535	0.0230	283.569	0.0035	6.5135	0.1535	6.0319
28	50.066	0.0200	327.104	0.0031	6.5335	0.1531	6.0960
29	57.575	0.0174	377.170	0.0027	6.5509	0.1527	6.1541
30	66.212	0.0151	434.745	0.0023	6.5660	0.1523	6.2066
31	76.144	0.0131	500.957	0.0020	6.5791	0.1520	6.2541
32	87.565	0.0114	577.100	0.0017	6.5905	0.1517	6.2970
33	100.700	0.0099	664.666	0.0015	6.6005	0.1515	6.3357
34	115.805	0.0086	765.365	0.0013	6.6091	0.1513	6.3705
35	133.176	0.0075	881.170	0.0011	6.6166	0.1511	6.4019
40	267.864	0.0037	1779.090	0.0006	6.6418	0.1506	6.5168
45	538.769	0.0019	3585.128	0.0003	6.6543	0.1503	6.5830
50	1083.657	0.0009	7217.716	0.0002	6.6605	0.1501	6.6205
n	$(1+i)^n$	$\dfrac{1}{(1+i)^n}$	$\dfrac{(1+i)^n-1}{i}$	$\dfrac{i}{(1+i)^n-1}$	$\dfrac{(1+i)^n-1}{i(1+i)^n}$	$\dfrac{i(1+i)^n}{(1+i)^n-1}$	$\dfrac{1}{i}-\dfrac{n}{(1+i)^n-1}$

TABLE F.6 20% Interest factors for discrete compounding.

	SINGLE PAYMENT		EQUAL PAYMENT SERIES				Uniform gradient-series factor
	Compound-amount factor	Present-worth factor	Compound-amount factor	Sinking-fund factor	Present-worth factor	Capital-recovery factor	
n	To find F Given P $F/P, i, n$	To find P Given F $P/F, i, n$	To find F Given A $F/A, i, n$	To find A Given F $A/F, i, n$	To find P Given A $P/A, i, n$	To find A Given P $A/P, i, n$	To find A Given G $A/G, i, n$
1	1.200	0.8333	1.000	1.0000	0.8333	1.2000	0.0000
2	1.440	0.6945	2.200	0.4546	1.5278	0.6546	0.4546
3	1.728	0.5787	3.640	0.2747	2.1065	0.4747	0.8791
4	2.074	0.4823	5.368	0.1863	2.5887	0.3863	1.2742
5	2.488	0.4019	7.442	0.1344	2.9906	0.3344	1.6405
6	2.986	0.3349	9.930	0.1007	3.3255	0.3007	1.9788
7	3.583	0.2791	12.916	0.0774	3.6046	0.2774	2.2902
8	4.300	0.2326	16.499	0.0606	3.8372	0.2606	2.5756
9	5.160	0.1938	20.799	0.0481	4.0310	0.2481	2.8364
10	6.192	0.1615	25.959	0.0385	4.1925	0.2385	3.0739
11	7.430	0.1346	32.150	0.0311	4.3271	0.2311	3.2893
12	8.916	0.1122	39.581	0.0253	4.4392	0.2253	3.4841
13	10.699	0.0935	48.497	0.0206	4.5327	0.2206	3.6597
14	12.839	0.0779	59.196	0.0169	4.6106	0.2169	3.8175
15	15.407	0.0649	72.035	0.0139	4.6755	0.2139	3.9589
16	18.488	0.0541	87.442	0.0114	4.7296	0.2114	4.0851
17	22.186	0.0451	105.931	0.0095	4.7746	0.2095	4.1976
18	26.623	0.0376	128.117	0.0078	4.8122	0.2078	4.2975
19	31.948	0.0313	154.740	0.0065	4.8435	0.2065	4.3861
20	38.338	0.0261	186.688	0.0054	4.8696	0.2054	4.4644
21	46.005	0.0217	225.026	0.0045	4.8913	0.2045	4.5334
22	55.206	0.0181	271.031	0.0037	4.9094	0.2037	4.5942
23	66.247	0.0151	326.237	0.0031	4.9245	0.2031	4.6475
24	79.497	0.0126	392.484	0.0026	4.9371	0.2026	4.6943
25	95.396	0.0105	471.981	0.0021	4.9476	0.2021	4.7352
26	114.475	0.0087	567.377	0.0018	4.9563	0.2018	4.7709
27	137.371	0.0073	681.853	0.0015	4.9636	0.2015	4.8020
28	164.845	0.0061	819.223	0.0012	4.9697	0.2012	4.8291
29	197.814	0.0051	984.068	0.0010	4.9747	0.2010	4.8527
30	237.376	0.0042	1181.882	0.0009	4.9789	0.2009	4.8731
31	284.852	0.0035	1419.258	0.0007	4.9825	0.2007	4.8908
32	341.822	0.0029	1704.109	0.0006	4.9854	0.2006	4.9061
33	410.186	0.0024	2045.931	0.0005	4.9878	0.2005	4.9194
34	492.224	0.0020	2456.118	0.0004	4.9899	0.2004	4.9308
35	590.668	0.0017	2948.341	0.0003	4.9915	0.2003	4.9407
40	1469.772	0.0007	7343.858	0.0002	4.9966	0.2001	4.9728
45	3657.262	0.0003	18281.310	0.0001	4.9986	0.2001	4.9877
50	9100.438	0.0001	45497.191	0.0000	4.9995	0.2000	4.9945
n	$(1+i)^n$	$\dfrac{1}{(1+i)^n}$	$\dfrac{(1+i)^n-1}{i}$	$\dfrac{i}{(1+i)^n-1}$	$\dfrac{(1+i)^n-1}{i(1+i)^n}$	$\dfrac{i(1+i)^n}{(1+i)^n-1}$	$\dfrac{1}{i}-\dfrac{n}{(1+i)^n-1}$

TABLE F.6 25% Interest factors for discrete compounding.

	SINGLE PAYMENT		EQUAL PAYMENT SERIES				Uniform gradient-series factor
	Compound-amount factor	Present-worth factor	Compound-amount factor	Sinking-fund factor	Present-worth factor	Capital-recovery factor	
	To find F Given P	To find P Given F	To find F Given A	To find A Given F	To find P Given A	To find A Given P	To find A Given G
n	$F/P, i, n$	$P/F, i, n$	$F/A, i, n$	$A/F, i, n$	$P/A, i, n$	$A/P, i, n$	$A/G, i, n$
1	1.250	0.8000	1.000	1.0000	0.8000	1.2500	0.0000
2	1.563	0.6400	2.250	0.4445	1.4400	0.6945	0.4445
3	1.953	0.5120	3.813	0.2623	1.9520	0.5123	0.8525
4	2.441	0.4096	5.766	0.1735	2.3616	0.4235	1.2249
5	3.052	0.3277	8.207	0.1219	2.6893	0.3719	1.5631
6	3.815	0.2622	11.259	0.0888	2.9514	0.3388	1.8683
7	4.768	0.2097	15.073	0.0664	3.1611	0.3164	2.1424
8	5.960	0.1678	19.842	0.0504	3.3289	0.3004	2.3873
9	7.451	0.1342	25.802	0.0388	3.4631	0.2888	2.6048
10	9.313	0.1074	33.253	0.0301	3.5705	0.2801	2.7971
11	11.642	0.0859	42.566	0.0235	3.6564	0.2735	2.9663
12	14.552	0.0687	54.208	0.0185	3.7251	0.2685	3.1145
13	18.190	0.0550	68.760	0.0146	3.7801	0.2646	3.2438
14	22.737	0.0440	86.949	0.0115	3.8241	0.2615	3.3560
15	28.422	0.0352	109.687	0.0091	3.8593	0.2591	3.4530
16	35.527	0.0282	138.109	0.0073	3.8874	0.2573	3.5366
17	44.409	0.0225	173.636	0.0058	3.9099	0.2558	3.6084
18	55.511	0.0180	218.045	0.0046	3.9280	0.2546	3.6698
19	69.389	0.0144	273.556	0.0037	3.9424	0.2537	3.7222
20	86.736	0.0115	342.945	0.0029	3.9539	0.2529	3.7667
21	108.420	0.0092	429.681	0.0023	3.9631	0.2523	3.8045
22	135.525	0.0074	538.101	0.0019	3.9705	0.2519	3.8365
23	169.407	0.0059	673.626	0.0015	3.9764	0.2515	3.8634
24	211.758	0.0047	843.033	0.0012	3.9811	0.2512	3.8861
25	264.698	0.0038	1054.791	0.0010	3.9849	0.2510	3.9052
26	330.872	0.0030	1319.489	0.0008	3.9879	0.2508	3.9212
27	413.590	0.0024	1650.361	0.0006	3.9903	0.2506	3.9346
28	516.988	0.0019	2063.952	0.0005	3.9923	0.2505	3.9457
29	646.235	0.0016	2580.939	0.0004	3.9938	0.2504	3.9551
30	807.794	0.0012	3227.174	0.0003	3.9951	0.2503	3.9628
31	1009.742	0.0010	4034.968	0.0003	3.9960	0.2503	3.9693
32	1262.177	0.0008	5044.710	0.0002	3.9968	0.2502	3.9746
33	1577.722	0.0006	6306.887	0.0002	3.9975	0.2502	3.9791
34	1972.152	0.0005	7884.609	0.0001	3.9980	0.2501	3.9828
35	2465.190	0.0004	9856.761	0.0001	3.9984	0.2501	3.9858
n	$(1+i)^n$	$\dfrac{1}{(1+i)^n}$	$\dfrac{(1+i)^n-1}{i}$	$\dfrac{i}{(1+i)^n-1}$	$\dfrac{(1+i)^n-1}{i(1+i)^n}$	$\dfrac{i(1+i)^n}{(1+i)^n-1}$	$\dfrac{1}{i}-\dfrac{n}{(1+i)^n-1}$

TABLE F.6 30% Interest factors for discrete compounding.

	SINGLE PAYMENT		EQUAL PAYMENT SERIES				Uniform gradient-series factor
	Compound-amount factor	Present-worth factor	Compound-amount factor	Sinking-fund factor	Present-worth factor	Capital-recovery factor	
n	To find F Given P $F/P, i, n$	To find P Given F $P/F, i, n$	To find F Given A $F/A, i, n$	To find A Given F $A/F, i, n$	To find P Given A $P/A, i, n$	To find A Given P $A/P, i, n$	To find A Given G $A/G, i, n$
1	1.300	0.7692	1.000	1.0000	0.7692	1.3000	0.0000
2	1.690	0.5917	2.300	0.4348	1.3610	0.7348	0.4348
3	2.197	0.4552	3.990	0.2506	1.8161	0.5506	0.8271
4	2.856	0.3501	6.187	0.1616	2.1663	0.4616	1.1783
5	3.713	0.2693	9.043	0.1106	2.4356	0.4106	1.4903
6	4.827	0.2072	12.756	0.0784	2.6428	0.3784	1.7655
7	6.275	0.1594	17.583	0.0569	2.8021	0.3569	2.0063
8	8.157	0.1226	23.858	0.0419	2.9247	0.3419	2.2156
9	10.605	0.0943	32.015	0.0312	3.0190	0.3312	2.3963
10	13.786	0.0725	42.620	0.0235	3.0915	0.3235	2.5512
11	17.922	0.0558	56.405	0.0177	3.1473	0.3177	2.6833
12	23.298	0.0429	74.327	0.0135	3.1903	0.3135	2.7952
13	30.288	0.0330	97.625	0.0103	3.2233	0.3103	2.8895
14	39.374	0.0254	127.913	0.0078	3.2487	0.3078	2.9685
15	51.186	0.0195	167.286	0.0060	3.2682	0.3060	3.0345
16	66.542	0.0150	218.472	0.0046	3.2832	0.3046	3.0892
17	86.504	0.0116	285.014	0.0035	3.2948	0.3035	3.1345
18	112.455	0.0089	371.518	0.0027	3.3037	0.3027	3.1718
19	146.192	0.0069	483.973	0.0021	3.3105	0.3021	3.2025
20	190.050	0.0053	630.165	0.0016	3.3158	0.3016	3.2276
21	247.065	0.0041	820.215	0.0012	3.3199	0.3012	3.2480
22	321.184	0.0031	1067.280	0.0009	3.3230	0.3009	3.2646
23	417.539	0.0024	1388.464	0.0007	3.3254	0.3007	3.2781
24	542.801	0.0019	1806.003	0.0006	3.3272	0.3006	3.2890
25	705.641	0.0014	2348.803	0.0004	3.3286	0.3004	3.2979
26	917.333	0.0011	3054.444	0.0003	3.3297	0.3003	3.3050
27	1192.533	0.0008	3971.778	0.0003	3.3305	0.3003	3.3107
28	1550.293	0.0007	5164.311	0.0002	3.3312	0.3002	3.3153
29	2015.381	0.0005	6714.604	0.0002	3.3317	0.3002	3.3189
30	2619.996	0.0004	8729.985	0.0001	3.3321	0.3001	3.3219
31	3405.994	0.0003	11349.981	0.0001	3.3324	0.3001	3.3242
32	4427.793	0.0002	14755.975	0.0001	3.3326	0.3001	3.3261
33	5756.130	0.0002	19183.768	0.0001	3.3328	0.3001	3.3276
34	7482.970	0.0001	24939.899	0.0001	3.3329	0.3001	3.3288
35	9727.860	0.0001	32422.868	0.0000	3.3330	0.3000	3.3297
n	$(1+i)^n$	$\dfrac{1}{(1+i)^n}$	$\dfrac{(1+i)^n-1}{i}$	$\dfrac{i}{(1+i)^n-1}$	$\dfrac{(1+i)^n-1}{i(1+i)^n}$	$\dfrac{i(1+i)^n}{(1+i)^n-1}$	$\dfrac{1}{i}-\dfrac{n}{(1+i)^n-1}$

TABLE F.6 40% Interest factors for discrete compounding.

	SINGLE PAYMENT		EQUAL PAYMENT SERIES				Uniform
	Compound-amount factor	Present-worth factor	Compound-amount factor	Sinking-fund factor	Present-worth factor	Capital-recovery factor	gradient-series factor
n	To find F Given P $F/P, i, n$	To find P Given F $P/F, i, n$	To find F Given A $F/A, i, n$	To find A Given F $A/F, i, n$	To find P Given A $P/A, i, n$	To find A Given P $A/P, i, n$	To find A Given G $A/G, i, n$
1	1.400	0.7143	1.000	1.0001	0.7143	1.4001	0.0000
2	1.960	0.5103	2.400	0.4167	1.2245	0.8167	0.4167
3	2.744	0.3645	4.360	0.2294	1.5890	0.6294	0.7799
4	3.842	0.2604	7.104	0.1408	1.8493	0.5408	1.0924
5	5.378	0.1860	10.946	0.0914	2.0352	0.4914	1.3580
6	7.530	0.1329	16.324	0.0613	2.1680	0.4613	1.5811
7	10.541	0.0949	23.853	0.0420	2.2629	0.4420	1.7664
8	14.758	0.0678	34.395	0.0291	2.3306	0.4291	1.9186
9	20.661	0.0485	49.153	0.0204	2.3790	0.4204	2.0423
10	28.925	0.0346	69.814	0.0144	2.4136	0.4144	2.1420
11	40.496	0.0247	98.739	0.0102	2.4383	0.4102	2.2215
12	56.694	0.0177	139.234	0.0072	2.4560	0.4072	2.2846
13	79.371	0.0126	195.928	0.0052	2.4686	0.4052	2.3342
14	111.120	0.0090	275.299	0.0037	2.4775	0.4037	2.3729
15	155.568	0.0065	386.419	0.0026	2.4840	0.4026	2.4030
16	217.794	0.0046	541.986	0.0019	2.4886	0.4019	2.4262
17	304.912	0.0033	759.780	0.0014	2.4918	0.4014	2.4441
18	426.877	0.0024	1064.691	0.0010	2.4942	0.4010	2.4578
19	597.627	0.0017	1491.567	0.0007	2.4959	0.4007	2.4682
20	836.678	0.0012	2089.195	0.0005	2.4971	0.4005	2.4761
21	1171.348	0.0009	2925.871	0.0004	2.4979	0.4004	2.4821
22	1639.887	0.0007	4097.218	0.0003	2.4985	0.4003	2.4866
23	2295.842	0.0005	5737.105	0.0002	2.4990	0.4002	2.4900
24	3214.178	0.0004	8032.945	0.0002	2.4993	0.4002	2.4926
25	4499.847	0.0003	11247.110	0.0001	2.4995	0.4001	2.4945
26	6299.785	0.0002	15746.960	0.0001	2.4997	0.4001	2.4959
27	8819.695	0.0002	22046.730	0.0001	2.4998	0.4001	2.4970
28	12347.570	0.0001	30866.430	0.0001	2.4998	0.4001	2.4978
29	17286.590	0.0001	43213.990	0.0001	2.4999	0.4001	2.4984
30	24201.230	0.0001	60500.580	0.0001	2.4999	0.4001	2.4988
n	$(1+i)^n$	$\dfrac{1}{(1+i)^n}$	$\dfrac{(1+i)^n-1}{i}$	$\dfrac{i}{(1+i)^n-1}$	$\dfrac{(1+i)^n-1}{i(1+i)^n}$	$\dfrac{i(1+i)^n}{(1+i)^n-1}$	$\dfrac{1}{i}-\dfrac{n}{(1+i)^n-1}$

TABLE F.6 50% Interest factors for discrete compounding.

	SINGLE PAYMENT		EQUAL PAYMENT SERIES				Uniform gradient-series factor
	Compound-amount factor	Present-worth factor	Compound-amount factor	Sinking-fund factor	Present-worth factor	Capital-recovery factor	
n	To find F Given P $F/P, i, n$	To find P Given F $P/F, i, n$	To find F Given A $F/A, i, n$	To find A Given F $A/F, i, n$	To find P Given A $P/A, i, n$	To find A Given P $A/P, i, n$	To find A Given G $A/G, i, n$
1	1.500	0.6667	1.000	1.0000	0.6667	1.5000	0.0001
2	2.250	0.4445	2.500	0.4000	1.1112	0.9001	0.4001
3	3.375	0.2963	4.750	0.2106	1.4075	0.7106	0.7368
4	5.063	0.1976	8.125	0.1231	1.6050	0.6231	1.0154
5	7.594	0.1317	13.188	0.0759	1.7367	0.5759	1.2418
6	11.391	0.0878	20.781	0.0482	1.8245	0.5482	1.4226
7	17.086	0.0586	32.172	0.0311	1.8830	0.5311	1.5649
8	25.629	0.0391	49.258	0.0204	1.9220	0.5204	1.6752
9	38.443	0.0261	74.887	0.0134	1.9480	0.5134	1.7597
10	57.665	0.0174	113.330	0.0089	1.9654	0.5089	1.8236
11	86.498	0.0116	170.995	0.0059	1.9769	0.5059	1.8714
12	129.746	0.0078	257.493	0.0039	1.9846	0.5039	1.9068
13	194.620	0.0052	387.239	0.0026	1.9898	0.5026	1.9329
14	291.929	0.0035	581.858	0.0018	1.9932	0.5018	1.9519
15	437.894	0.0023	873.788	0.0012	1.9955	0.5012	1.9657
16	656.841	0.0016	1311.681	0.0008	1.9970	0.5008	1.9757
17	985.261	0.0011	1968.522	0.0006	1.9980	0.5006	1.9828
18	1477.891	0.0007	2953.783	0.0004	1.9987	0.5004	1.9879
19	2216.837	0.0005	4431.671	0.0003	1.9991	0.5003	1.9915
20	3325.256	0.0004	6648.511	0.0002	1.9994	0.5002	1.9940
21	4987.882	0.0003	9973.765	0.0002	1.9996	0.5002	1.9958
22	7481.824	0.0002	14961.640	0.0001	1.9998	0.5001	1.9971
23	11222.730	0.0001	22443.470	0.0001	1.9999	0.5001	1.9980
24	16834.100	0.0001	33666.210	0.0001	1.9999	0.5001	1.9986
25	25251.160	0.0001	50500.330	0.0001	2.0000	0.5001	1.9991
n	$(1+i)^n$	$\dfrac{1}{(1+i)^n}$	$\dfrac{(1+i)^n-1}{i}$	$\dfrac{i}{(1+i)^n-1}$	$\dfrac{(1+i)^n-1}{i(1+i)^n}$	$\dfrac{i(1+i)^n}{(1+i)^n-1}$	$\dfrac{1}{i}-\dfrac{n}{(1+i)^n-1}$

Index

MAR 4 1994

HUMBER COLLEGE
LIBRARY